Towards Environmental Innovation Systems

Matthias Weber
Jens Hemmelskamp
(Editors)

Towards Environmental Innovation Systems

With 35 Figures and 33 Tables

Springer

Dr. Matthias Weber
ARC Systems Research GmbH
Department of Technology Policy
2444 Seibersdorf
Austria
matthias.weber@arcs.ac.at

Dr. Jens Hemmelskamp
University of Heidelberg
Directorate Research and Project Management
Seminarstraße 2
69117 Heidelberg
Germany
hemmelskamp@zuv.uni-heidelberg.de

Cataloging-in-Publication Data

ISBN 978-3-642-06100-4 e-ISBN 978-3-540-27298-4

Springer is a part of Springer Science+Business Media

springeronline.com

Printed in Germany

Hardcover-Design: Erich Kirchner, Heidelberg

Printed on acid-free paper

In Memoriam
Vicki Norberg-Bohm
died March 21, 2004 at the age of 48

Acknowledgements

This book forms part of the work of a growing network of researcher dealing with sustainability and innovation. We thank the contributors for investing time in preparing their chapters and helping us to realise this book. In particular we would like to acknowledge the financial support for this book project from the German Federal Ministry of Education and Research (BMBF) and the Austrian Ministry of Transport, Innovation and Technology (BMVIT). In particular, we would like to thank Alexander Grablowitz (BMBF) and Hans-Günther Schwarz (BMVIT). We also thank our research assistant Ilja Karabanow who was responsible for the technical edition and supported us in proof reading of the papers. And last but not least a big thank you to Kathrin Eggs for her support in organizing the conference.

Contents

PART THREE: STRATEGIES FOR ENVIRONMENT AND INNOVATION

PART FOUR: POLICY ASSESSMENT (ISSUES) AND NEW REQUIREMENTS

Merging Research Perspectives on Innovation Systems and Environmental Innovation: An Introduction

Matthias Weber and Jens Hemmelskamp

This book is about innovation systems and the possibilities and limitations of shaping their evolution and transformation. It is also about innovation and the environment, or, perhaps more precisely, innovations and sustainability. As such, it addresses the question of the direction of innovative activities in social systems. It is widely recognised nowadays that giving innovation a direction that takes it beyond the contribution it can make to economic growth represents a major challenge for modern societies, a challenge that requires more than technological innovations. In addition, major changes are required along the entire production consumption chain, its flows, its multi-level architecture, its institutions and structures, and – not least – the behaviour of the actors involved in it, from resource extraction to the final consumption of goods and services. Innovation systems are thus required to deliver a new quality of outcomes. This new quality can be captured by the term environmental system innovations, which can be defined by five characteristics: functional changes with a jump in eco-efficiency; a combination of technological, organisational and institutional innovations;the involvement of a multitude of actors; the existence of new guiding principles and sets of goals; and long-term changes at micro- and meso-level (Butter 2002). Innovation systems expected to generate environmental system innovations require new policies and governance approaches operating at and coordinating between different levels and realms of policy-making.

The objective of this book is to contribute to the shaping of a multi-disciplinary research field that helps address these policy challenges. Over the last decade two main streams of research work have emerged that are of major relevance to this issue. First of all, the work on national, regional and sectoral systems of innovation, based mainly on an evolutionary perspective of technological change, has widened the spectrum of innovation determinants that are regarded as relevant. Institutional and organisational frameworks, cumulative learning processes between users and producers and the importance of spatial and technological characteristics are now regarded as indispensable elements in innovation research (Lundvall 1992; Nelson 1993; Braczyk et al. 1997; Edquist 1997; Malerba 2002). However, most research work in this tradition has concentrated on the economic consequences of innovation and hardly taken environmental considerations into account.

Secondly, research work on the impact of environmental regulation on innovation emerging in environmental economics has been a major building block in closing this gap (i.e. Hemmelskamp, Rennings & Leone 2000; Licht et al. 1995; Kemp 1994; Green et al. 1994). Considering environmental regulations as push and pull factors and thus as part of a comprehensive system of framework conditions influencing innovation, this research has broadened the debate on the dynamic effects of environmental policy instruments. By combining elements of innovation research, environmental economics and policy analysis a new, applied perspective on environmental innovation has emerged. It improves our understanding of the impact of environmental regulation on innovation behaviour and provides an analytical framework for explaining the first-mover hypothesis. Subsequently, new policy models like transition management, lead market concepts or time strategies for policy actions were developed that strongly influence current debates about policy for innovation and the environment.

While both streams provide complementary insights into innovation processes in systems, technological change and their potential for improving the environment, little interaction has taken place between the two in terms of building a common conceptual and theoretical framework. This is not to say that no research work had been conducted at the interface between the two streams of research. But there are few forums where this interaction has been deepened.

Since the end of the nineties, however, the linkages between innovation and the environment have become a major issue in national and international research programmes. Moreover, the focus on environmental aspects has been broadened by taking sustainability as an overarching frame of reference. For instance, in 2000, a new research programme was initiated by the German Federal Ministry of Education and Research (BMBF) that aims to advance our knowledge about the role of "framework conditions for innovations towards sustainability (RIW)". It builds on a preceding programme dealing with innovation and the environment (Klemmer 2000) but its scope is broader in that it also includes other dimensions of sustainability beyond the environmental. Merging the two aforementioned streams of research is regarded as a promising approach to providing a theoretical and conceptual foundation for a better understanding of the role of framework conditions for innovations towards sustainability and in particular the possibilities of inducing system innovations by means of policy.

This interest in system innovations for sustainability is shared by research communities in several other countries and corresponding research programmes have been implemented that deal with innovation, the environment and sustainability. For instance, the concept of transition management has become very influential in the Netherlands as a guiding framework for informing policy-makers about long-term strategies for system transformations towards sustainability (Rotmans et al. 2001). At a European level, both within the Fifth and the Sixth Framework Programme, sustainability and innovation has become a major research issue (CEC 2002).

Therefore, one of the aspirations of this book is to bring the research communities from different disciplines and countries together and enable an exchange of experience and a review of the frameworks used to guide research and policy "towards environmental innovation systems". It brings together a selection of contributions exploring new directions of work at the borderline between innovation systems and environmental innovations research. It comprises empirical and conceptual as well as policy-oriented contributions. It is structured along the lines of four main areas of interest with respect to environmental innovation systems: conceptual foundations, empirical experiences, strategic approaches and experience with policy instruments. Accordingly , the first part of the book brings together contributions with a conceptual and ground-laying research interest, dealing in the first instance with key concepts and approaches that promise to represent a useful foundation for thinking about system innovations geared towards sustainability and in particular towards improving the environment.

In his article, Paolo Saviotti discusses one of the fundamental relationships underpinning the transformation of innovation systems towards sustainability, namely the co-evolution of technologies and institutions. After introducing some basic interpretations of the term co-evolution, as well as its use in economics, he shows how institutions and technologies have co-evolved in the past, taking motor vehicles and biotechnology as examples. These illustrations then serve to elaborate some generalised hypotheses about the co-evolution of technologies and institutions – including a taxonomy of institutions - and the extent to which there is scope for inducing innovation by means of introducing "appropriate" institutional changes. This brings him finally to discuss the "win-win" argument that is often applied with respect to environmental innovations, i.e. the hypothesis that innovation can be beneficial to both the innovating firm and the environment. He suggests that while it may apply in selected cases, it is – to say the least – very difficult to disentangle whether the net environmental effect is positive or outweighed by second-order effects that may occur. For instance, efficiency-enhancing technologies may well induce demand effects that over-compensate the direct efficiency gains achieved. As a consequence, designing appropriate institutions to induce technological innovations for improving on the (co-evolving) environment is an extremely challenging task which needs to be informed by further work on the co-evolution of technology, institutions and the environment.

System innovation offers a route for achieving sustainability benefits. This requires the management of evolution processes and the orientation of private and public actors to transition goals. New policy concepts and instruments are therefore needed. Transition management attempts to gear existing dynamics to transition goals chosen by society. Through its focus on long term goals of sustainability and its attention to dynamics it aims at overcoming the conflict between long-term ambition and short-term concerns. In the article by Rene Kemp and Jan Rotmans, transition management is described and exemplified in connection with a low-emission energy supply system.

Frans Berkhout's contribution addresses the core issue of this book, namely system-level changes that promise to bring about significant reductions in environmental impacts. Using the concept of regime shifts, he is interested in the question of whether and how such regime shifts could be induced and in particular what role they might (or might not) play in contributing to such an inducement. More specifically, he examines evidence on the question of whether environmental pressure does indeed have an impact on the rate and direction of technical change, and if so , whether more environmentally friendly technology trajectories or regimes can be expected to inform policy. By looking at two rigid PVC products used for construction purposes and at coated printing and writing paper, he identifies three main interdependent channels through which environmental pressure exerts an influence on innovation: product design, process changes and abatement technology. In both cases, a sudden reversal at the regime level was observed , rather than a smooth transformation and bottom-up improvement of the incumbent technological regime. In this respect, he differs from the proponents of transition management who favour a "soft" transformation approach to achieving a regime shift. This has important implications for policy. Key factors in inducing a regime shift are, in his view, competition and the encouragement of new incipient regimes, rather than complicated adjustments of the dominant regime.

The second part of the book contains a selection of empirical case studies based on a combination of an innovation systems perspective and a keen interest in the transformation of these systems to achieve higher environmental performance. National, regional and sectoral perspectives are brought together.

Halina Brown discusses the transformation of the Polish industry and innovation system that brought about a significant improvement in environmental quality during the 1990s. Evidence of these improvements is given by drawing upon different indicators of environmental performance. The modernisation of the regulatory system is recognised as one of the key factors that contributed to this success story. The strengths of the regulatory system are traced back to a high degree of continuity in institutions, policies and modes of societal transaction, to widely shared values and attitudes among the key societal actors and to a broad support for the rule of law and due process. In other words, a culture of compliance was successfully established. For the future, the ability of the Polish innovation systems to generate technology and system innovations that allow for a reduction of the environmental intensity of production and consumption is seen as crucial in coping with the consequences of current economic growth, especially as Poland is not performing particularly well in terms of innovation.

Gerd Schienstock addresses the issue of innovation systems and sustainable development from a regional perspective. He argues that the changing nature of innovation processes, reflected in the greater emphasis now being placed on tacit knowledge and informal cooperation, strengthens the importance of the spatial vicinity of the actors involved in networked innovation processes. Integrated environmental technologies represent a new development pathway to which regional policy can contribute, for instance by shaping appropriate *Leitbilder* and by acting as a facilitator of regional cooperation for sustainability-oriented innovation.

Based on six case studies of the Scandinavian energy system, Atle Midttun and Anne Louise Koefoed deal with the dynamics of innovation and discuss the interplay between processes within a broad set of institutional contexts, ranging from politics to markets, and their importance for successful innovation.

Frank Becker and Frank Englmann investigate the role of public policy and voluntary initiatives for promoting water-benign process innovations in the chemical industry. In an empirical analysis they analysed the West German chemical industry during the mid-1990s. Their results show that compliance with environmental regulations seems to have been by far the most important reason for implementing both end-of-pipe and integrated environmental innovations. These findings are of particular interest with respect to the implementation of the new EU Water Framework Directive and also for EU chemicals policy.

The third part of the book focuses on future strategies favouring the move towards environmental innovation systems.

Nicolas Ashford challenges certain tenets of the theories of reflexive law and ecological modernization. While far-sighted prevention-oriented and structural changes are needed, some proponents of these theories argue that the very industries and firms that create environmental problems can be transformed - through continuous institutional learning, the application of life-cycle analysis, dialogue and networking with stakeholders and the implementation of environmental management systems - into sustainable industries and firms. However, while useful, these improvements are inadequate . It is not marginal or incremental changes that are needed for sustainability but rather major product, process and system innovations – which are often beyond the capacity of the dominant firms and industries. Ashford also questions the alleged failure of regulation to stimulate necessary technological changes and identifies the conditions under which innovation for sustainability can occur. Finally, he discusses differences in policies for innovation and the environment in industrialized and developing countries.

A national lead market is often the geographical starting point for a global diffusion of products or processes. Lead markets for environmental technologies also depend on various kinds of policy action. Martin Jänicke and Klaus Jacob discuss the potential role of lead markets in the context of global economic modernisation. In particular, they go into the interplay between the diffusion of environmental policy innovation and environmental technology.

Georg Erdmann's article focuses on the time-dependence of innovation pathways and the fact that solutions to a current problem may represent the problems of the future. As a consequence, he argues that sustainability should be understood as a continuous process rather than as a stable end-state. Moreover, critical factors for the success of a specific instance of innovation are not just its inherent characteristics but also whether it emerges at a time when it can benefit from reinforcing effects such as economies of scale, learning and network effects or specific regulatory conditions. There are windows of opportunity when innovations have a high success probability , while at other points in time they may be of negligible importance. He underpins his argument by discussing the example of how different chlorine production technologies have spread in Europe and Japan, showing that Japanese technology policy applied an appropriate time strategy for establishing an advanced technology in the chemical industry, whereas European regulation was rather counterproductive by imposing rigid emission standards too early on, thus preventing a superior technology from establishing itself once it became available. In his article , Erdmann points to some general lessons to be learned for innovation (and diffusion) policy, based on his insights on time strategies for policy intervention.

Remco Hoogma, Matthias Weber and Boelie Elzen present one possible approach to the induction and management of transition processes towards sustainable innovation systems. This approach is based on bottom-up processes of niche development that can potentially shift the dominant technological regime. Past examples of how regime shifts have occurred are used to illustrate the momentum that small niches can develop. Strategic niche management is suggested as a kind of modulation policy capable of inducing regime shifts. They pay particular attention to the role of niche managers, i.e. companies or public authorities that take a lead function in the niche development process.

Nigel Roome argues that the real challenge is to develop a form of continuous iteration between the policy framework and the demands of local environmental and specific socio-technical systems. He develops a conceptual model of the transition from innovation in environmental compliance-driven industry to innovation in more sustainable forms of enterprise. Proceeding from this model, he describes principles and processes as key elements of a "design guide for sufficiency" which is the outcome of the EC Expert Group on "Policies and Actions for Sustainable and Competitive European Production Systems". Finally, Nigel Roome formulates key points for policies and conditions for innovation improving competitiveness within the framework of sustainability.

The fourth and final part of the book deals with the assessment of recent policy initiatives aimed at improving the integration of environmental considerations in technology and innovation policy. On this basis, new requirements for policy and research are derived.

Yukiko Fukasaku discusses the particularities of environmental innovation, the issues the policy makers need to address in designing effective environmental policies and research and innovation policies that optimise environmental innovation. She discusses the specific information needs of policy makers enabling them to contribute to optimising environmental innovations. Fukasaku demands information on public and business expenditure in environmental R&D and on how and where the funds are spent. She also demands better knowledge about the determinants of environmental innovations in firms, how firms assess the costs and benefits involved and how they can acquire relevant information.

Vicki Norberg-Bohm and Theo de Bruijn concentrate on the emergence processes characteristic of policy innovations within the context of environmental innovation systems. Drawing on new institutional theory and the literature on technology policy and management, they discuss the role of voluntary, collaborative and information-based strategies in technological innovation. They compare three U.S. programs to three Dutch programs and investigate why comparable approaches are successful in one context and fail in another. They also recommend ways in which policy innovation can either work within or change the existing regulatory structure.

The IPPC Directive lays down a framework requiring EU member states to issue operating permits that contain conditions based on best available techniques (BAT). It requires the European Commission to organize an exchange of information between member states and the industries concerned with best available techniques. David Hitches, Frank Farrell, Josefina Lindblom and Ursula Triebswetter discuss the impact of the implementation of BAT on the competitiveness of existing plants. Focusing on three industries, the paper answers various questions: Are BAT plants viable? Do they suffer disadvantages in the face of international competition? What are the implications for the economic viability of the sectors concerned? The principal methodology adopted is a case study approach contrasting the economic performance of plants that have adopted most of the elements of BAT with the performance of other 'non-BAT' plants in the various industries.

In the attempt to achieve system innovation, Philip Vergragt recommends that the role of the government should be to formulate and legitimise the direction to be taken in connection with sustainable development. He describes the concept of back-casting and reviews the important "Dutch Sustainable Technological Development" program and the "Strategies towards the Sustainable Household" project. In particular, he reflects on the role of government, business and societal actors in transitions and system innovations.

Finally, Ken Green gives a synthesis of the outlook on issues for the future. Building on his assessment of emerging key challenges to achieving the shift from 'environmental' towards 'sustainability' innovation policy, he identifies both new research needs and policy requirements. In particular, finding the right division of labour between national, sub-national and supra-national entities with respect to the evolutionary process of innovation for sustainability is seen as a crucial issue for the future.

On the Co-Evolution of Technologies and Institutions

Pier-Paolo Saviotti

1. Introduction

To say that technologies and institutions evolve together means that they evolve in interaction. This ought to be hardly surprising if looked at from the vantage point of a systems approach. Such an approach has lately acquired a considerable weight both at a theoretical and at a policy level, in particular by means of the concept of innovation system. Innovation systems have been studied at different levels, ranging from national innovation systems (Lundvall 1992; Freeman 1987; Nelson 1992; Edquist 1997) to regional innovation systems (Braczyk et al. 1998) to sectoral innovation systems (Carlsson and Stankiewicz 1991; Carlsson and Jacobsson 1997; Breschi and Malerba 1997). The common assumption underlying all these approaches is that innovation is not created simply by pouring money into an R&D bucket from which innovations come out. On the contrary, innovations are created by a system, constituted by different parts or components interacting and determining the final outcome. Amongst the evidence that led to the concepts of innovation systems mentioned above is the strong specificity, or persistent asymmetry, shown by several of these systems in the course of time. Thus, for example, some national innovation systems acquire a particular pattern of specialisation, or areas of strength, that show a high persistence in the course of time. The strength of the German chemical and pharmaceutical industry dates from the second half of the XIXth century. Likewise Japanese specialisation in electronics, motor cars and photographic equipment, or Italian industrial clusters specialising in ceramic tiles, leather products or optical equipment, show a considerable persistence. Furthermore, each of these innovation systems shows an institutional specificity that is even stronger than the output asymmetries mentioned above, with patterns of institutions and institutional interaction differing sharply amongst countries, regions or sectors.

These persistent features of innovation systems are very difficult to explain in terms of a framework based on the assumption that the economic system is in equilibrium most of the time and that, if there are any deviations from equilibrium, the system quickly returns there. The persistence of the above asymmetries means that the equilibrium of the system is not unique, or that the system spends a very large part of the time in states that are out of equilibrium, even if not necessarily to a very large extent. The first case would be obtained if the innovation system were to display multistability, a feature known in complex systems. It would then be possible for different institutional configurations or compositions of the economic system to produce comparable if not identical outcomes. In these circumstances we could expect innovation systems to evolve in such a way to produce a comparable performance, as judged by indicators such as GDP per head, productivity, trade balances, etc., while preserving some specific structural features.

The present contribution is not about innovation systems, but about the co-evolution of technologies and institutions. Innovation systems have been mentioned here as an example of systems, that is of collections of parts (or components) interacting in such a way that, in presence of environmental variations the system adapts while preserving its structural identity. That is, for environmental variations within a given range the system reacts by means of internal adjustments (e.g. feedback loops) in order to preserve its adaptation to the external environment. This state of self-regulation or homeostasis is stable up to the point where environmental variations exceed those boundaries beyond which the system collapses. Self-regulation is a subset of the wider problem of self-organisation, which analyses the conditions under which the system can be formed, starting from its constituent parts and the conditions under which it remains stable. It is to be observed that the external environment considered so far is that of a particular system, that is the collection of external entities influencing the behaviour of the system and of the variables representing them. For example, the external environment of a firm in an industrial sector contains the variables influencing the supply and demand conditions for the sector, which may include the external physical environment. Such external environment will be different for each sector, and at a very micro level for each firm. Thus the environment of a particular firm or organisation is a subset of the physical environment but it contains also some man-made features, for example institutions, or man-made resources.

Even if we recognise that the study of subjects like innovation or the environment amounts to studying complex systems, the approach is analytically difficult to apply. To single out a small number of the components of a system and to study their interacting evolution gives us an analytically more tractable problem. Thus we can consider the co-evolution of technology and institutions, or for that matter of any pair of important components of a social system, as a limited subset of the same social system that is elected for analytical convenience. This means that by stressing mainly the interactions between technologies and institutions we are assuming that these interactions are much stronger and more important in determining the behaviour of the system than the interactions between either technology or institutions and other components of the social system. This may be a very reasonable approximation, but it does not amount to an exemption from the study of other interactions within the system. Thus the study of co-evolution must be considered an intermediate step towards the understanding of the socio-economic system, or, at least of parts of it more closely related to the evolution of technology.

In the following section of this paper the use of co-evolution by economists and biologists will be reviewed in order to provide a basis for a more detailed discussion of the interaction between technology and institutions.

2. Meaning and Uses of Co-Evolution

2.1 On the Meaning of Co-Evolution

Let us begin this section with a few definitions. At least technology and institutions have to be defined. This does not imply that the definitions given will be absolutely correct and applicable to all possible cases. Yet they will serve as a guide for the subsequent discussion. Technology is in a very general sense the set of activities by means of which human beings modify their external environment. In a fundamental sense it may be said that human beings, like other biological species, adapt to their external environment. Such external environment has both given (i.e. physical), and man-made components. Human beings need some resources (food etc.) in order to survive. Technology is the means whereby they improve their access to resources, as compared to obtaining them based only on their natural endowments (strength, speed, etc.). Technology widens the range of resources that human beings can have access to. Thus agriculture increases the output of those vegetables that can constitute resources (food, building materials etc.) for human beings. The environment-modifying function of technology is exemplified by the importance of tools. Anthropologists think that the capacity to use tools is one of the most important characteristics distinguishing human beings from other animal species. According to Georgescu-Roegen (1971) tools can be considered exo-somatic instruments. Tools are of course a way in which human beings can improve their ability to modify their external environment with respect to using their arms, nails, etc.. Thus technology is intimately related to the external environment of mankind. However, it is to be noted again that the external environment as customarily understood in environmental studies is not necessarily equal to the one that was referred to in the above considerations on technology. The external environment used in environmental studies is mostly physical and biological, while the one that technology is expected to modify contains also man-made components.

Science is the set of activities by means of which human beings explore and try to understand their external environment. Some forms of knowledge are involved in both science and technology. Yet scientific knowledge is considered to be the most complete and reliable form of knowledge (Ziman 1978). The close relationship that exists between the two in modern societies is due to the fact that it is easier to modify our external environment if we know the way it is constituted. Thus science enhances our ability to modify our external environment. Again, science should in principle help us to modify both our natural and man-made environment.

The relationship between science and technology can be better appreciated by referring to some considerations about knowledge. Knowledge can be considered both as a correlational and a retrieval/interpretative structure (Saviotti 1999). The first property of knowledge can be understood if we think that science establishes correlations between variables representing observables of the external environment. Not only well known formulas and theorems do that, but also more qualitative types of theories correlate different entities or their representative variables.

The second property means that those who already know a part of a theory have greater chances of being able to learn or retrieve another more advanced part of the same theory. For what concerns R&D, the retrieval/interpretative function of knowledge leads naturally to the concept of absorptive capacity (Cohen and Levinthal 1989, 1990). To perform R&D on given topics not only may lead to new goods and services, but it improves the capacity of the performing firm to learn/retrieve external knowledge similar to the one on which it has previously performed R&D. All these considerations about knowledge help us to understand the links between science and technology. Let us assume that our welfare would be improved by modifying certain aspects of the external environment. It may be very difficult or impossible to modify directly those aspects. However, if we know that such aspects are related to others, we may be able to develop alternative and easier routes to modify the intended aspects of the external environment. Alternatively, we may say that to know the external environment reduces the number of trial and error experiments that we have to perform in order to modify the external environment. Thus there are very close relationships between science and technology. These relationships become even closer if we take into account that many problems begin as technological problems and are subsequently explained by science. The scientific contribution improves the subsequent development of the technology. The steam engine is an often quoted example. However, according to Popper (1972) scientific developments always begin with some sort of practical problem.

To summarise our considerations so far we can say that technology is the set of activities by means of which human beings modify their external environment, and that such modification is made easier by science. Thus at is very roots technology is intimately related to the external environment of mankind. Again, the external environment that technology can modify has both a natural and a man-made component. Of course, while the modifications of the external environment carried out by technology were intended to improve human beings adaptation to the external environment they do not always do so, but sometimes they produce unintended side effects (externalities) in addition to the desired modifications. These considerations amount to say that technology is not an activity that by accident creates environmental problems, but an activity that by definition is intended to modify the external environment and that together with intended modifications can create undesirable ones.

Institutions are obviously very important for economic life, yet they have had alternate fortunes in economic thought. They were given a very considerable importance by American institutionalist economists, only to be largely abandoned, except for an a-institutionalist version of the market by neo-classical economists. Recently they have come back to the forefront of attention by many economists (see for example Hodgson 1988). In spite of their acknowledged importance institutions are not particularly easy to define and we have for them many acceptable definitions rather than a very good one that is accepted by everyone. One of the clearest definitions of institutions, even if not necessarily one that is analytically more accurate than many others, has been given by North (1990) "(I)nstitutions are the rules of the game and organisations are the players. Institutions are usually credited with establishing patterns of human action, by excluding some types of behaviour and encouraging others". In alternative words, institutions are "...ways of thought or action of some prevalence or permanence, which is embedded in the habits of a group or the customs of people" (Hamilton 1932, cited in Edquist 1997). That is, institutions facilitate and maintain patterns of habitual behaviour (Hodgson 1988), as opposed to rational behaviour, that would be decided in any particular set of circumstances based on the conditions of the external environment. Institutions are 'settled habits of thought common to the generality of man' (Veblen 1919). In turn institutions can also be considered both a response to the extremely high information and computation costs that a so-called rational behaviour would have and a co-ordinating device for human actions (Loasby 1999). A discussion of the nature of institutions would lead us very far and it is not compatible with the objective of this contribution. The previous definitions and considerations should suffice for our purpose here.

To go back to the main objective of this paper, we need to discuss the co-evolution of a series of activities attempting to modify the external environment of mankind (technology) and a set of institutions that influence and are influenced by the evolution of technology. First of all, it must be observed that technology can hardly ever have been a-institutional. By establishing particular patterns of interaction with the external environment technology was identifying patterns of human behaviour which, if adopted on a large scale, would automatically have become institutions. However, in a sense the type of institutions that are directly modeled on particular technologies are hardly an interesting type of institution. To say that the development of electronics requires firms that are different from food processing firms is hardly something that will teach us very much about the co-evolution of technologies and institutions. The most interesting part of the question is: what institutions other than those which directly produce the new technology can influence the mode and level of development of the technology itself?

In order to try and answer this question we now go back to the basic meaning of the term co-evolution. Biologists have been the first to use it and have used it to a much greater extent and more systematically than social scientists. Although biological theories cannot be transposed unchanged to economics, some interesting lessons can be learnt. Biologists have classified the types of possible interactions between the interacting populations in co-evolution. These types are (Roughgarden 1996):

- Parasitic: one party benefits at the clear expense of the other
- Commensal: one party, the guest, benefits at the clear expense of the other, the host, in a way that brings negligible harm to the host
- Mutualistic: both parties clearly benefit each other

It seems somewhat difficult to imagine a lasting parasitic relationship between technologies and institutions, although such a relationship cannot be excluded for a short period. Even a commensal relationship, although possible, does not seem the type of relationship that would be capable of determining the development of an important new technology. Amongst the three types mentioned above a mutualistic relationship is the most likely to lead to the emergence of a new and important technology. Moving to a more economic terminology, we can expect that co-evolving technologies and institutions are likely to be complementary. The word important is stressed here to distinguish between common and uneventful technologies that follow the development of the economic system and those technologies that create and renew economic development. The latter would be those technologies that for Schumpeter (1934) can rescue an economy from a recession and induce a new period of growth, or, those technologies that underlie the leading sectors in economic growth (Rostow 1960). The concentration on these technologies does not mean that other technologies, those that were called before uneventful, are not economically important. All technologies are important for economic development. The concepts required here are rather those of technology or industry life cycle (see for example Abernathy and Utterback 1975; Klepper and Simons 1996). The development of the institutions that are required to underpin the development of the technology itself has to be studied along the technology or industry life cycle. Typically theories of industry life cycle include different stages of the life of a technology, going from very early ones, often characterised by high if uncertain growth rates, to periods of maturity in which the rates of growth of the technology slow down. The creation of the institutions required for the further development of the technology occurs at particular phases of the life cycle. To the extent that such institutions are required, their creation is necessary in order for the diffusion of the technology to proceed beyond a given extent. Thus institutions will typically be created in the early stages of the diffusion of the technology and the extent of diffusion will be determined by the presence (or absence) of such institutions.

2.2 Uses of Co-Evolution in Economics

Perez (1983, 1985) and Freeman, Perez (1988) introduced the concept of techno-economic paradigm. Such a concept relies on those of technological paradigm (Dosi 1982) and of long waves of economic development. A technological paradigm can be considered a set of routines, technological practices shared by a large number of firms in the production and utilisation of particular technologies. It is conceived at the level of aggregation of one technology. On the other hand Freeman et al (1982), following a basic Schumpeterian intuition, maintained that clusters of several interacting technologies were required in order to create the long term economic cycles predicted by Kondratieff. Freeman et al (1982) called the set of technologies that could do that new technology systems. Perez observed that the development of institutions was typically slower than that of technologies. Once it is accepted that institutions are required for the full and complete development of a technology, or a set of, it follows that the duration of the life cycle of a set of connected technologies depends crucially on the time needed to build the required institutions. While co-evolution is not specifically mentioned in this concept, a techno-economic paradigm corresponds clearly to the co-evolution of technologies and institutions. However, within this concept the mechanisms of co-evolution are not analysed. We do not know what types of institutions are required nor to what extent they can change the dynamics of creation and diffusion.

A more detailed analysis of the processes leading to the creation of institutions as a new technology develops is done by Nelson (1994). He points out that there is historical evidence that various features of the institutional environment tend to adapt to the emergence and evolution of new technologies. For example, people in new industries become conscious that there is a new industry and that it has collective interests and needs, thus forming industry or trade associations. Furthermore, this joint evolution of technologies and institutions gives rise to a particular relationship between science and technology. Quite often technology comes into existence with a limited theoretical understanding. This induces scientific research to understand it and establishes the basis for further technological progress. New disciplines, such as metallurgy or chemical engineering, sometimes arise to understand the problems linked to particular technologies. These technology-oriented sciences tend to tie industry to universities. Intellectual property rights are sometimes required to enable the further development of the industry. Thus it is not only new firms that accompany the development of new technologies, but several other layers of institutions are usually created in this process.

For all their usefulness as facilitators of technological evolution, institutions can sometimes act as inertial forces, delaying the required adaptation of an industrial system. Institutions required to favour the development of a particular technology in the early phases may constitute a hindrance to their subsequent change or replacement by newer technologies. For example Lazonick (1990) argued that the same methods of training and work organisation that had worked so well in the XIXth century in Britain became a handicap in the twentieth century.

McKelvey (1997) studied the co-evolution of science and technology within the commercial development of genetic engineering and of their related institutions. Her study was concentrated on the early years of genetic engineering and on the USA. The institutions that she studied were research institutes, Dedicated Biotechnology Firms (DBFs) and large pharmaceutical firms. The activities carried out by these firms straddle the boundaries between science and technology. Thus DBFs carry out what could be called basic research yet this basic research is not identical to that carried out in Universities at the same time. When trying to express insulin or growth hormone, firms tend to emphasise usefulness while Universities try to advance general knowledge. According to McKelvey, to understand the relationship between contemporary science and technology we need to think in terms of four selection environments: the techno-economic, the scientific-economic, the techno-government, and the basic scientific. The hybrid status of the scientific economic seems particularly well suited to the activities of DBFs. McKelvey's study concentrates on firms and raises the question of the specificity of the institutional structure studied to the economy of the USA. At the time of her study very few DBFs existed in Europe and virtually none in Japan. This raises the question of whether the institutional structure appropriate to a given technology is unique and necessary or whether it can vary amongst different countries, a problem also raised by Nelson (1994).

Rosenkopf and Tushman (1998) study a specialised subsector, that of the flight simulation industry, and focus on the co-evolution of community networks and technology. In particular, they concentrate on the so called 'cooperative technical organisations' (CTOs). Examples of CTOs are various types of standards and technical committees that are usually created at the very beginning of a technology, but that often persist beyond the emergence phase. CTOs are involved in community networks and contribute to shape the co-evolution of new technologies and of other related institutions. For example, CTOs help to define technical options, thus influencing the nature of the dominant design. They examine the proposition that the networks in CTOs are involved evolve differently depending on the stage of the life cycle of the technology considered and on the consequent level of uncertainty. They find that while CTOs are continuously founded, those founded during eras of ferment have more lasting potential to resolve uncertainty and to advance technological development. Also, the density of the networks in which CTOs are involved increases regularly with the age of the technology.

3. Considerations about some Sectors/Fields

Let us now use some examples in order to try and understand the problem of the co-evolution of technologies and institutions.

3.1 Motorcars

No technology has been marking the XXth century as deeply as motor cars. Thus we have a large number of observations about this sector. First, different types of institutions are involved in its development. In addition to firms, that were always new, there are many other institutions that accompanied and co-determined the development of this sector. For example, regulating institutions determining the side of the road on which people are going to drive and other rules of behaviour, or certifying institutions defining the competencies required to drive a car and confirming that people have in fact attained such competencies are of fundamental importance. These institutions perform the basic function of co-ordination. If we imagine the results of choosing randomly the side of the road on which to drive we can easily realise that co-ordination faults of this type would largely deny the advantages of the use of motor cars. Thus we can expect that one of the basic functions of co-evolving with a technology is to solve the co-ordination problems created by the emergence of the technology itself.

A second important class of institutions that can influence the evolution of a technology are those related to infrastructures. Infrastructures are common inputs required for the use of particular technologies, but usually not provided by the producers of the technologies due to either their size or to the competencies required. Often infrastructures are supplied by the state. In the case of motor cars infrastructures are, for example, roads, petrol pumps etc.. Here we can see how the presence and quantity of infrastructures determines both the pace and the scope of the evolution of a technology. If in a virtual experiment we were to place the cars that exist today in an industrialised country on the roads that the same country had in the year 1900, we can easily imagine how the resultant congestion would raise the negative externalities created by motor cars greatly above their benefits. Thus, the pace at which infrastructures are built determines the pace at which the technology can diffuse. Conversely, if we imagine the technology to evolve through a life cycle, the size of the population of potential adopters is determined by the final stock of infrastructures built after the technology has attained maturity.

A further class of institutions influencing co-evolution are those related to complementary technologies, that is technologies that supply complementary inputs and maintenance services. In the case of motor cars examples of such institutions can be found in the petroleum refining and in the tyre-making industries. Both industries supply inputs without which a motor car could not deliver the required services. However, they not only shaped the evolution of motor cars, but were themselves shaped by it. The output of the petroleum-refining industry and its evolution has been largely influenced by that of motor cars. The output produced by petroleum refining changed from the fuels required for heating and lighting in the second half of the XIXth century to the lighter fractions required by motor cars starting from the 1930s. Recent developments, requiring refiners to supply unleaded petrol, follow the same trend. Likewise, the tyres we use today are very different from those made at the beginning of the XXth century not only because they last longer, but because they are adapted to drive on the paved roads and motorways that we have today.

The stability of the system constituted by a technology and by its co-evolving institutions, like that of any complex system, is limited by a range of environmental variables. The motor car system underwent a considerable challenge, that it was capable of overcoming, during the oil price shocks of the 1970s. A sudden and very drastic increase in the price of the main input for motor cars risked destabilising profoundly the system. Although subsequently the price of oil fell to very low prices, the reaction of the system to the crisis would have probably stabilised it even if oil prices had remained high. The immediate reaction of the system was a form of induced innovation, that rapidly increased the efficiency of fuel consumption of motor cars (see for example Ohta and Griliches (1976, 1986) and other references cited in Jaffé et al 2001). Of course, other complementary sectors adapted as well. It is enough to think about oil exploration, that rapidly increased the supply of it, thus counteracting the effect of the oil price shock.

Without going into further details an important general point can be made here. Both infrastructures and complementary technologies are stabilising the technology that is considered here. Thus, if we start generalising, we can see that a very important role can be played in the co-evolution of technologies by complementarity. This would not be surprising at all for a biologist, in view of the types of interactions that are usually considered to be involved in co-evolution there. The complementary relationship envisaged in this example seems rather a mutualistic one, in which all co-evolving entities benefit from the interaction. While parasitic or commensal types of relationships cannot be excluded, the admission of an important complementary role would be more difficult to accommodate in economics, where competition is the only interaction studied seriously, and where complementary goods are just mentioned at the beginning of introductory textbooks.

3.2 Biotechnology

Biotechnology is a collections of technologies that share the common feature to use biological methods to produce their outputs. Biotechnology is really a very heterogeneous set of technologies, some of which are very old. However, a part of the biotechnologies are at the moment both some of the most dynamic technologies driving economic evolution and at the same time raising considerable controversy. The parts of biotechnology that some commentators call third generation biotechnology are those based on molecular biology, a discipline developed starting from the 1930s mainly in the USA. Molecular biology was born of the attempt to apply to biology the methods of physics. Until the 1970s molecular biology was essentially a theoretical if promising speciality. At the beginning of the 1970s some important discoveries opened the way to a stream of industrial applications. Biotechnology is now by some considered the technology of the XXIst century. Its scope is immense and even the remarkable achievements obtained since then probably do not represent more than a small part of what we can expect to see developing in future. Yet for all its potential biotechnology raises a series of ethical problems that slow down its development. Is this just a case? Is biotechnology just a victim of a wave of irrationality or anti scientism? The connections between the intimate nature of biotechnology and the ethical preoccupations and controversies raised are far from being casual.

Biotechnology shares with all other technologies the property that it can produce disadvantages as well as benefits. Thus some people are rightly worried about the risks involved in using biotechnology for example to produce food or to create new plant varieties. Critics of biotechnology argue that genetically modified organisms are not safe and that to create genetically modified plants is against nature and can create serious risks for the environment. Such preoccupations are having an important impact on R&D policies, public and private, and on firm strategy world-wide. Towards the middle of the 1990s the concept of the life science company emerged as the company that using a common biotechnology knowledge base could supply several differentiated markets. This strategy was essentially based on the primacy of knowledge as the component defining the firm, as opposed to the nature of its output. Although such a concept seemed to make sense, especially as we are expected to be moving towards the knowledge-based economy, it became prematurely obsolescent due to a drastic change in the selection environment of biotechnology based firms and sectors. The barriers to the acceptance of the products of pharmaceutical and agro-food firms changed dramatically and differentially. While new drugs and new types of treatments, such as those promised by gene therapy, are still eagerly awaited, the thought of using GMOs in food seems at present unacceptable to a large part of Europeans. Equally, genetically modified plants are seen as a threat to both the environment and to the well-being of less developed countries. As a result the large firms that were beginning to adopt the concept of the life science firm rapidly changed their strategy and separated sharply their pharmaceutical from their agro-food activities. This reversal of strategy is not due to any failure of the life science firm but rather to a sharply increasing heterogeneity of the selection environment for the biotechnology based sectors. Pharmaceuticals, showing no significantly increased barrier to acceptance by consumers, keep receiving the lion's share of the investment, while agro-food, for which no one predicts a quick return to profitability by using modern biotechnology, is kept on perfusion. Thus the institutional structure of the biotechnology based sectors has been heavily influenced by some features of the selection environment.

Let us observe here that the selection environment is not given, but it is created during the evolution of the technology, or it co-evolves with it. If fact, the selection environment is constituted by a combination of factors, amongst the most important of which there are the institutions influencing the technology. Such institutions are not there before the technology emerges, but they need to be created during the co-evolution process. Ethical committees, regulating bodies, citizens' organisations, etc. either still need to be created or at best are at an emergent phase. We still do not have a generally accepted legislation to deal with cloning, stem cell research, etc.. Thus a whole institutional infrastructure that will determine the progress of biotechnology is still being put in place but its development is now slower than that of science and technology in the same field. The creation of this institutional infrastructure is particularly complicated in biotechnology, given the ethical problems that it raises. Thus the development of a technology that can determine that of several interconnected sectors is highly dependent on the setting up of a series of institutional infrastructures that have to solve, among others, ethical problems. The separability of homo economicus does not seem to be assured all along the life cycle of broadly based industries or technologies.

3.3 The Co-Evolution of Demand and Supply

Textbook theories of demand stress that the task of economists is to study the behaviour of consumers following from a given set of preferences. To enquire about preference formation is considered to be the task of other social science disciplines. However, it can be argued that no preferences and no demand can exist for goods and services that are so radically new that the consumer cannot even imagine their properties (Saviotti 2001). In these cases preferences and demand are created gradually as an innovation diffuses and as various forms of learning take place, both on the consumer and of the producer side. These forms of learning are mutual, in the sense that at the very beginning producers have to inform consumers about the innovation, but then producers themselves gradually learn how to evaluate demand as the innovation diffuses. Thus we can say that demand and supply co-evolve during the life cycle of a technology. Of course, the creation of demand and of supply cannot take place in a vacuum, but it requires institutions; firms and regulating institutions on the supply side, shops, supermarkets consumer organisations etc on the demand side. Thus, to say that demand and supply co-evolve means that their institutions co-evolve.

4. Some Generalisations

In the previous sections we examined the nature of co-evolution, the use made of the concept by some economists and the applications of the same concept to some examples. We now proceed to see if any generalisations about the co-evolution of technologies and institutions can be derived from the previous analysis.

4.1 Separability and Co-Evolution

The possibility to study the co-evolution of different entities is limited by their separability. It would not make much sense to study the co-evolution of an entity with itself. The entities whose co-evolution we intend to study must be separable to a certain extent. For example, to analyse the co-evolution of science and technology is complicated by the fact that their component activities are largely overlapping. The goals of science and technology are clearly different, but many of the actual projects carried out by them are very similar. The institutions of science and technology have traditionally provided separation, but in the present circumstances their boundaries are becoming increasingly fuzzy.

4.2 Division of Labour and Co-Ordination

Division of labour and co-ordination are two of the most fundamental phenomena in economic development. There are essentially two mechanisms of division of labour, both recognised by Adam Smith, even if one of them was analysed in far greater detail than the other. Division of labour can increase by subdividing existing production processes into a greater number of steps, each one carried out by a different worker. The division of labour created by means of this mechanism depends on the extent of the market: the larger the market, the greater the number of steps into which a given process can be subdivided. Also, new types of work functions can be created by completely new goods and services, requiring new processes of production, new skills and competencies. The extent of division of labour created by the second mechanisms can increase by means of the first mechanism as the market for a new technology increases. Thus new activities, new goods and services give rise to an increasing division of labour, but in the meantime the create new co-ordination problems. One of the main roles of the new and augmented institutions that co-evolve with the technologies is to solve the co-ordination problems created by the technologies themselves. A consequence of this symbiosis of technologies and institutions is that if the output variety of the economic system increases (Saviotti 1996), we can expect institutional variety to increase as well.

4.3 Institutional Taxonomy

Technologies cannot exist without institutions. If we follow the definition of institutions as patterns of habitual behaviour, we can see that these patterns are required in the existence of any productive activity. No organisation can survive without rules. Although routines are usually distinguished from institutions, in the sense of being internal to a firm, rules determining the interactions of firms with their selection environment are always required. Thus technologies cannot exist without institutions. The problem of the co-evolution of technologies and institutions may then seem trivial. However, the important part of it is: what types of institutions can we expect to be required for and to favour the development of a technology?

- Firms are the obvious starting point, we can call them the direct institutions of a given technology. However, other types of institutions are generally required.
- Regulating institutions provide rules for the co-ordination of economic activities and are required to reduce the level of uncertainty. In the case of motorcars, institutions were required to decide the side of the road on which to drive, while for biotechnology they should decide what applications are legitimate, what should be forbidden and exercise a quality control on outputs (e.g. FDA in the USA).
- Infrastructures are common inputs required by all users of a given technology and whose presence and size determines the possible number of users and the benefits the technology can give them.
- Institutions providing complementary inputs. These are of course the firms producing such inputs, but also the other institutions required for firms to function properly, as discussed above. Financial institutions are an example of a generally required complementary institutions, although they may take different forms in different industries. The extent to which a given technology requires complementary inputs is specific to the technology. We can expect very pervasive technologies to require many complementary inputs and to produce may outputs used by other technologies as inputs. Thus a very pervasive technology could be expected to have a high degree of connectedness within the economic system. It is to be noted that such connectedness is not limited to links with other firms but it includes links to the various types of institutions co-evolving with the technology. Also, we can expect the size of the network in which a technology takes part to influence the time required for the construction of the network, and thus the life time of the technology life cycle. For example, we could expect a more connected technology to have a longer life cycle. Conversely, we could expect that if a technological system, including a technology and its co-evolving institutions, takes longer than another to develop, it is likely to have a greater inertia. Of course, such hypotheses are just examples and would need to be structured by more empirical and theoretical work.

The taxonomy proposed here is not intended to be exhaustive, but to indicate the work required to give a more systematic structure to work on co-evolution of technologies and institutions and on innovation systems in general.

4.4 Complementarity

Amongst the institutions co-evolving with a new technology firms compete against one another, but many of the other institutions are complementary with respect to firms and sometimes amongst themselves. While this may seem a trivial statement, it points to a very important problem. Competition is a very well studied relationship in economics, but complementarity is rarely mentioned. Complementarity is usually found in introductory textbooks, where the distinction between substitute and complementary goods is introduced. After that it disappears without traces. Yet, if the previous considerations about co-evolution are right, a technology during the course of its development becomes included in a network of institutional links, some of which are competitive but others are complementary. In fact, the number of complementary links is likely to be a large percentage of the total number of links. For example, regulating institutions that provide co-ordination and/or infrastructures are clearly complementary with respect to the main technology. Of course, technologies providing complementary inputs, such as petroleum refining or tyres in the case of motor car, are complementary. Thus we can expect complementary interactions between institutional actors to play a very important role in the development of technologies. For all this importance we find very little help in economics for what concerns complementary interactions.

Goods, services or activities can be considered complementary if they are jointly required to produce a given output. Furthermore, all complementary activities and the institutions in which they are embodied can benefit from their complementary relationship. These definitions of complementarity are based on the results of such a relationship. But, can we predict whether two activities are likely to be complementary without actually making them work together on the basis of their internal characteristics? Substitute goods can compete with one another. Their competition will be more intense the more similar they are. We can then imagine to evaluate the degree of substitutability of two different goods by comparing their service characteristics. On the contrary complementary goods cannot be similar. They have to be different, but not all different goods are complementary. Thus we do not have an ex-ante definition of complementarity that allows us to tell whether two activities are complementary without placing them in a situation in which their complementarity can be tested. The only possible generalisations about complementary activities are the following:

1. If several activities are generated by means of a process of division of labour then all these activities are likely to be complementary. All the activities are jointly required to produce their common output.
2. Parts of a complex system are likely to be complementary, since the system would collapse if the parts were removed. Thus the parts of a complex artefact, such as a car or a computer, are complementary.

Both of these definitions, that do not necessarily encompass all the types of existing complementary activities, are based on the way in which the activities are generated rather than on their internal characteristics.

We can conclude this discussion by saying that complementary activities play a very important role in the co-evolution of technologies and institutions and altogether in economic development. However, our general understanding of complementary relationships is far more limited than that of competitive relationships.

4.5 The Scope and Pace of Development of a Technology

The scope of a technology can be considered as the number of users it can have. This does not necessarily include only the number of people directly using the technology but also those using the complementary technologies and involved in the co-evolving institutions. Thus, the scope of a technology does not depend only on the technology itself, but also on the co-evolving institutions. Of course, to the extent that institutions are co-determinants of technological evolution the rate at which they are put in place may determine the pace at which technologies are adopted at particular places. A clear example of this situation is given by regulations on biotechnology, allowing or not certain types of field tests or of scientific experimentation, and varying depending on the countries. Countries with easier regulations are likely to see a more rapid development of the activities concerned, of course other conditions being equal.

We can obtain an intuitive representation of this situation by reference to diffusion curves. A typical diffusion curve is sigmoidal, with a slow start, followed by a period of acceleration and leading finally to saturation when all potential adopters have adopted. A diffusion curve is defined by three parameters: the time at which the whole process starts (e.g. the invention time), the slope of the accelerating part, and the percentage of the population at saturation. The first parameter is independent of the institutional structure because it usually corresponds to a pre-institutional phase. The slope of the curve during the accelerating part is likely to depend on the rate at which the relevant institutional infrastructure is put in place, while the saturation level is likely to depend on the size of the institutional infrastructure at saturation.

4.6 System Stability

Complex systems are stable in presence of variations in their external environment, provided that these changes do not exceed a given range. This same principle can be applied to technological systems. They are adapted to a given environment, but adaptation may disappear if the environment changes too much. A typical example was given before by the oil price shock for motor cars. However, we know very little about the actual range of stability of given technological systems. In biology the diversity of the system is expected to influence its stability, but we have no knowledge about the stability of technologies.

5. Induced Innovation and the Co-Evolution of Technologies and Institutions

The induced innovation hypothesis was formulated in the 1930s mostly to deal with distribution issues. To many observers of economic life it seemed as if technical progress was inherently labour saving and thus likely to cause necessarily unemployment. The first systematic formulation of the induced innovation hypothesis was due to Hicks (1932), according to whom a change in the relative price of any factor of production is 'a spur to invention directed to economising the use of the factor that has become relatively more expensive'. The subsequent use of the induced innovation (Binswanger and Ruttan 1978; Jaffe et al. 2001) hypothesis took place in the production function framework, thus taking into account the effect of the relative prices of the factors of production on factor intensities at an aggregate level. Both a micro-economic and an institutional dimension were lacking. In this essay we are particularly interested in the latter missing dimension. To incorporate it, the induced innovation hypothesis has to be reformulated as follows: Any change in the economic environment induces innovations aimed at compensating that change, that is at developing a technology and a set of institutions that are appropriate to the new environment. The term appropriate technology has been used in the context of the economic development literature, but we are using it here in a more general sense. It is important to point out that the changes of the environment that can induce a change in technologies and in institutions are not necessarily exogenous: for example, during its evolution a technology can create a growing environmental impact that induces innovations aimed at reducing such environmental impact. In this case the change would be largely endogenous to the economic system. It is also important to point out that in this generalised formulation the induced innovation hypothesis acquires a systems dimension. Thus we are studying a system, that is a subset of the whole environment that is constituted for example by a technology, by all the firms using it and by those institutions that influence more directly the technology considered. If we wanted to find out what types of institutions were involved and by what mechanisms it happened we should return to the early phases of the technology and reconstruct the process whereby the formation of particular institutions was 'induced'. The correspondence between a particular technology and its related institutions would constitute a pattern of inducement. The previous considerations do not imply that a technology once created remains unchanged and that in this rigid form it induces the formation of particular institutions. On the contrary, there is clear evidence that technologies created in an initial form cannot diffuse beyond a very limited extent and that in order for their diffusion to proceed further appropriate institutions have to be created. Even after these institutions are created the pattern of inducement does not go uniquely from technologies to institutions. The institutions appropriate to a given technology constrain its future development by excluding particular paths and by encouraging others. During the life cycle of a technology there are inducements both from the technology to its appropriate institutions and from these institutions to the technology. Of course, we can expect that as a technology matures a stable set of institutions is created that remains unchanged during the subsequent development of the technology. During this maturity period only incremental innovations are likely to develop. When formulated in this generalised way induced innovation becomes equivalent to the co-evolution of technologies and institutions. In both cases the important question that arises and to which we do not have any clear answer is: what are the characteristics of the institutions appropriate to a particular technology?

6. Environmental Impact

If the environment were a particular technology the previous considerations would in principle be applicable to it. We could say that the environment co-evolves with a set of appropriate institutions. Of course, a priori we do not know what institutions would be appropriate to the changes in the environment that we are experiencing, but our level of knowledge, or of uncertainty, would not be different from that related to any particular technology. In fact, the environment is far more general than a specific technology, because it can be affected by all technologies, although each technology has a specific way of acting upon it. Thus the environmental impact of any technology can be considered as a change in the external environment of all technologies, change that 'induces' a corresponding change in the technologies themselves and in their related institutions. This generality of the environment implies that it is likely to induce both changes in each individual technology and in its specific institutions, and the creation of general 'environmental' institutions. This situation complicates the problem of working out the patterns of inducement and of co-evolution of technologies, of institutions and of the environment, but does not involve the need to abandon either the induced innovation hypothesis or the co-evolution of technologies and institutions both as relevant metaphors and as general conceptual models. Admittedly, the task to trace the co-evolution of a technology and of its institutions is already complex and the addition of the environmental dimension is likely to complicate it further. In what follows of this essay we begin to scratch the surface of the problem of the co-evolution of technologies, of their specific institutions and of the environment.

For what concerns the purposes of this essay the changes in technology that took place since the beginning of the industrial revolution are of three types:

1. Existing technologies underwent a considerable increase in productivity
2. Qualitatively new technologies were created, producing goods and services completely different from anything that was previously available.
3. Up to a certain time very limited attention was paid to the environmental impact of technologies.

We can then ask ourselves whether a different development path would not have been possible, in which greater attention had been paid to the environmental impact of technologies since the industrial revolution, and whether such alternative development path would have been superior (or not) to the one actually followed. Let us start by observing that according to the so called win-win theory (Jaffé et al 2001) an innovation that increases the efficiency of a given technology can also reduce its environmental impact. This may happen because if the innovation considered reduces the quantities of inputs required to produce one unit of output (and thus also reduce waste) it will simultaneously and automatically reduce the environmental impact of the same technology. If the win-win theory were to apply all the time the problem of reducing the environmental impact of technologies would be greatly simplified. For example, all the policies that supported innovations aimed at increasing the efficiency of given technologies would automatically reduce the environmental impact of the same technologies. A contradiction between efficiency increasing innovations and the reduction of environmental impact would still exist to the extent that the above innovations reduced the unit environmental impact of the technologies considered but raised their level of diffusion by lowering their production costs. Thus a reduction in unit environmental impact could be more than compensated by the concomitant diffusion of the same technologies. The previous story finds a different interpretation when admittimg that productive efficiency and environmental impact are two different and separate dimensions of the same technology. That both improved in the same proportions without causing any distortions in the development of the technologies considered would be the exception rather than the rule. According to Sahal (1985) the ratios of the different characteristics of a given technology are unlikely to remain constant when each of the characteristics changes over very wide ranges of values. If such ratios were to remain constant the technology would become progressively unstable. In other words, technologies do not develop 'proportionally', keeping the ratios of their characteristics constant. The win-win theory would involve a certain form of development and it cannot be expected to be valid all the time, even if we cannot exclude that it can apply to particular situations or periods of the life cycle of given technologies. A further reason for which the win-win theory would not lead to an automatic improvement in environmental impact is the complementarity of increasing efficiency in pre-existing sectors and of the rate of creation of new sectors, that is the rate of growth of variety (Saviotti 1996). To the extent that this complementarity holds, even if any increase in productive efficiency were to reduce the corresponding environmental impact, the growing output variety could easily raise the costs involved in cleaning up the pollution created by the given technologies. For example, the information and co-ordination costs involved in dealing with the waste products of a growing variety of technologies can be expected to grow with this variety. In summary, the win-win theory cannot be expected to apply systematically for at least three reasons:

- Even if the unit environmental impact of a technology were to be reduced by efficiency enhancing innovations the total environmental impact of the same technology might actually worsen due to the increased diffusion following from the higher productive efficiency.
- A simultaneous and proportional improvement of productive efficiency and of environmental impact for given technologies is unlikely to be a general phenomenon, since it would constitute an example of proportional development, a type of development which can only be observed over limited ranges of the variables representing a technology.
- Even if the total environmental impact of a (wide) set of technologies were to be reduced by efficiency increasing innovations, it would still be possible for the increased variety of the economic system to create higher costs of cleaning up the pollution generated by the set of technologies, due to enhanced information and co-ordination costs.

Again, this implies that the win-win theory cannot apply systematically to all technologies and efficiency increasing innovations, but does not exclude that simultaneous improvements in productive efficiency and in environmental impact can be achieved by means of the same innovations in a wide range of circumstances.

From the previous reasoning it follows that innovations that improve other characteristics of a technology do not automatically lead to a reduction in the environmental impact of the same technology. Thus it would have been possible for other characteristics of technologies, for example those linked to productive efficiency or to output quality, to improve while simultaneously the environmental impact worsened. This limited validity of the win-win theory would have been compounded by the fact that up to a point producers were selectively induced not to innovate because the costs of pollution were not borne by the polluter. In those circumstances it would have been rational to reduce other costs more than the costs of polluting. Thus until recently there were reverse inducements to reduce the environmental impact of technologies. Furthermore, environmental impact is a density dependent phenomenon. If the environmental impact of a given technology were concentrated in one or few places where it was not immediately felt and where it could be diluted, then it might not worry anyone except for those who live in the immediate neighbourhood or work in the factories where the pollution is produced. However, as the density of polluting plants increases, the average environmental impact can be expected to rise and to affect the majority of the population, thus inducing the creation of 'general' environmental institutions. These would be institutions aimed at protecting the environment in its entirety, not a particular subset of it. On the contrary, in the early phases of industrialisation only the creation of 'specific' environmental institutions, designed to protect a particular subset of the environment, was induced. Thus the protection of health and safety at work started to be a priority much earlier than general environmental impact. We can see that the pattern of inducement to the creation of environment protecting institutions is linked to the nature of the phenomenon that institutions are expected to control, and in particular here to its density dependent character.

In summary:

- The win-win theory cannot apply all the times. Thus environmental improvement can only be achieved if resources are explicitly allocated to it, and not as the unintended outcome of the allocation of resources to other objectives. However, the win-win theory can be profitably used to facilitate environmental innovation.
- In the past producers were selectively encouraged to neglect environmental innovation relative to other aspects of technologies by the lack of internalisation of environmental impacts. The fact that the polluter did not pay amounted to a reverse induced innovation, slowing down environmental improvement relative to other dimensions of technological performance.
- Environmental impact is a density dependent phenomenon, affecting only limited subsets of the population in its early phases, and beginning to be generally felt only after the diffusion of plants raises their density above a value that affects the majority of the population. The pattern of institutional inducements, leading to the creation of institutions protecting specific subsets of the environment and groups linked to these subsets in the early phases of industrialisation and proceeding to the creation of 'general' environmental institutions in later phases, reflects the basic nature of the phenomenon.

The combination of these three factors explains why we could not have expected the environmental impact of technologies to improve automatically as other features of the same technologies progressed.

If in spite of these circumstances producers had been persuaded to reduce environmental impact more than they actually did in the course of time the likely outcome would not have been a comparable rate of growth with a cleaner environment, but a possibly cleaner environment with a lower rate of growth. Environmental innovation would have been achieved by allocating resources to it, thus reducing progress in other dimensions, for example production cost. A lower rate of growth of productive efficiency could easily have entailed a lower rate of output growth. The existence of a trade-off between environmental improvement and other dimensions of technological performance does not imply that the industrialisation path actually followed was optimal. It simply means that an earlier reduction in environmental impact could have been achieved only by subtracting resources from other objectives and thus slowing down rates of output growth. These considerations do not take into account the learning and synergistic effects involved in economic development. The existence of these effects can lead to both the rapid emergence of new sectors and phenomena and to their premature extinction. Thus alternative industrialisation paths do not involve only slower growth but also the failure of industrialisation to take off. A judgement about the relative desirability of the actual and of alternative paths followed towards industrialisation would require considerations about welfare and about its distribution among members of society. The considerations in this paper simply amount to saying that the actual industrialisation path followed was created by a set of inducements and that as a result of these it was not wholly irrational. The same considerations do not exclude that an alternative path more respectful of the environment would have been possible, but they imply that a path preserving the same rates of growth while providing a better environment would not have been easy to achieve.

If we were to take into account the environment as a further dimension of economic development in addition to technology and institutions, then we should say that the environment did not co-evolve with technologies and institutions. For a long time, due to the combination of the fact that no one was responsible for it and that and that environmental impact is a density dependent phenomenon, limited attention was paid to the environment. The first and early exception, which confirms the density dependent nature of environmental impact, was constituted by health and safety at work. General institutions emerged later.

Several institutional development paths are in principle possible. For what concerns direct institutions, we can expect all existing firms to be affected but also new specialised firms to emerge. Furthermore, although some factors affecting the formation of environment-related institutions are likely to be similar in various regions or countries, other factors are likely to differ. The local mixture of activities, population density, political and cultural factors are likely to be additional determinants of environment related institutions.

7. Summary and Conclusions

The co-evolution of technologies and institutions received lately a considerable attention. The meaning of the problem, however, is not entirely clear. Technologies cannot exist without institutions, except possibly in their very early and emergent stages. Thus it does not make any sense to think about the co-evolution of two entities, one of which, technology, is a-institutional. In this sense the co-evolution of technologies and institutions might seem a trivial problem. The important dimensions of it are the types of institutions that are required for the development of a technology and their dynamics. In this paper some taxonomic considerations about the institutions co-evolving with technologies are developed. Thus in addition to firms, the direct institutions of a technology, there are regulating institutions, providing co-ordination, infrastructures, and the institutions providing complementary inputs. This list is not complete, but it provides an example of the type of taxonomic work required to place work on co-evolution on a sound analytical basis. Technologies can then be expected to evolve within institutional networks whose composition is to be determined more systematically. The dynamics of these networks is essentially initiated by the emergence of new technologies, but the subsequent development is jointly determined by technologies and institutions.

Thus, if the output variety of the economic system increases in the course of time (Saviotti 1996) we can expect the institutional variety will have to increase as well, although not necessarily by the same amount.

Two important general points that can be made about the co-evolution of technologies and institutions. First, the emergence of new technologies increases the division of labour in the economy, but in the meantime creates new co-ordination problems. One of the roles of co-evolving institutions is to provide co-ordination. Second, although the firms producing and using the new technologies compete, other co-evolving institutions are in a complementary relationship with the main technology. Infrastructures and institutions providing complementary inputs are examples of this type of relationship. If we consider the system of co-evolving technologies and institutions a network, we can see that the number of complementary links within the network is likely to be a large percentage of the total links. While competition is widely studied in economics, the attention paid to complementary interactions is much more limited. These considerations on the co-evolution of technologies and institutions suggest that greater attention should be paid to complementary relationships.

Co-evolving institutions determine the pace and scope of the development of technologies. Of course, the potential of the technology itself is also an important determinant, but at constant potential institutions can indeed play a very important role. We can imagine their effect by reference to a diffusion curve: institutions determine the slope of the rising part of the curve and the percentage of the population of potential adopters that adopt at saturation.

In the end the co-evolution of technologies and institutions needs to be studied in a systems framework. The dynamics of the networks of co-evolving institutions and technologies is very poorly understood, although it can have very important consequences. Examples such as motor cars or biotechnology prove how such networks have been important in the past and how they are likely to influence future development.

What can be said on the basis of these considerations on the environment? We could think of self-sustaining development as the co-evolution of technologies and the environment. In the past, let us say since the beginning of the industrial revolution, there was no co-evolution. The environment was considered a bottomless sink. Of course, environmental consciousness has greatly increased ever since and no-one doubts the need to take care of the impact of technology on the environment. In terms of the previous analysis this co-evolution will necessarily be the co-evolution of several types of institutions, firms, regulating institutions etc. With respect to previous historical experience we could expect this to be a more general process than even the construction of a techno-economic paradigm, since the integration of environmental preoccupations is a more general process, not limited to some sector of the economy. At least a part of the institutions required will have to provide co-ordination over many or all the sectors of the economy and will have to straddle international boundaries. Thus we can expect it to be a long lasting process. A number of possible paths are in principle open for the development of environment-related institutions. For what concerns firms, either existing firms could incorporate pollution control devices in their products and processes or new, specialised firms could emerge. The responsibility for the regulation and control of environmental impact is likely to be shared amongst sectoral, general, regional, national and international institutions. Institutional development paths in this field as elsewhere cannot be expected to be unique. Although the physical nature of the environmental impact is likely to provide similar inducements in different countries and regions, there are other determinants of the creation of environment related institutions. The local mixture of activities, population density, cultural and political factors are important additional determinants of institutional evolution in this field. Thus we can expect patterns of development of environment-related institutions to differ amongst countries and regions. The study of these patterns could be very useful to improve our understanding of the co-evolution of technology, institutions and the environment.

The Management of the Co-Evolution of Technical, Environmental and Social Systems

René Kemp and Jan Rotmans[1]

1. Introduction

Environmental problems of pollution have been countered quite successfully through the use of control technology and cleaner processes at the production side. At the consumer side little has changed. People still engage in the individual use of motorised transport and energy-intensive life styles. The common explanation for this is that people *want* automobility, cheap energy and cheap food. Such an explanation assumes that people preferences are fixed and that the system is geared toward satisfying these. It fails to see that people want many things; that consumer choices are restricted by supply choices, and that user benefits may conflict with societal benefits. Supply and demand not only interact but also interlock. Their interaction gives rise to particular trajectories, which are sustained by industrial interests vested in it, assumptions about user needs and high costs of making a system change, both for the actors concerned and society of large. The above helps to understand why most change is incremental, aimed at exploitation rather than exploration (March 1991).

Environmental policy has been unsuccessful in changing behaviour and bringing about societal transformations, involving a change in both technology and behaviour. There is a consensus that the existing trajectories in transport, energy, and agriculture are not sustainable, but the alternatives are not clear or deemed unsatisfactory by experts. There is a conflict between short-term goals of policy and the long-term change needed for sustainability. Whilst the goal of sustainable development has been accepted there is a paucity of concepts and tools to work towards it. This paper offers an approach to further sustainability goals. We have called this approach transition management because it aims at managing the processes of co-evolution that make up a transition. Transition management consists of the management of phases of a transition in a reflexive, iterative and stepwise manner. Dutch policy makers in the new national environmental policy plan *Een wereld en een wil* (A world and a willing) have embraced it and made it official policy. The plan uses 2030 as a time horizon. In this paper we explain the notion of transition management and explain why it is a useful model for managing processes of co-evolution and transitions.

---

[1] The paper draws on joint work of the authors with Marjolein van Asselt, Frank Geels, Geert Verbong and Kirsten Molendijk for the 4th Dutch National Environmental Policy plan (NMP-4).

2. The Need for Transformation or System Innovation

The accumulation of stock pollutants and ever-increasing scale of economic activity, undoing environmental improvements per unit of output, call for system innovation. End-of-pipe solutions and other types of changes at the supply side will not suffice. We need more comprehensive responses, involving a change in production chains, in product-service systems, and the ways in which we consume and live (Kemp and Soete 1992; Kemp 1995; Weterings et al. 1997; Weaver et al. 1999; Vellinga and Herb 1999; Ashford et al. 2001). In the vocabulary of innovation studies, we need system innovation besides system improvement. System innovation in the sociotechnical realm involves changes in sociotechnical systems beyond a change in (technical) components. It is associated with new linkages, new knowledge, different rules and roles, a new 'logic of appropriateness', and sometimes new organisations.[2] System innovation usually consists of a combination of new and old components and may even consist of a novel combination of old components, as in the case of industrial ecology – the closing of material streams through the use of waste output from one company by another.

Two other examples of system innovation offering environmental benefits are: the hydrogen economy (with the hydrogen generated in clean ways, for instance through the use of renewables); and integrated mobility (or chain mobility). In the latter case, people are using different transport modes (collective ones and individual ones such as a car and bicycle) using information services from mobility agencies that offer them travel plans and make reservations. Chain mobility involves a wide range of changes, in infrastructure (in the form of P+R stations and special bus lanes), in technology (such as light rail in conurbations) but also an array of social and organisational changes: the collective ownership and use of cars (car-sharing and riding), the creation of mobility agencies offering and selling intermodal transport services, the integration of collective transport schemes, and the introduction of transport management system for employees by companies.[3]

System innovation transcends a single country or a single continent and goes beyond the use of more efficient manufacturing processes and green products (Vellinga and Herb 1999). The transformation may be beyond those that the dominant industries and firms are capable of developing easily, at least by themselves (Ashford et al. 2001). The time scale for system innovation, one generation or more, is long from a policy point of view. An indication of the time scale and geographic scale of system innovation (vis-à-vis the scales for other types of change) is given in figure 1.

---

2 A related distinction is that between sustaining innovations and disrupting innovation (Christensen, 2000).

3 Three other examples, described in Ashford et al. (2001) are: biomass-based chemistry, multiple sustainable land-use (the integration of the agricultural function with other functions in rural areas) and flexible, modular manufactured construction.

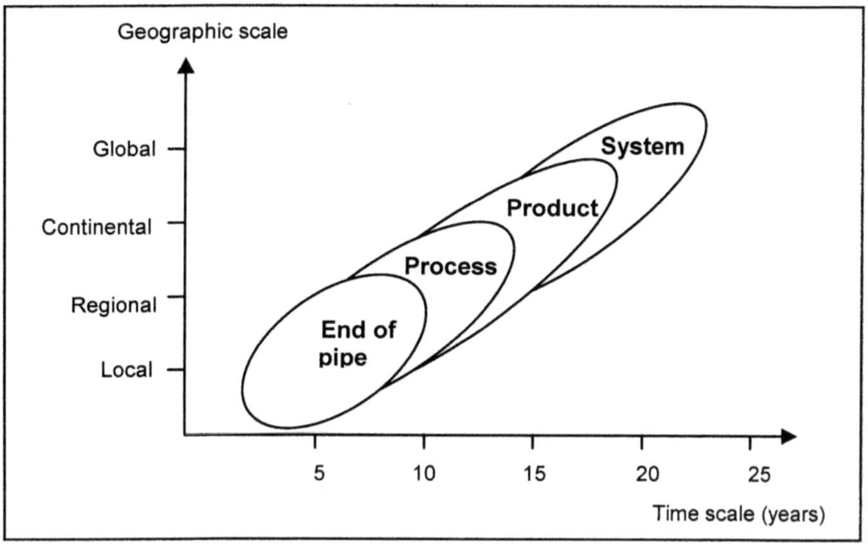

Fig. 1. The time Scale and Geographic Scale of Societal Responses to the Issue of Environment. Vellinga and Herb 1999

System innovation may consist of the development of a new system (such as the development of the grid-based electric system) and the transformation of an existing system, such as the emergence of a regime of chain mobility out of the existing regimes of individual and public transport.

The distinction between system innovation and system optimisation is useful because it forces one to think about the long-term consequences of innovations: whether they give rise to or contribute to system innovation or do not alter the current path of development. An example of such a mapping of innovations (and corresponding policy measures) is provided in Figure 2.

Figure 2 depicts whether the innovations and policy measures to counteract transport problems contribute to system optimization or to system innovation. Personalized public transport such as dial-a-ride services and CO_2 policies are believed to contribute to system innovation in the form of chain mobility with people combining individual means of transport and collective means for their travel needs. Anti-congestion policies are believed to sustain the current trajectory of motorized passenger transport based on the individual use of cars. Some innovations may be part of system optimization and of system innovation. An example is urban cars, which may be used in combination with collective forms of transport or as a 2nd or 3rd household car. The fact that innovations may be used both within an existing system and within a new system is not uncommon for innovations. Such innovations may be called *two-world* innovations and may play an important bridging function within a transition, together with hybrid technologies (Geels and Kemp 2000).

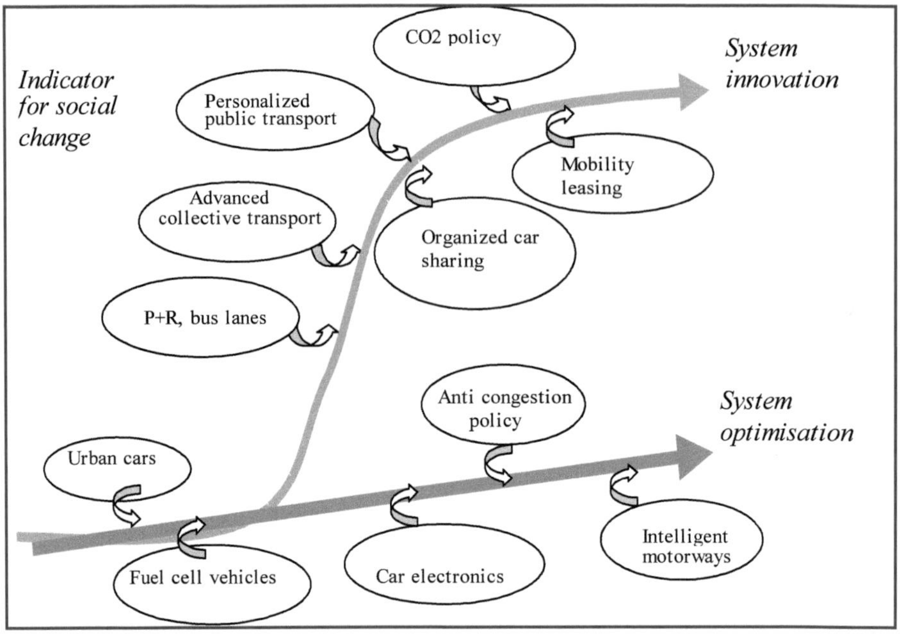

Fig. 2. Systemoptimazation versus System Innovations in Land-based Passenger Transport

3. Transition and Co-Evolution

For the purposes of managing change processes to sustainability it is useful to use the concept of a *transition* rather than system innovation because it brings into focus four things:

1. The end state (new equilibrium);
2. The path towards the end state, made up of different stages;
3. The transition problems that dog the transition process;
4. The wide range of developments internal and external to a particular system that shape the outcomes.

A transition is the confluence of developments that span various systems and domains. A transition consists of a set of connected changes in technology, the economy, institutions, behaviour, culture, ecology and belief systems that reinforce each other. Within a transition there is multiple causality and co-evolution of independent developments (Rotmans et al. 2000, 2001)

Although transitions are characterized by non-linear behaviour, the process itself is a gradual one. The nature and speed of change differs within each of the four stages (see figure 3):

In the *predevelopment* phase there is very little visible change but there is a lot of experimentation

- In the *take-off* phase the process of change gets under way and the state of the system begins to shift.
- In the *acceleration* (breakthrough) phase structural changes take place in a visible way through an accumulation of socio-cultural, economic, ecological and institutional changes that react to each other; during the acceleration phase, there are collective learning processes, diffusion and embedding processes.
- In the *stabilisation* phase the speed of social change decreases and a new dynamic equilibrium is reached.

During a transition there are changes in:

- The speed of change;
- The size of change; and
- The time period of change

Being the three system dimensions of a transition.[4]

It should be noted that the concepts of speed and acceleration are relative. All transitions contain periods of slow and fast development, caused by processes of positive and negative feedback. Within a transition there are no great jumps and it does not occur quickly. A transition consists of a gradual, mostly continuous process typically spanning at least one generation (25 years). This can be accelerated by unexpected or one-time events, for example, war, large accidents (e.g. Chernobyl) or an oil crisis, but not be caused by such single events.

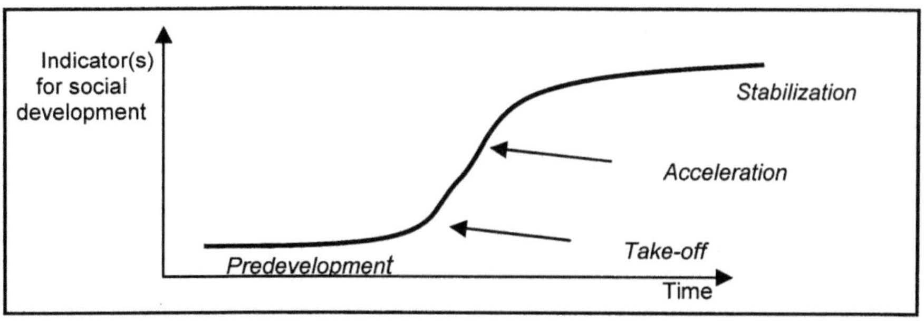

Fig. 3. Four Phases of Transition, Rotmans et al. 2000, 2001

Transitions involve structural change but not everything changes. What changes most fundamentally are the assumptions, practices and rules. Technological changes may be secondary, which is a different way of looking at transitions than most people do; especially engineers are inclined to view technology changes as

4 The *nature* of change can be viewed as a fourth dimension.

primary, and institutional changes as secondary, which are often seen as being forced by technology, overlooking the fact that the technologies are made by people guided by new ideas, a new outlook and a new set of assumptions.

A transition

... is the shift from an initial dynamic equilibrium to a new dynamic equilibrium

... is characterised by fast and slow developments as a result of interacting processes

... involves innovation in an important part of a societal subsystem

A transition is the result of long-term developments in stocks and short-term developments in flows (Figure 4).

Since stocks change slowly, the dynamic pathway of a transition is characterized by an S curve (for example a logistic curve). The developments occur in various domains: technology, economy, social life, culture, nature. Every domain has its own dynamics. Cultures only change slowly, just like ecological systems. Economic changes, may occur very rapidly, price fluctuation being an example. Institutional and technological changes are somewhere in between. The whole picture, therefore, forms a hybrid mixture of fast and slow dynamics. The various time axes shift over each other and constantly influence each other. The slowest processes to a great extent, determine the tempo and the direction of the entire dynamics, i.e. by the developments in stocks.

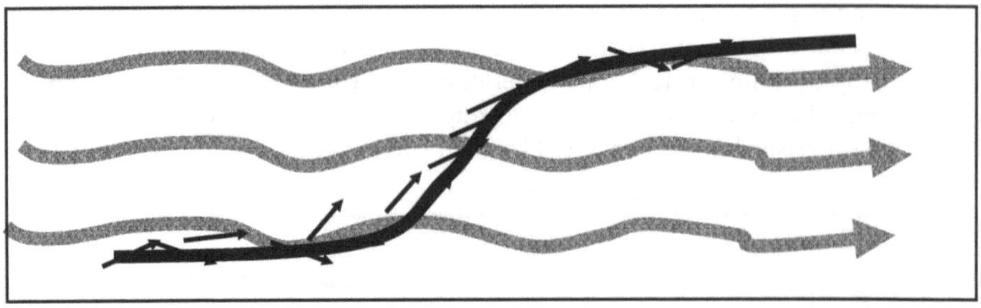

Fig. 4. A Transition is the Result of Long-term Developments in Stocks and Short-term Developments in Flows, Rotmans et al. 2000, 2001

The concept of transition can be used at different aggregation levels. When analysing sociotechnical systems, it is useful to use the multilevel scheme of Rip and Kemp (1998), which makes a distinction between niches, regimes and the sociotechnical landscape. The advantage of this scheme is that it is not based on a concentric view, with the existing system in the middle, but pays attention to the wider context and the dynamics in it: the evolution of macro-variables such as globalization, the evolution of prices and incomes, political changes, changes in policy belief systems and values, regime changes and microscopic changes: the development and use of new technologies in niches, local initiatives, leading to social learning processes that in due course may transform an existing system.

We can't go much deeply into this scheme. For those interested in it we refer to Rip and Kemp (1998) and to Kemp and Geels (2000). An important level is the meso level of *regimes*. A regime refers to the dominant practices, rules and shared assumptions that guide private action and public policy in a field, structuring the behaviour of actors which tend to be geared towards optimizing the system rather than transformation. The distinction between niches, regimes and sociotechnical landscape helps to understand change processes, for example why radical innovations often come from outsiders: because the regime actors are locked into old ways of thinking and old technologies, which leads them to improve existing technologies and strategic action to fight off a new development. But once the regime rules change – for example when regime actors start to see a new development as an opportunity rather than a threat – a reversal of strategy may occur; the regime actors may give a new development momentum through the application of large amounts of capital and organizational and marketing power. When this happens, that is when the mental models that guide key actors change, new developments get momentum and things may change very rapidly.

Niches are the local domains in which new or non-standard technologies are used. The niches may be market niches or technological niches, protected places. Military demand often afforded a niche for radical technologies. Companies may also create a niche for new products for strategic reasons, as a springboard to mass markets (Lynn et al. 1997), but niches may be created by all kind of actors or simply be the result of the heterogeneity of demand and local circumstances.

The third level is the macro level of the *sociotechnical landscape*. This relates to background variables such as the material infrastructure, political culture and coalitions, social values, worldviews and paradigms, the macro economy, demography and the natural environment, which channel transition processes, and change themselves slowly in an autonomous way. The term landscape refers to the lay of the land with its gradients. The macro landscape channels both micro and meso developments. In imagery terms, changes in worldviews (belief systems) and macro policies (such as agreements in WTO rounds or CFC control policy) may rain down upon the macro landscape, but its contours still dictate their convergence into rivers (figure 5).

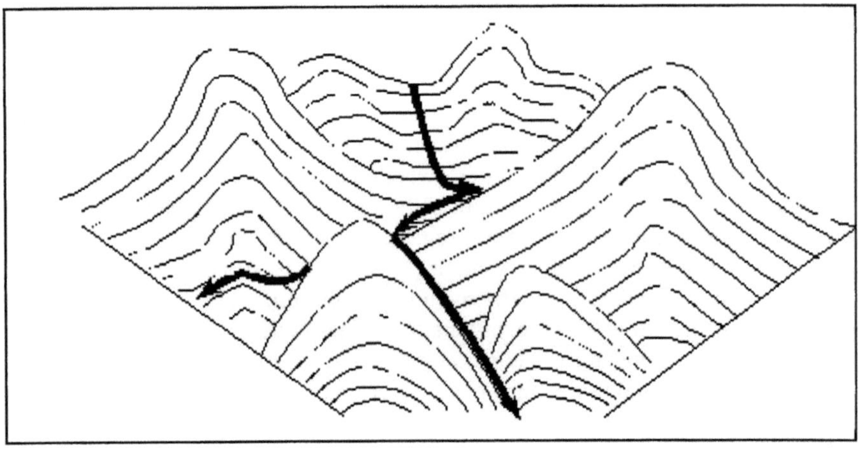

Fig. 5. The Macro Landscape Channels Micro and Meso Developments (from Sahal 1985)

4. Policy Programmes for System Innovation

After the reliance on end-of-pipe solutions and clean technology to deal with environmental problems, some countries have accepted the need for transformation of functional systems (especially agriculture, but also transport and energy). The Netherlands is such a country, acting as a forerunner. The term system innovation is used as a policy concept and several policies exist for it. Before we turn to transition management it is useful to describe these policy initiatives, focussing on what is missing in the programmes. The policy initiatives are: the DTO programme (a research programme for sustainable technologies), the white paper Environment and Economy, NIDO and EET.

DTO is an interdepartmental research programme for sustainable technologies, which ran from 1993-1997. The goal of the programme was to identify and work towards technology options offering a factor 20 environmental efficiency improvements in broad areas of need such as nutrition, transport, housing, and water supply and protection. Industry was an important actor in the programme. Industrial opinion leaders were asked to think about long-term technological solutions offering magnitude environmental benefits. They were selected for their imagination and their position within industry, because the programme wanted to influence the industrial research agenda. Many of the industry people were research directors. In total 25 million guilders (11.3 mln €) was spent under the programme by the Dutch government. The financial contribution from industry was low, about 10% of the costs of the illustration projects, in the form of money and time. The DTO programme led to the development and articulation of 14 illustration processes for sustainability. The 14 illustration processes were:

- Sustainable multifunctional land use (nutrition)
- High-tech agro production (nutrition)
- Integrated crops utilisation (nutrition)
- Novel protein foods (nutrition)
- Underground tube systems for transport (transport)
- Automatic demand and supply management of transport streams (transport)
- Hybrid electric propulsion (transport)
- Fuel cells in mobile applications (transport)
- Mainport Rotterdam (transport)
- Sustainable village renewal (housing)
- Sustainable office building (housing)
- Sustainable chemistry (chemistry)
- Integrated water chains (water)
- Sustainable washing (water)

The project was successful in tapping people's mind and imagination and led to ideas for system innovation and networks of collaboration but failed to influence industries' research agenda in an important way for the reason that the technologies were far from economical. Their use would require a change in the frame conditions giving the sustainable technologies a competitive edge. A 5 million guilder (2.3 mln €) programme of knowledge transfer called DTO-KOV followed the programme but like the original programme this programme did not address the root problem of unfavourable frame conditions. The absence of a pull mechanism frustrated the further development of these technologies and the occurrence of processes of co-evolution resulting in transformations and the creation of new systems.

The second policy initiative is a government white paper about the role of technology in environmental policy. The paper called Environment and Economy came out in 1997 and contained a large number of examples of system innovation offering environmental benefits together with private user benefits. The paper articulated the policy belief that economic growth and environmental protection can be reconciled through the use of innovative technology. It made a call upon localized actors, market actors and local government, to develop these options. An evaluation of the paper and the initiatives in its wake by the CPB said that the paper was successful in giving a sense of direction through the use of 'figureheads' (boegbeelden) but that one also needs generic policies that internalize the environmental costs.

The 3rd initiative is the EET programme, a research programme for breakthrough innovations offering economic and environmental benefits in a time space of 5-20 years.[5] So far 70 projects have been funded (plus 38 KIEM projects, technical feasibility studies). An average EET project has a size of 8 million guilders (3.6 million €) of which half is funded by the government (Willems and van den Wildenberg 2000). The minimal size is 1 mln guilders (0.45 mln €). The total size

[5] EET stands for Economie, Ecologie, Technologie.

of the EET projects funded since its start in 1995 is 529 million guilders of which the government paid 280 million guilders. It is a very large and perhaps unique subsidy programme through its focus on both economic and environmental benefits.

EET complements environmental technology programmes that have a more narrow focus on environmental benefits and that offer little opportunities for system innovation. The focus on radical innovation is good, given the long development times for such innovation and positive spillover effects. A less good aspect of it is that little attention is given to the societal boundary conditions that are needed for the use of the innovations that are under development: the price of energy, whether a energy tax is needed for its use (many projects aim to develop energy efficient innovations), the systems aspects (of complementary technology, infrastructure, skills and so on needed for its use) and the social acceptability. Applicants could have been asked to think about and write about in the application. The programme and selection of projects could also have been linked to transition agendas and to road maps made by industrial actors. As it stands, the EET programme is not really aligned with environmental policy and oriented towards one aspect of system innovation: which is technology.

NIDO (Nationaal Initiatief Duurzame Ontwikkeling) is a programme which for a period of two years supports 'jump projects', initiatives which offer sustainability benefits. It is less technology focused than DTO and EET and more oriented towards practical implementation. The NIDO budget for 2001 is 8.5 million guilders (3.9 million €), which is used to support 4 programmes: *van financieel naar duurzaam rendement about* the coupling of companies' financial performance indicators with companies' ecological and social performance indicators; *duurzame logistiek* which is about sustainable logistic chains; *wonen, leven, werken* about sustainable living and livings, and *waarden van water* about integrated and sustainable urban water management. The private contribution to these projects is 3.5 million guilders (1.6 million €). Apart from supporting the programmes financially NIDO helps participating parties with obtaining additional funds and the dissemination of knowledge. The small size of the projects and short period of support means that for some type of changes (such as the shift to an emission-low energy system or a different type of transport systems) the support from NIDO will be too little to have much of an impact. Like the other programmes, NIDO does not deal with the overall frame conditions. It does explore visions for system innovation.

5. Transition Management

The experiences with the above Dutch programmes (especially DTO) led Dutch policy makers to look for a more integrated and comprehensive approach to work towards transitions. They asked the authors of this chapter to analyse the possibilities for managing transitions, and to come up with a model for transition management. The below model is the result of this project in which we worked in close interaction with the working group responsible for the 4th National Environmental

Policy plan (NMP-4) and a larger group of policy officials.[6] Several considerations informed the model. The most important of these were:

- The need to orient myopia of actors, both business actors and government actors, towards the future and to societal goals;
- The existence of barriers to system innovation, having to do with interests, costs, beliefs and standard assumptions favouring incremental change;
- The need for coordination of fragmented policy fields: Science & Technology policy, economic policy, innovation policy, environmental policy, transport policy and agriculture policy, all of which have a role to play in the transition to a low-emission energy system;
- The need for legitimising policies towards structural change and democratically setting goals;
- The need for opting for an approach of gradual change and learning about a variety of options;
- The need for flexibility both with respect to the goals and paths towards the goals.

Transition management consists of a deliberate attempt to bring about structural change in a stepwise manner. It does not attempt to achieve a particular transition goal at all cost but tries to utilise existing dynamics and orient these dynamics to transition goals that are chosen by society. The goals and policies to further the goals are not set into stone but constantly assessed and periodically adjusted in development rounds.[7] Existing and possible policy actions are evaluated against two criteria: first, the immediate contribution to policy goals (for example in terms of kilotons of CO2 reduction and reduced vulnerability through climate change adaptation measures), and second, the contribution of the policies to the overall transition process. Policies thus have a *content goal* and a *process goal*. Learning, maintaining variety and institutional change are important policy aims and policy goals are used as means. The use of development rounds brings flexibility to the process, without losing a long-term focus.
A schematic view of transition management is given in figure 6.

---

6 The project team consisted of Jan Rotmans, Marjolein van Asselt and Kirsten Molendijk from ICIS, René Kemp from MERIT, Frank Geels from the University of Twente and Geert Verbong from TUE.

7 The idea of development rounds comes from Teisman (2000).

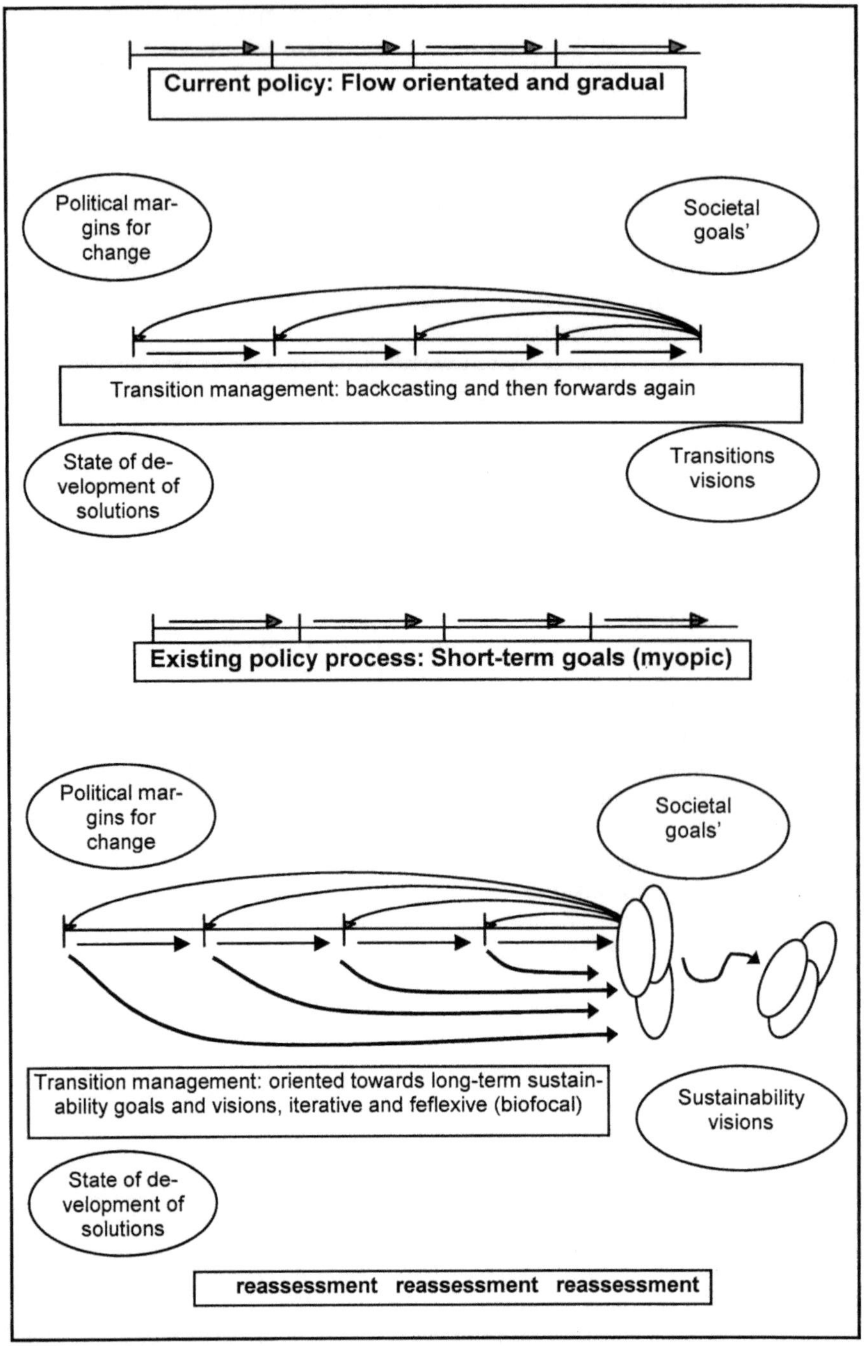

Fig. 6. Short-term versus Long-term Policy

Transition management is based on a two-pronged strategy. It is oriented towards both system improvement (improvement of an existing trajectory) and system innovation (representing a new trajectory of development or transformation). The role of government differs per transition phase. For example, in the predevelopment stages there is a need for social experimentation and creating support for a transition programme, the details of which should evolve with experience. In the acceleration phase there is a special need for controlling the side effects of large-scale application of new technologies. Throughout the entire transition the external costs of technologies should be reflected in prices. The changing nature of policy is shown in figure 7.

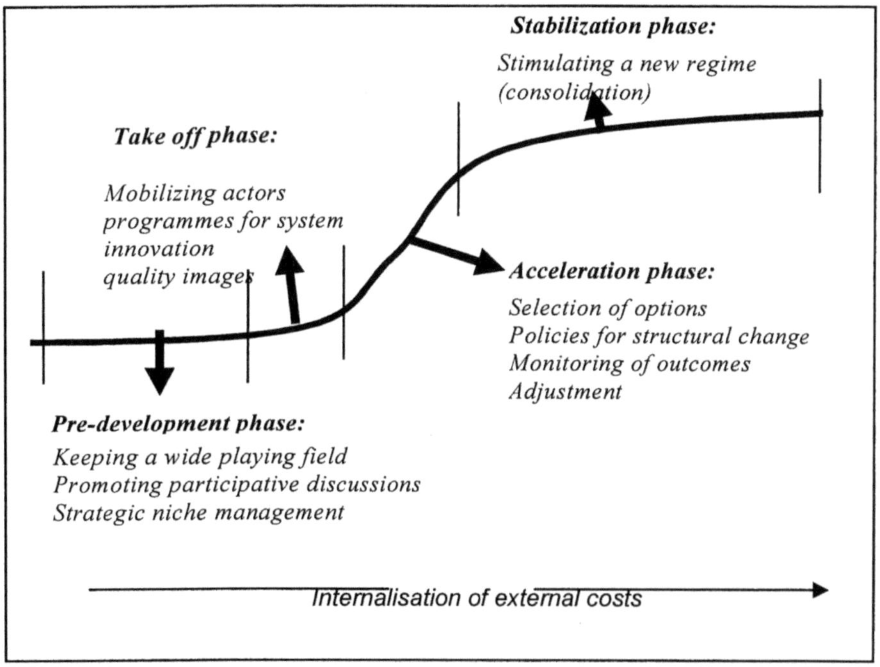

Fig. 7. Role of the Government in various Phases of a Transition Process[8]

Transition management breaks with the planning and implementation model and policies aimed at achieving particular outcomes. It is based on a different, more process-oriented philosophy. This helps to deal with complexity and uncertainty in a constructive way. Transition management is a form of process management

[8] Strategic niche management is the creation and management of a niche for an innovation with the aim of promoting processes of co-evolution. The innovation is used by real users. This helps to promote interactive learning (between suppliers and users) and helps to build product constituencies (which include policy actors). The approach of SNM is described in Kemp et al. (1998a), Kemp et al. (1998b), Kemp et al. (2001) and Hoogma et al. (2001).

against a set of goals set by society whose problem solving capabilities are mobilised and translated into a transition programme, which is legitimised through the political process.

Key elements of transition management are:

- Long-term thinking (at least 25 years) as a framework for shaping short-term policy
- Thinking in terms of more than one domain (multi-domain) and different actors (multi-actor) at different scale levels (multi-level); how developments at one level with one type of actors gel with developments in other domains
- A focus on learning and a special learning philosophy (learning-by-doing and doing-by-learning)
- An orientation towards system innovation
- Learning about a variety of options (which requires a wide playing field).

Transition management does not aim to realize a particular path. It may be enough to improve existing systems, it may also be that the problems turn out to be less severe than at first thought.

Transition management is not an instrumental activity. The actual policies are the outcome of political negotiations and processes of co-evolution which inform further steps, but the basis steps are:

The transition goal

This consists of a basket of images, not a societal blueprint. The transition goal is multi-dimensional and should not be defined in a narrowly technological sense. The goals should be democratically chosen and based on integrated risk analysis.

This will constitute a radical break with current practice in environmental policy where quantitative standards are set on the basis of studies of social risk, and adjusted for political expediency. Risk-based target setting is doomed to fail when many issues are at stake and when the associated risks cannot easily be expressed in fixed, purely quantitative objectives. This holds true for climate change but also for sustainable transport.

Transition management relies on integrated risk analysis and the setting of minimum levels for certain stocks (e.g. health, ecosystem diversity and capital) and aspiration levels. The estimates of various types of risk are subjective, since the risks are surrounded by structural uncertainties, legitimating the incorporation of various perspectives (van Asselt 2000). The net result is a policy corridor for key variables, indicating the margins within which the risks are considered acceptable.[9]

The use of transition visions

Transition management is based on long-term visions that function as a framework and a frame for formulating short-term and long-term objectives and evaluating existing policy. To adumbrate transitional pathways, these visions must be

9 The idea of a policy corridor is described and applied in Rotmans and den Elzen (1993).

appealing and imaginative and be supported by a broad range of actors. Inspiring final visions are useful for mobilizing social actors (such as 'underground transport' and 'multifunctional land use'), although they should also be realistic about innovation levels within the social subsystem in question.

The 'basket' of visions can be adjusted as a result of what has been learned by the players in the various transition experiments. The participatory transition process is thus a goal-seeking process, where both the transition goals and visions change over time. This differs from so-called 'blueprint' thinking, which operates from a fixed notion of final goals and corresponding visions.

Interim objectives

Figure 8 shows the similarities and differences between current policy-making and transition management. In each case, interim objectives are used. However, in transition management these are derived from the long-term objectives (through so-called 'backcasting'), and contain qualitative as well as semi-quantitative measures. In other words, the interim transition objectives contain *content* objectives (which at the start can look like the current policy objectives, but later will increasing appear to be different), *process* objectives (quality of the transition process, perspectives and behaviour of the actors concerned, unexpected developments) and *learning* objectives (what has been learned from the experiments carried out, have more options been kept open, re-adjusting options and learning objectives).

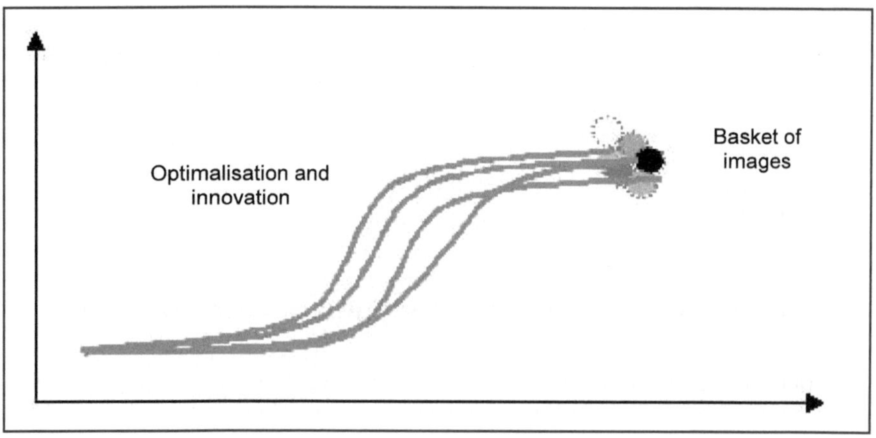

Fig. 8. Multi-dimensional Transition

Evaluating and learning

Transition management involves the use of so-called 'development rounds', where what has been achieved in terms of content, process dynamics and knowledge is evaluated. The actors who take part in the transition process evaluate in each interim round the set interim transition objectives, the transition process itself and the transition experiments.

The set interim objectives are evaluated to see whether they have been achieved; if this is not the case, they are analysed to see why not. Have there been any unexpected social developments or external factors that were not taken into account? Have the actors involved not complied with the agreements that were made?

The second aspect of the evaluation concerns the transition process itself. The set-up and implementation of the transition process is put under the microscope. How do the actors concerned experience the participation process? Is it dominated by certain parties (vested interests)? Is it too consensual (too cosy), or is there too little commitment? Are there other actors who should be involved in the transition process? Are there other forms of participation that must be tried out?

The final issue for evaluation is the amount of learning or 'enrichment' that has taken place in the previous period. A special point of attention is what has been learned from the experiments carried out to stimulate the transition. What have been the most important learning moments and experiences? Have these led to new knowledge and new circumstances? And what does this means for future policies?

Creating public support

A continuing concern is the creation and maintenance of public support. This is important for the process to keep going and preventing a backlash, which may occur when quick results do not materialize and setbacks are encountered. One way to achieve this is through participatory decision-making and the societal choice of goals. But societal support can also be created in a bottom-up manner, by engaging in experiences with technologies in areas in which there is local support. The experience may take away fears elsewhere and give proponents a weapon. With time solutions may be found for the problems that limit wider application. Education too can allay fears but real experience is probably a more effective strategy. Through the prudent use of new technologies in niches, societal opposition may be circumvented.

6. Transition Management in Relation to Current Policy

Transition management should be seen as complementing rather than conflicting with current policy. The concept of transition places short-term policy within a time frame of one, two or even three generations (25-75 years) rather than the maximum of 5-10 years, which is typical of current policy. It is also oriented towards system innovation. Unfortunately, the fruits of technical fixes will contribute more quickly to policy objectives in the short term. An example of this is CO2 collection and storage. Another example is the catalytic converter which helped to achieve reductions in automobile NOx emissions but increased energy use and that did not deal with the many social and economic problems related to car use. Technical fixes are no solution for complex social problems.

This does not mean that transition management rejects the improvement of existing systems as a route towards sustainability. It says that you must aim for both system optimisation and system innovation instead of one of the two. The two strategies are not necessary mutually exclusive: cleaner cars can go hand-in-hand with innovative public transport systems. System improvements may thus act as a stepping-stone for system innovation. Another example is organized car sharing, which facilitates intermodal travel.

A characteristic of transition management when successful is that structural change is achieved in gradually, without too much destructive friction in the form of social resistance or high costs. This is done through the use of hybrid technologies and two-world technologies and exploitation of niches, attractive domains of application. You do not need centralised comprehensive planning for the creation of a new system. It can also be achieved through in a gradual way, by adding for example new elements to an existing system, which facilitate further change.

Transition management tries to utilize the opportunities for transformation that are present in an existing system. It joins in with ongoing dynamics instead of forcing changes. Transition management also implies refraining from large-scale investment in improvement options that only fit into the existing system and which, as a result, stimulate a 'lock-in' situation.

The role of government in transition management is a plural one: facilitator-stimulator-controller-director, depending on the stage of the transition. The most effective (but least visible) is the guidance in the pre-development phase, and to a lesser extent, in the take-off phase. Much more difficult is the guidance in the acceleration phase, because the direction of development in this phase is mainly determined by reactions which reinforce (or weaken) each other and cause autonomous dynamics. It is still possible at this stage to adjust the direction of development, but it is almost impossible to reverse it.

7. The Transition to a Low-Emission Energy Infrastructure

This section applies the idea of transition management to energy supply. It examines the possibilities for managing the transition to a low-emission energy supply system. The development of a low-emission energy supply in the Netherlands makes a good case for transition management. The production, transport and distribution of energy represents an important societal sub-system, of which the services extend into social life. As with any transition, a number of important boundary conditions are set by other domains, which can either slow down or strengthen the transition. The economic domain demands affordability and sufficient economic returns; the socio-cultural domain values health, safety and asks for reliability of delivery; while from the ecological point of view, the risks for nature and the environment are important. Global and European 'landscape' developments have a major influence on the Netherlands' future energy supply.

From a transitional perspective, the transition in energy is still in its pre-development phase. The main unsustainability aspects are: the CO2 emissions contributing to climate change causing rivers overflows and increased sea levels, and the dependence on fossil fuels, making it vulnerable to price changes, which may cause economic problem but also political problems, as was demonstrated by the unrest over high diesel prices in 2000. Alternatives are expensive at their current level of development and seen as longterm options. But deferment of the transition to new energy sources only shifts the problems to later generations, because future options for the energy supply are, to a large extent, determined by current investment in R&D (IIASA-WEC). The SER, an influential advisory board in the Netherlands has stated that the energy infrastructure must change fundamentally in the long-term.

The perceived unsustainability of the existing energy system by all the policy actors and the Dutch commitment to the Kyoto protocol are drivers for change, but there are many obstacles to an actual transition. One important hindrance is the overproduction of fossil fuels, leading to low energy prices. A second obstacle are the interests of the oil companies in oil and gas, a powerful policy actor with great financial resources. Although they claim to be investing in alternative sources of energy, they fear a lock-in, and are scared of placing all their eggs in one basket (i.e. choosing the 'wrong' energy technology). As a result, the current energy producers and users causing the CO2 emissions, have no real incentive for change. Finally, there is no groundswell of popular support for a change in sources of energy. In these circumstances, how can a low-emission energy system be developed through transition management, what kind of difference does it make?

8. Energy Transition Management

An essential element of transition management is the selection of a collective transition objective. This objective needs to be multi-dimensional, and not only quantitative. From the socio-cultural viewpoint, safety and reliability of delivery are important requirements. The ecological risks might be specified in CO2 concentrations. A low-emission energy supply is often translated in terms of CO2 reductions, of the order of 50% of 1990 levels, to be realized over a period of 50 to 100 years.

The second step concerns final visions of energy transition. A recent study by the Dutch Energy Centre, ECN, articulated three visions for the future of the Dutch energy supply:

1. Status quo: In this vision the current energy infrastructure remains intact, but final energy fuels are made from renewable energy resources (solar, wind and biomass). Oil, methane and electricity remain the final energy fuels. There will be more conversion steps, particularly for biomass and coal, where the primary energy fuels are both renewable and 'clean' fossil fuels (use of fossil fuels, with storage of CO2 in empty natural gas fields or coastal seas).
2. The hydrogen economy: In this vision, hydrogen is the dominant final energy fuel, particularly for industry, transport and built-up areas. This requires a thor-

ough adaptation of the current natural gas network, so that, for example, cars are able to run on hydrogen.
3. The all-electric society: Here, the role of electricity as the final energy fuel is dominant in all sectors of society. This requires a fundamental transformation of the current energy infrastructure, including a large-scale electricity network in order to allow cars to run on electricity, for example.

These three final energy visions are not mutually exclusive, and each combines centralized with local systems of power generation. They are, however, purely technological in their perspective. Real transition final visions must have a social dimension. The social, cultural, institutional and environmental contexts of a transition must be considered carefully if the process is to attract the support of actors involved.

The ECN analysis suggests that all three final energy visions may lead to the desired 50% reduction in CO2 emissions, but only if they are followed scrupulously. The roles of renewable energy sources (solar, wind and biomass) and clean fossil fuel energy in each final vision are clear; what is not so obvious is how much all the visions continue to rely on nuclear power and the parallel development of energy-saving technology. One thing that is clear is that the biomass for energy cannot be produced in the Netherlands. To produce the biomass alone for the first vision would require the entire land of the Netherlands to be used for growing energy crops.

It is difficult to make judgements about the viability of the various options, as costs were not estimated. At first sight, the status quo final vision offers a lot of advantages, since the existing infrastructure can be preserved, although an exorbitant quantity of biomass is required. The hydrogen society final vision has the advantage that it can be entirely CO2-free. Furthermore, there is considerable enthusiasm for such advanced technology. On the other hand, such a fundamental changeover would require a great deal of time and effort. The electrical society final vision opens up the prospect of a gradual transfer to low CO2 emissions, or possibly even a CO2-free energy supply in the long-term. There is, however, not a great deal of enthusiasm for this, partly as a result of the risks (breakdowns, disasters) and the way in which it could sideline a number of innovative technologies presently in development.

Formulating interim objectives is the third step of transition management. This allows us to describe the various transition paths behind the final energy visions. Linking the chosen final energy visions to the various transition paths can outline a transition management strategy. If we look at the characteristics of an energy path, a couple of things catch the eye. Firstly, there is no one-to-one relationship between the transition path and the final transition vision. Secondly, the energy transition is not a series of jumps, but a process of gradual development.

Given the present uncertainty about which option is best, all final visions must be kept open, at least for the time being. It may take decades for a technology winner to emerge (see figure 9).

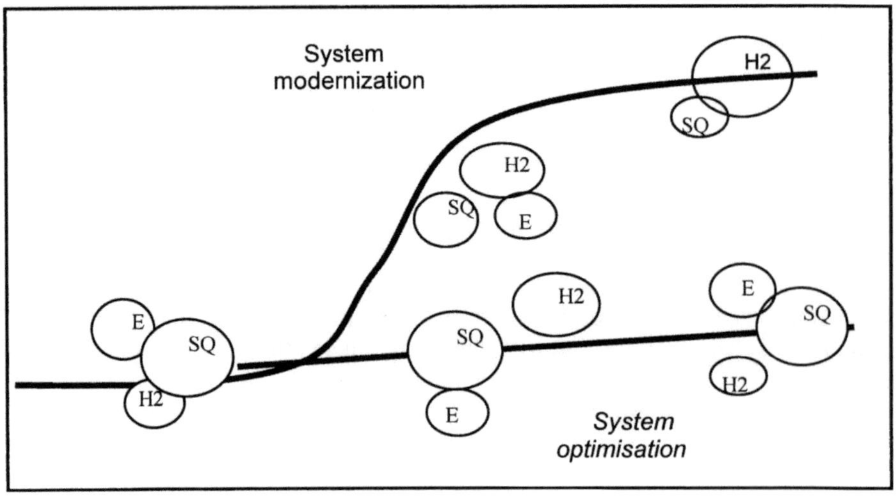

Fig. 9. Keeping open Transition Images in the Course of Time, Rotmans et al. 2000, 2001

The other options then gradually disappear from the picture, although a hybrid always remains possible. Though the rise and fall of options is evolutionary and largely autonomous, it is not outside the control of government. Even within a continuously changing economic, technological, environmental and institutional context, a strategic policy towards system innovation can refocus or redirect the transition.

The Netherlands' current policy is orientated towards observing agreements such as the Kyoto Protocol in 2010. But neither the Kyoto policy nor the proposed, tighter Kyoto+ policy is an example of energy transition management. A great deal of the CO2 reductions will be achieved abroad through low-cost options that do not contribute to system innovation.

The Netherlands could achieve CO2 reductions of approximately 13% in the period 2010-2020, according to the ECN report (making final CO2 emissions in 2020 approximately 6% lower than in 1998), but only by a Herculean effort. Unless accompanied by structural change in the energy infrastructure, it would require massive use of renewable energy and enormous investment in energy saving. Yet this seems to be the way the country is headed. With the focus on the medium term (reaching no further than 2020), there is little sign of change to the current energy infrastructure, based on oil, gas and electricity.

Not only does this reduce the time available to real change from 50 to 30 years, it effectively locks out two of the three transition visions: the hydrogen and electricity societies (see figure 10). Nothing is turned upside down, there is no forced change to the energy infrastructure. Promising alternative energy options are locked out. A transition may still be possible but one does not really prepare for it.

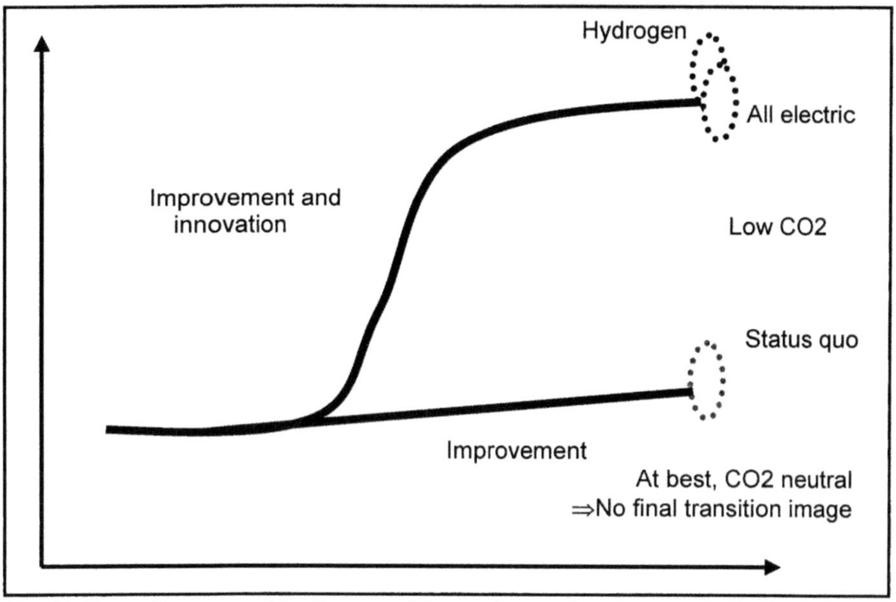

Fig. 10. Kyoto Process and the Process of System Innovation, Rotmans et al. 2000, 2001

The value of transition management is that it does not choose for one solution and also does not let time choose. Transition management does not attempt to choice the best path but attempts to learn about various options and to modulate dynamics towards societal goals. An energy transition policy contains the current climate policy, but adds three things to it: a long-term vision, an impulse for system innovation, and a framework for aligning short-term goals and policies to long-term goals.

However, our analysis also shows that it won't be easy to realise such an energy transition. Apart from the overall frame conditions that should change, it requires a double role of the government. In process terms the government has to facilitate the transition process, whereas in terms of contents, the government has to inspire the other social actors, by giving direction. The guidance for the process of a transition will require a different form of participation, however, with new actors. Via a process of so-called *niche participation*, new players who are as yet insignificant but who may become important in the future should become involved in the process. These actors may be brokers for renewable energy, communities for sustainable energy lifestyles, or producers of new energy technologies. In organizing the transition process, the government can form an interdepartmental body or create an external entity of private and public decision makers responsible for transition management. The details of this need to be further worked out.

9. Summary and Concluding Remarks

There is a convergent view that several of the present trajectories of development are not sustainable and require fundamental change. This chapter has described a method for managing the change process. We have called this method transition management because the challenge of sustainability involves the management of transition problems: the costs of adaptation, resistance of vested interests, and uncertainty about the best option. Through transitions environmental benefits may be achieved, by shifting to new systems that are inherently more environmental benign, but transitions may also produce wider sustainability benefits in the form of preservation of natural capital, health protection and social well-being.

Transition management

... is a collective, cooperative effort to work towards a transition in a flexible, stepwide manner, utilising dynamics and visions

... involves a wide range of policies with their choice and timing gauged to the particular circumstances of a transition

... involves system innovation and system improvement

Although transitions cannot be managed, one can work towards them. This is what transition management attempts to do. Transition management consists of a deliberate attempt to bring about structural change in a stepwise manner. It tries to utilise existing socio-technical dynamics and orient these dynamics to transition goals that are chosen by society. The goals and policies to further the goals are constantly assessed and periodically adjusted in development round. Through its focus on long term ambition and its attention to dynamics it aims to overcome the conflict between long-term ambition and short-term concerns.

Transition management is based on a two-pronged strategy. It is oriented towards improving existing functional systems (system improvement) and towards system innovation to meet the transition goals. Policies for system innovation are adaptive and time-limeted. The role of government in transition management is a plural one: facilitator-stimulator-controller-director, depending on the stage of the transition.

The value added of transitions management is that it orients myopic actors to the future and to societal goals, that it creates societal support for a transition (resulting in a transition programme which is politically legitimised) and commits societal actors to change. It provides a basis for coordination of public and private action. It does not fix a path but explores various options.

In our view, transition management offers a promising alternative for a planning and control approach and the use of economic incentives that both suffer from serious problems: economic incentives are likely to be too weak and probably too general to promote system innovation whereas a planning and implementation approach is likely to be disruptive, by failing to include the multitude of microconcerns at the decentralized level. It is a different type of governance model, not an instrument.

Transition management involves a change in policy making, which is oriented toward long-term goals of sustainability (instead of short-term goals), to system innovation and to new actors. Transition management is not something consensual. Transition management does not exclude the use of control policies, such as the use of standards and emission trading. We need corrective policies besides push policies. The policies can be chosen and legitimised as part of the transition endeavour or independently from it. For example the use of CO2 taxes and other types of economic incentives can be legitimised by the economic principle that one should internalise external costs. The introduction of corrective policies will not be easy. Perhaps the commitment to a transition facilitates their introduction. We don't know. Perhaps it will forestall the introduction of taxes. We have to see. Transition management is not a panacea for every problem but a promising perspective. Two of ist great advantages are that it may be used to achieve a greater coherence in policy and in societal actions for sustainability and that it also is *do-able* at least in a country such as the Netherlands.

Technological Regimes, Environmental Performance and Innovation Systems: Tracing the Links

Frans Berkhout

1. Introduction

There has been a growth in academic and policy interest in the notion of systems-level technological changes (regime shifts, systems innovations) that promise to bring about radical improvements in environmental efficiency (cf. Frosch and Gallapoulos 1989; Vellinga et al 1998). Perhaps the clearest example is the debate about shifts to low-carbon energy economies as a way of mitigating climate change (Royal Commission 2000; Grübler et al 1999). This debate stems from three kinds of conviction: that current patterns of economic development are environmentally unsustainable; that these patterns of development are nevertheless deeply entrenched by technological, economic, institutional and cultural commitments; and that alternative technological and institutional configurations can be designed that will deliver both environmental and economic benefits over the longer term.

The notion of regime shifts raises a number of profound questions. Can regime shifts be induced or stimulated? Is it possible to have foresight about their outcomes (economically, socially and environmentally)? Are they governable, or do they possess some inherent and autonomous inertia? If regime shifts can be induced, foreseen and managed, then the problem for policy is to formulate and implement a strategy that will encourage the innovation of new and known technologies, and to create around these technologies institutional frameworks that will enable their broad and effective diffusion. With this characterisation of the problem in mind, a regime shift is a goal-oriented system innovation carried out at a large-scale. Almost by definition, such processes of innovation are unlikely to emerge from existing market conditions and relationships. An innovation policy would therefore be central to a regime shift.

However, if regime shifts cannot be induced, if their outcomes are uncertain, and if, where they occur, they are substantially autonomous in their dynamics, the set of policy prescriptions would be very different. Instead of formulating and implementing strategy, the aim would be to seek to adapt and adapt to emergent features of new technological and institutional forms as they unfold. That is, rather than moving along a planned route towards a predetermined destination, the aim would be to incrementally follow a path the direction of which is only vaguely known and which may be subject to revision. The debate about technological regime shifts therefore mirrors a much older debate about innovation and strategic management (compare the 'rationalist' school of Ansoff (1965) with the 'emergent' school of Mintzberg (1987)).

This paper is concerned with innovation and environmental performance in technological regimes. The aim is to understand how the full range of technical changes in large, integrated technological systems interact, and how they influence the shape of these systems' environmental profiles. We also investigate the way in which expectations about future alternative trajectories of change may influence environmental profiles. We are concerned with two interlinked questions:

- How does technical change occur in technological regimes, and in particular, how far can 'environmental' factors be seen to have induced these changes? Where can we find evidence that 'environmental' pressures have had a significant impact on the rate and direction of technical change? And, following the arguments of some commentators, are these environmental pressures leading to a more preventive approach to environmental management?
- Is it possible to distinguish ex ante between more and less environmentally desirable trajectories of regime change? Can we employ technology foresight and environmental assessment techniques to describe a clear route of transition for technological regimes?

Drawing on what has been learnt from the combination of qualitative innovation studies and quantitative environmental assessment studies in two regimes (paper and PVC production), we describe a conceptual model that sets out the relationships between different forms of technical change and a range of economic and institutional factors that appear to determine them. We characterise innovation (and environmental performance) in technological regimes as unfolding dynamically out of the interaction of four types of innovation: abatement innovations; process innovations; product innovations; and infrastructural changes. Each of these forms of innovation link to distinct components of innovation systems and have distinct environmental outcomes. We argue that the opportunities and pressures for each form of innovation (and their interaction) are specific to the technological regime and sector. We conclude by reflecting on what might be termed the 'paradox of commitment' (Walker 2000). This is the observation that in order for innovation to take place there is a need for some degree of technological, economic and institutional commitment, but that at the level of the technological regime (as at lower levels of the system) the outcomes of such commitments are emergent and highly uncertain.

2. Framing Environment-Innovation Studies

Innovation studies concerned with the environment are interested in capturing environmentally relevant changes in technology, institutions and the behaviour of market actors. A previous generation of environment-innovation studies were primarily interested in the generation and diffusion of specific 'environmental technologies'. Skea (1995) provides a typical classification of environmental technologies (table 1). For these studies the appropriate frame of analysis was the technological artefact and the management procedure. For these studies the critical problem was to understand how to induce change in these artefacts and procedures. The empirical evidence shows that pressures that induce these types of change come from many sources, including regulators, customers and management routines within the firm (Irwin and Vergragt 1989; Dorfman et al 1992; Groenewegen and Vergragt 1991; Jackson 1993). Few generalisations emerged from this work. To a large extent the neo-classical literature on innovation and the environment has retained this focus on discrete techniques and their innovation and diffusion (cf. Jaffe et al. 2000; Ruttan 2000). Analysis tends to stress the importance of price as an efficient means to induce the innovation and diffusion of specific techniques and processes.

Table 1. Categories of Environmental Technology

Class of technology	Definition
Pollution control	pollution abatement; effluent removal (classic end of pipe techniques)
Waste management	handling, treatment and disposal of wastes
Recycling	waste minimisation through reuse of materials recovered from waste streams
Waste minimisation	production processes and techniques to minimise waste
Clean technology	production processes that give rise to low levels of environmental impact
Measurement and monitoring	sampling, measurement and data analysis
Clean products	products that give rise to low levels of environmental impact through their life cycles

Skea 1995

The environmental technology literature has tended to stress the distinction between abatement (end of pipe) technology, and 'clean' technology (typically seen as novel process innovations). A strong and highly influential argument for a more *preventive* approach to environmental management builds on this distinction[1]. According to this argument, the prevention of waste and emissions is preferable, from an environmental and an economic perspective, to their abatement. By reformulating products, changing inputs and operating production processes, it is possible to avoid the generation of wastes, so avoiding the need for investments in abatement technology. Emphasis is therefore placed on the need for process and product innovation that is oriented at eliminating waste and reducing emissions. If this form of innovation can be induced, or if it is motivated by the competitive advantage that may be gained through associated cost savings, so goes the argument,

[1] The 1996 EC integrated pollution prevention and control directive (IPPC) regime can be seen as a direct policy response to this 'preventive' approach to environmental management.

environmentally-driven process innovations will substitute for innovation in abatement technologies.

The final expression of this position suggests that innovation in environmental technologies will follow a number of phases, recalling earlier 'stages models' of environmental management (cf. Hunt and Auster 1990; Roome 1992). These phases of innovation will be defined by the interplay of regulation, growing innovative capabilities in industry, and more sharply defined incentives in the market (cf. ACOST 1992). In the first phase, starting from a basis of low environmental pressure and low technical capabilities amongst technology suppliers, the primary response of industry to environmental problems is well-established end-of-pipe controls (electrostatic precipitators in cooling towers, for instance). In a second phase, with growing environmental pressure and a greater emphasis on waste reduction and management, specialist suppliers of abatement technology emerge and suppliers of process technologies also begin to compete on environmental performance. In a third phase, integrated, strong environmental pressures from regulation and the market lead to preventive approaches to environmental management becoming a key focus of innovation amongst capital goods suppliers and the market for abatement technologies declines. A clear path from reactive to preventive management approaches enabled by a transition from end of pipe to 'clean' technology is laid out in this 'model' of environmental technology innovation.

While intuitively appealing, there are a number of limitations to this framing of environment-innovation analysis. First, a focus on atomised, micro-level changes in technology is liable to miss dynamics across the wider technological system that may be more significant. For instance, the substitution of a cleaner way of synthesising chlorine may be less significant than the rise in overall efficiency through the growth in the scale of production. Second, studies are faced with the non-trivial definitional problem of distinguishing between a 'clean' and a 'dirty' technology. The definition usually appears to rest on claims made by technology suppliers about how much environmental 'effort' went into the design and configuration of a new technology. What sets an environmental technology apart is therefore the strength of the regulatory or other pressure that can be claimed to have influenced its development, rather than an objective measure of its environmental performance. Where environmental outcomes are measured, a single dimension of performance is typically highlighted (SO_2 abatement in flue gas desulpurisation equipment, for instance). Little account is taken of the broader systems impacts that a new 'clean' technology may have. Indeed, a common feature of early environment-innovation studies was their lack of attention to the quantification of environmental performance. Third, the emphasis on discrete technologies, leads to a focus on new investment and substitution of one technology for another, and a lack of attention to processes of incremental innovation. In many technological systems incremental change is extremely significant, especially since capital turn-over rates are slow. To give an example from one of the case studies discussed later, a survey in 1997/98 of paper machines in the EU revealed that their median age (not accounting for rebuilds) was 32 years (Berkhout et al. 2000). But this slow rate of substitution did not mean that production processes were static, or that their environmental performance remained unchanged.

Numerous small adjustments and adaptations are made to industrial processes which, over time, have a significant influence on the environmental performance of a plant and an industry.

In response to these problems, more recent environment-innovation studies have broadened the scope of analysis. In the 'ecological modernisation' literature, which has emphasised the importance of technological innovation in reconciling economic development with ecological sustainability, there has been a demand for 'meso-level' explanations. In particular, there has been a drive to include institutional contexts and processes into the picture, arguing that the correct focus should be on the co-evolution of technical and institutional innovations (for recent reviews, Mol and Sonnenfeld 2000; Murphy 2000; Anderson and Massa 2000; Saviotti 2003 in this volume).

In more technically-informed 'industrial ecology' literature there has likewise been a shift towards 'systems studies' that aim to understand the resource and environmental profiles of technological systems in the round, typically along supply chains from cradle to grave (Socolow et al. 1994; Graedel and Allenby 1995; Ayres and Ayres 1996). This may be seen as an attempt at an analysis of the co-evolution of environmental systems (services and sinks) and industrial systems - the aim being to understand the total environmental consequences of a given product or service delivered through technological activities. A series of more normative objectives for technological transformation have emerged from this work, with a vision being painted of highly cyclical, solar-powered industrial systems achieved over the longer term.

The more recent concern with systems innovations and technological regime shifts can be seen as emerging from this intellectual context (Kemp et al. 1998). This institutionalist analysis of technical change is concerned with linking between several levels of change - micro-, meso- and the macro - what Geels (2002) term niches, regimes and socio-technical landscapes. Again, the stress is on the co-evolution of technical and institutional systems, the primary difference being goal-orientation. Here clear socio-technical goals are defined through a process of deliberation and systems innovation is managed by integrating adjustments and changes across multiple levels.

3. Change in Technological Regimes

The greater emphasis on technological regimes has changed the terms of the analysis of environment and innovation. The nature, rate and direction of change in a technological regime differs from change in discrete technological artefacts. Regimes are composed of stable assemblages of technical artefacts, organised in co-evolving market and regulatory frameworks. Because of the inter-related and interlocking nature of technological regimes, change is both slower and may be seen as following more predictable trajectories. For a regime change to occur it must be recognised as necessary, feasible and advantageous by a broader range of actors and institutions than would be the case for a discrete technological change.

In general, analysis of change in regimes has therefore tended to emphasise stability and continuity, seeking to explain why competing technological regimes only rarely emerge.

A range of explanations for these processes of technological channelling, path dependence, 'lock in' and 'lock out' have been proposed. Dosi (1988), using the term 'technological paradigm', argued that technological regimes were defined as '...a pattern for solution of selected techno-economic problems based on highly selected principles…'. In this analysis the choice of technical problems is defined by prevailing knowledge and problem-solving heurstics that '…restrict the actual combinations in a notional characteristics space to a certain number of prototypical bundles.' Arthur (1989) argued that learning effects and increasing returns to economic scale would lead to a process of technological 'lock in' that would systematically exclude competing and possibly superior technologies. David (1985) in his famous, though controversial, example of the QWERTY keyboard argued for three factors leading to path dependency in technological change: technical interrelatedness; economies of scale; and quasi-irreversibility. The first and the last of these relate to the 'switching costs' involved in moving from one technological regime to another (Berkhout 2002). A number of other well-known studies use different cases to make similar arguments (Cowan and Gunby 1996; Islas 1997; Leibowitz and Margolis 1999). Finally, Walker (2000) in analysing the persistence of nuclear reprocessing technology in the UK stresses the importance of embedded institutional, political and economic commitments to a particular technological regime identified with a long-term need. He argues that this process of institutional 'entrapment' is ubiquitous in large technical systems where infrastructures contain large and lumpy blocks of capital. Without heavy commitments by key interests defection would be too easy and technological regimes too fragile to develop.

In sum, the literature on change in technological regimes places emphasis on persistence of change along well-defined pathways because the generation of novelty is bounded by working assumptions and procedures inherent to that regime, or because there are a range of institutional and technical barriers to switching away from one regime to another (Berkhout 2002). In the following section, the links between innovation and environmental performance in two technological regimes will be analysed and compared.

4. Innovation and Environmental Performance in two Technological Regimes

The manufacture of two products – rigid polyvinylchloride (PVC) used in construction and coated printing and writing papers – was analysed and compared using a common methodology[2]. These two materials were chosen because they represent commodities that had remained relatively stable over a significant period, so permitting a longitudinal study of innovation and environmental performance.

Two sets of analysis were carried out. The first analysis aimed at understanding innovation and technology dynamics within each sector, disaggregated into major process steps (for example, forestry, pulping, paper milling, paper recovery, deinking and fibre recycling for the paper sector). These studies focused on changes in technology and productivity, but also dealt with changes in industrial organisation, the strategies of firms organised across production systems, and relationships between technology suppliers and users. The second set of analysis used life cycle analysis (LCA), covering identical production systems, to explore the system-wide environmental effects of infrastructural, process, product and abatement innovations that had been documented in the innovation studies. Life cycle models segment production systems into stages that were analysed in the context of the whole technological system. No *a priori* distinctions were made between innovations judged to be 'environmental' and those that were not – all relevant technical changes were included in the analysis[3]. The study took a dynamic perspective with both a 'back-casting' historical review of trends in innovation and environmental performance, and a forward-looking scenario-based analysis of alternative portfolios of technical change. The time-frame for the study was 1980-2010, using 1995 as a base year. The two case studies were carried out in parallel according to a common research design. A matching level of analysis was adopted for the study of innovation and environmental performance in these technological regimes so that the innovation studies and the environmental assessments were tightly coupled. An analysis of the impacts on competitiveness of each of the identified innovations was also carried out, but this is not discussed here.

4.1 The Innovation Studies

The innovation studies had two objectives. First, to establish the drivers, sources, rates and direction of technical and organisational changes in production processes. 'Backcasting' over the period 1980-95 was designed to provide an under-

---

2 These case studies are drawn from the Sustainability, Competitiveness and Technical Change study (1997-2000) funded under the EC's F4 Environment and Climate Programme. The study was coordinated at SPRU, University of Sussex, and partners included the Department of Economics, Technical University of Berlin, and the Institutet for Vatten- och Luftvardsforskning (IVL), Stockholm and Gothenberg.

3 'Incremental' and 'radical' innovations were included, although no attempt was made to classify innovations according to these categories.

standing of underlying technological and industrial dynamics in the two sectors. Second, the studies aimed to develop futures scenarios for production processes (archetypal process routes) to serve as the basis for environmental and policy analysis. Forecasting covered the period 1995 to the 2010s. Four alternative scenarios were elaborated in both the case studies (dynamics as usual, recovery and recycle, eco-efficiency, and pollution prevention, see below). Different sets of technological options were bundled to reflect the specific objectives of each scenario allowing alternative models of process routes to be built up. The main sources of evidence for the innovation studies were primary and secondary literature, and interviews with technologists and researchers in the paper and chemicals industries[4].

4.2 Life Cycle Analysis Models of Environmental Performance

A formal modelling approach was taken to environmental performance assessment. Models of 'archetypal' process routes were developed for both case studies on the basis of existing LCA software (KCL). These model process routes were taken to be representative of EU production systems in the two sectors. Slightly different approaches were taken to model construction and selection of parameter values in the two cases. For the paper case, existing KCL data (developed by a Finnish forest industries research organisation) was modified using new data from Swedish, Finnish and German pulp and paper mills. For the PVC case, no mature and parameterised LCA model was available, and a new model was constructed on the basis of data from a single Swedish facility. This data was compared with data available in the literature from other plants, and modified where appropriate to improve its representativeness and consistency.

In choosing parameter values, a balance had to be struck between the competing considerations of representativeness, policy relevance and data availability. A *hybrid* approach was adopted. For background modules (primarily of energy and electricity production) EU-averages were used. For foreground data (those relating to the production processes themselves), the aim was to use information from existing validated databases and plants that would represent 'good' productivity and environmental performance. Parameters chosen were peer reviewed by industry experts throughout the model development process.

The life cycle inventory models were used as research tools to investigate the environmental impact of technical and structural changes in the two industries. For the PVC case, six model configurations were developed: benchmark processes for 1980 and 1995, and four alternative scenarios for 2010. For the paper case, eleven configurations were generated, including the six listed above. Five additional runs were conducted to take account of different energy and fibre contexts in Scandinavia and west-central Europe.

---

4 In all some 150 interviews were conducted in the period 1997/98.

4.3 The Technology Scenarios

Scenario analysis is a well-established approach for dealing with uncertainty about the future (cf. Ringland 1998). Scenarios were used in the study as a way of providing clear principles for identifying technical changes that might be expected under different future policy contexts. In order to make claims about the impacts on environmental performance and competitiveness of alternative trajectories of technological and organisational change, it was necessary to begin with a 'dynamics as usual' scenario. Under this scenario, a common approach was taken by using existing Best Available Technology (BAT) standards for 2010. In the paper industry case this was defined in recently published IPPC Best Available Technology Reference (BREF) note. In the PVC case these were derived from definitions provided by the European trade association EUROCHLOR in support of the BREF Note, and from specifications produced under OSPARCOM[5].

In each of the three alternative scenarios (eco-efficiency, pollution prevention, recovery and recycling), the aim was to test the technological and environmental implications of pursuing different policy goals. A wide-ranging technology foresight exercise was undertaken. This generated inventories of innovations for each of the major process steps for both paper and PVC. The scenarios were framed by assuming specific technology choices (represented as specific parameter values in the LCA models) that matched a given policy objective, with other objectives given less prominence. Bundles of technologies were clustered according to how appropriate they appeared to address a specific policy goal. The scenarios chosen met three basic criteria: 1) they were applicable across the two case studies; 2) they illuminated current policy choices and 3) they pointed up theoretical debates about the relationship between technical change and environmental performance. The three scenarios were characterised as follows:

1. Eco-Efficiency: Maximising resource productivity was a key goal of technological changes in this scenario. Inputs of materials and energy were assumed to be minimised, regardless of the impacts on emissions and on recycling (in fact, recycling is often consistent with resource productivity). This involved process changes, as well as changed assumptions about the composition of final products.
2. Pollution Prevention: Minimisation of emissions to the environment was the main goal of technological changes modelled in this scenario. Some process changes and full adoption of available abatement techniques were included under this scenario. Input and product composition were retained from the baseline scenario.
3. Recovery and Recycling: Maximum reuse of materials and energy resources was the key goal under this scenario. The main focus was on post-consumer wastes, in-process recovery and recycling being integrated into the eco-efficiency scenario.

---

[5] The Oslo and Paris Commissions under which marine pollution in the NE Atlantic is regulated.

4.4. Technological and Market Characteristics of the Two Sectors

4.4.1 Polyvinyl Chloride

Polyvinyl chloride is probably one of the oldest polymers in modern use. Regnault in France first produced vinyl chloride monomer in 1835 and Baumann first recorded its polymerisation in 1872 when he exposed sealed tubes containing the monomer to sunlight. The earliest patents for PVC production were taken out in the USA in 1912, and pilot plant production of PVC began in Germany and the USA in the early 1930's. A production site was first started in Schkopau in 1938.

PVC is a chlorinated hydrocarbon polymer. In contrast to many other plastics, it is not exclusively based on crude oil/natural gas resources, but contains a considerable amount of chlorine produced by chlor-alkali electrolysis of rock salt or brine. However, while PVC uses less oil, the production and use of chlorine causes a number of environmental burdens. An early draw-back of PVC was its tendency to de-hydrochlorinate at higher temperatures. Not until the discovery of suitable stabilisers could processing technology advance to the point where the full potential of polymer could be realised. Today, by choosing suitable stabilisers and plasticisers, the polymer can be converted into a wide variety of different products. Some of these additives, principally plasticisers (phthalates) and stabilisers (often based on heavy metals like cadmium or lead) may cause environmental burdens during the production and conversion of PVC. Problems may also arise during use phase of PVC and during waste management. The incineration of PVC wastes causes additional environmental hazards, including the generation of dioxin and hydrogen chloride (HCl) during the incineration of PVC containing waste.

PVC production follows a standard production route. Hydrocarbon feedstock is converted by cracking to ethylene (ethene). Sodium chloride is electrolysed as an aqueous solution to produce chlorine with sodium hydroxide and hydrogen as co-products in a process known as *chlor-alkali electrolysis*. The ethylene and chlorine are then reacted to produce 1,2-dichloroethane (ethylene dichloride, EDC) in a process called *direct chlorination* (DC). The EDC is then decomposed by heating in a high temperature furnace (cracking) to produce vinyl chloride and hydrogen chloride in a process called *pyrolysis*. If the process were stopped at this stage, 50% of the input of chlorine would be lost from the system, representing a significant loss of raw materials. In practice, the hydrogen chloride is reacted with further ethylene in the presence of oxygen to produce more EDC in a process called *oxy-chlorination* (oxy). The EDC produced by oxy-chlorination is also decomposed by pyrolysis.

PVC resin can be made by three different processes: suspension, emulsion and bulk (mass) polymerisation. The resins obtained from these processes possess somewhat different physical properties and are generally used in different applications. Suspension PVC (S-PVC) is general-purpose grade and is used for most rigid PVC applications such as pipes, profiles and other building materials. It is also used for most flexible applications such as cable isolation, foils, and various products made by injection moulding. Emulsion PVC (E-PVC) is primarily used for coating applications such as PVC coated fabrics. Bulk or mass PVC (M-PVC) is used for specific types of hard sheets and bottles. Finally, the PVC resin is compounded with different additives (related to the final application) and converted by different processes to the desired application.

Production of PVC is highly concentrated in a few global chemical companies, although the degree of vertical integration varies. Producers of PVC do not all carry out the complete sequence of operations; some buy commodities such as sodium chloride, chlorine, hydrogen chloride, ethylene dichloride (EDC) and even vinyl chloride monomer on the open market and operate only the later stages of the process. Some PVC producers are also engaged in PVC compounding and manufacturing of PVC product.

4.4.2 Paper

Early European manufacture of paper (15th century onwards) was based on non-wood fibres, such as hemp, flax, linen and cotton rags. Industrialisation in the 19th century amplified demand for a non-seasonal raw material, and wood fibres became the dominant raw material for European paper production. First, wood is processed so that the fibre raw material is separated out or *defibrated.* This 'pulp' is then mixed into a water suspension which is sprayed onto a 'wire' conveyor belt, such that the water drains away to leave a 'web' of interconnected fibres. This web is squeezed between rollers and dried so that yet more of the water is removed to leave the final paper product: a web of cellulose fibres. Following use, the paper can be recovered, the fibres separated and cleaned, after which they can be re-used. Modern paper machines are improved derivatives of the first fourdrinier machine built by Donkin in 1807.

In practice, the industrial production of paper and board is complex and highly capital-intensive. Although the main raw material is wood, there are competing process options along the route from wood to paper. Different pulping technologies exist (mechanical and chemical). The fibre suspension which enters the paper making machine can be modified with additives and chemicals, and can be made up from a mixture of different pulp types. The pulp can be bleached by various methods in order to improve brightness of the final paper product. Apart from wood (or non-wood) fibres, paper may contain 'fillers' (kaolin and calcium carbonate) and colours. Fillers are used as a cheaper substitute for fibre and to impart opacity to the paper. Once the base paper has been produced, it may be coated or polished (calendering).

Post-use there are alternative techniques for de-inking fibres in commercial operation (calendaring). The precise mix of raw materials and processes chosen will determine the type and quality of the paper being produced (the main paper grades are: graphic paper; sanitary and household paper; packaging; and others (from cigarette paper to roofing materials)). Even within a given paper grade there is considerable flexibility in the configuration of pulp types and process options which may be available. Figure 1 is a schematic of the main process steps in the paper life cycle.

The structure of paper and board production follows two distinct patterns: integrated or non-integrated. In integrated production, pulp and paper and board production are co-located, usually near to forest resources. This structure is evident in Scandinavian paper production. Pulp production based on virgin fibre may also be separated from paper production that is located closer to final markets for paper. Most commonly, in this non-integrated structure mechanical pulp production remains co-located with paper making, while 'market' chemical pulp is purchased from other producers. Non-integrated mills are more typical in the central and southern EU. Proximity to markets also tends to improve the potential for using recycled fibres. De-inking plants are typically co-located with paper production.

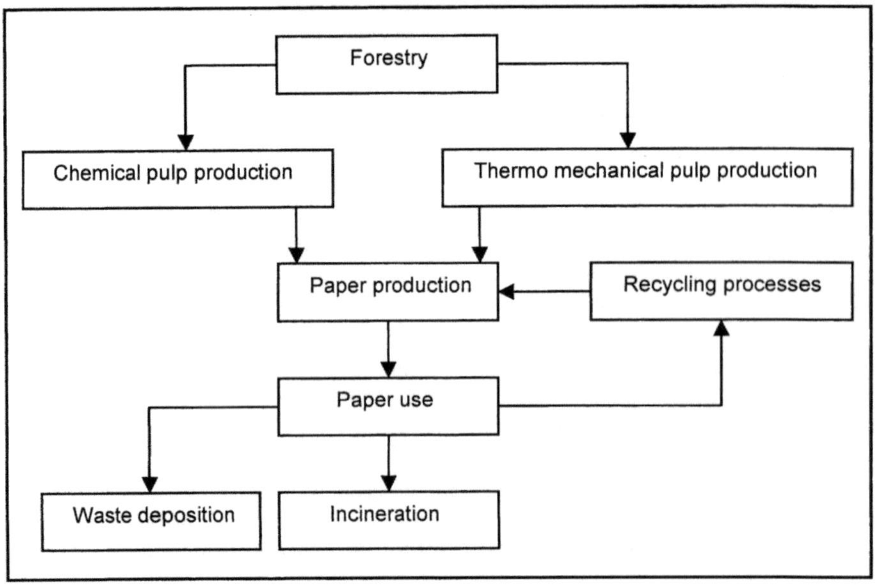

Fig. 1. Schematic of Paper Production System

4.5 Dynamics of Innovation and Environmental Performance: Principal Findings

In describing and analysing the co-evolution of technological innovation and environmental performance in two mature process industries, we were interested in two questions, both related to the question of technological transitions under the influence of 'environmental' pressures. First, we were interested in whether we could find evidence of a transition in the pattern of innovation from 'end of pipe' to clean technology (novel process innovation) as predicted by 'stages' models of environmental technology. Second, we were interested, using scenario analysis, in whether it was possible to distinguish between the environmental performance of alternative bundles of existing or forecast technologies grouped together in pursuit of alternative policy goals: eco-efficiency; waste minimisation and recycling. As explained, these alternative process routes were designed to respond to prevailing policy debates, but they also serve as a more generic 'thought experiment' to test the assertion that it is possible to define an 'end state' towards which alternative (more sustainable) trajectories of technological change may be directed. The discussion of results of the study is organised under a series of six synthetic headings.

Innovation in mature industries takes many forms: The pulp and paper and PVC case studies demonstrate that a wide range of inter-related technological changes occur continuously and in parallel within mature process industries. Most of these are incremental and cumulative changes made to processes, although there are clear gradations between different forms of process change. Some involve more radical changes – the development of commercial-scale de-inking in the paper sector, for instance – while others are more limited procedural changes. Table 2 provides a categorisation of different forms of process change, including an assessment of the types of knowledge required for firms to carry them out. Clearly, the more profound the change, the more costly and risky it will become for producers, and the less frequently they will be made. Some more radical and novel process changes may be avoided by some producers altogether.

But process changes alone do not tell the whole story. The study showed that to a varying extent, product innovation continues to be an important factor in mature price-competitive industries. These product changes involve both quality improvements (better strength and printability characteristics for paper), as well as more significant reformulations that force process changes. A good example here is the demand for elemental chlorine-free paper in the early 1990s that required a new bleaching process to be adopted and a whole set of other incremental adjustments along the production process. Product innovation is generic and continuous, rather than rare and discontinuous. In both sectors a complex and dynamic interaction between innovation in products and processes is observed, with multiple and location-specific changes being made. The prediction of a decline in product innovation made in the classical Abernathy-Utterback model is therefore not confirmed (Abernathy and Utterback 1978).

Table 2. Categories of Process-based Innovation[6]

Category	Knowledge/Vintage	Example
Radically novel	New and comprehensive/total replacement	De-inking technology (paper)
Novel A	Substantial advance/modular replacement	Elemental chlorine-free bleaching (pulp), membrane technology (PVC)
Novel B	Substantial advance/rebuild, retrofit or re-design	Instrumentation and digital control systems (PVC and paper)
Incremental A	Minor advance/replacement	Upscaling of paper machines (paper)
Incremental B	Existing or minor advance/rebuild, retrofit or redesign	Fibre stock recirculation (pulp), system integration (paper)
Incremental C	Existing or minor advance/continuous	New materials in naphtha furnace (PVC)
Incremental D	Procedural or organisational	Environmental management systems

Two further sources of innovation were also highly influential for environmental performance. First, the development of progressively more efficient abatement technologies, and second the background development of energy infrastructures. In energy-intensive industries the configuration of the electricity grid has a profound influence on the overall environmental performance of the final product. As these electricity infrastructures are modernised and periodically transformed through the introduction of new technologies (the rapid diffusion of gas-fired power stations in the UK in the 1990s, for instance), so the environmental profile of products in the whole economy can also be reshaped.

Environmental performance of production systems is influenced by a mix of technological changes. The prevailing conceptual framework of environmental change in industry is a 'pressure-response' model - environmental pressures (whether regulatory or market-driven) influence firms leading to changes in technological choices and management routines. These, in turn, lead to relative improvements in environmental performance. Systematic modelling of two production systems and their environmental impacts has shown that technological changes underpinning changes in environmental performance frequently are not shaped by environmental factors. While there are cases (investments in abatement or novel chlorine-producing processes, for instance) where environmental factors are clearly dominant, there are a range of technological changes with major environmental performance impacts that are motivated by cost-saving or quality changes. Attempts to segregate those changes that are environmentally-driven from those that are not in the manner implied by the pressure-response model is both theoretically and empirically unjustified. All innovation, no matter what the factors shaping it are, must be considered as having potentially positive or nega-

6 Adapted from M. Bell, 'Cleaner Technology: Where does it come from?', SPRU, University of Sussex, March 2000.

tive environmental impacts. There is a need to consider the relationship between innovation and environment in a more integrated way, beginning with an analysis of innovation, rather than with an attempt to isolate 'environmental' or 'clean' technological changes *ex ante*.

Tables 3 and 4 show that, seen from the perspective of the technological system, the key changes underlying the dynamics of environmental performance include investments in abatement, process change, product changes and changes in the background energy mix. Some of these changes were made with environmental performance being a consideration, while many others were not. The results show that the 'green lens' often applied by analysts of the relationship between innovation and environment is unhelpful. By integrating innovation studies with life cycle analysis, methods now exist for moving beyond this framing of the problem.

Table 3. Technology Changes Underlying Environmental Performance Dynamics: Pulp and Paper Production, 1980-95

Indicator	Key technology drivers of environmental performance change
CO2	Background energy mix change
Timber use	Product change (higher filler and recycled fibre content in paper), Process change (fibre stock recirculation)
NOX	Engine efficiency (transport), process change (higher energy efficiency in pulping), background energy mix change
SO2	Sulphur dioxide abatement (pulping)
BOD	Abatement (waste water treatment), process change (heat recovery from organic pulping wastes in mechanical pulp), product reformulation (higher recycled fibre use)
COD	Waste water treatment
AOX	Process change (elemental chlorine-free bleaching)

Table 4. Technology Changes Underlying Environmental Performance Dynamics: PVC Production, 1980-95

Indicator	Key technology drivers of environmental performance change
CO2	Background energy mix, process change (VCM production)
NOX	Process change (naphtha production), abatement
SO2	Sulphur dioxide abatement (low sulphur fuels)
Dioxin	Process change (polymerisation)
Chlorinated hydrocarbons	Process change (closed reactor polymerisation)
Hydrogen chloride	Waste management (on-site incineration of chlorinated VCM wastes)
Mercury	Process change (phase-out of mercury cells in chlorine production)
Cadmium (air)	Product reformulation (phase-out of cadmium stabilisers)
Lead (air)	Product reformulation (introduction of lead stabilisers)

If it is not possible to attach an environmental 'flag' to innovations leading to changes in environmental performance, then the 'environmental pressures-technological response' model proves inadequate. This poses both a theoretical and a policy challenge. The theoretical challenge is to develop a better model. The policy challenge stems from the recognition that even apparently direct instruments like technology-based emission standards are only part of what lies behind the reshaping of the environmental profiles of industries. Other policies, influencing industrial structures and the general technological performance in firms also play a significant role.

Novel process technologies (clean technologies) do not play a dominant role in defining the environmental profile of production systems. The 'clean technology' model implies that, through time, technological opportunities for pollution abatement will decline. In this account, and assuming a constant level of 'environmental' pressure, we would expect to find evidence of a falling significance of investments in abatement and a growth in significant process innovations. In other words, we would expect to find a transition from a focus on abatement innovations towards process innovations.

The study tracked technological changes over 15 years (between one and two investment cycles in both the paper and integrated chemicals sectors). We did not observe a transition in the nature of innovation from abatement to process innovation, even though process-based industries represent a good case for such a transition. Moreover, we found no evidence that opportunities in abatement are being exhausted. New abatement techniques and continuous incremental changes in existing technologies are likely to play a significant role in changing industrial environmental performance in the future, in tandem with process and product changes. Novel process changes do occur: chlorine-free bleaching of Kraft pulp; and new chlor-alkali electrolysis technologies in vinyl production are clear examples. It is also clear that environmental pressures can be the principal drivers of these more profound technological changes. Nevertheless, the fundamental economic constraints on the introduction of novelty remain and this suggests that there is unlikely to be an acceleration of this type of innovation, except in cases where a sudden technological breakthrough is achieved, or a powerful exogenous shock is applied to the industry which brings to market viability new technologies that are currently uneconomic or not viewed as being mature.

Another aspect of novel process innovations is that they appear to provide very specific environmental benefits. Chlorine-related emissions may have been radically reduced through the introduction of new bleaching and electrolysis techniques, but they appear to have had little impact on other dimensions of environmental performance. Integrated environmental performance improvements appear to be achieved through more continuous, incremental technical change. Novelty therefore needs to be seen in the context of incremental change when considering the complete environmental profile of a technological regime. Encouraging novel or radical technologies may be at the expense of broader gains through smaller steps. These results suggest that a more balanced picture needs to be drawn in which all forms of technological change continue to play an often mutually-reinforcing role in achieving environmental performance improvements.

In mature process industries, improved resource productivity lies at the heart of competitiveness. The argument that there can be a correlation between improved environmental performance and competitiveness (the 'Porter hypothesis') rests primarily on the assertion that greater resource productivity brings appropriable economic benefits to producers primarily through lower input and waste management costs. This position suggests that by accelerating the rate at which resource productivity improvements are achieved by, for instance, introducing more novel resource-saving techniques, producers will be able to achieve greater competitive advantage.

While the basic premise of the argument is certainly correct, the inference that there are generally available opportunities for increasing the rate at which innovations leading to improved resource productivity are introduced does not appear to be supported by the evidence. Resource productivity is already a major focus for innovative activity in resource-intensive industries. The scaling-up of production capacity (the growth of paper machines, increased throughput in naphtha crackers and so on) and the adoption of yield and efficiency improving techniques are central to the normal technological activity of all producer firms.

The rate at which these changes are made is determined by the availability of new technologies from suppliers, the length of investment cycles, and the balancing of risk with expected returns from new investments. Although there are different technology strategies available to firms (some choosing a more risky, innovative strategy and others taking a more secure follower strategy), the increments in resource productivity that can be achieved at any time tend to be limited and predictable. This is borne out in the rates of change in some key input parameters for the PVC life cycle, as shown in Table 5.

Table 5. Annualised Rates of Change of Resource Productivity: PVC Production

Time period	Energy	Crude oil	Rock salt
1980-1995	2.07%	0.23%	-2.08%
1995-2010	1.09%	0.15%	-1.67%

Note: Positive rates of change signify improvements in resource productivity over time, negative rates signify declines.

Table 5 also shows another general trend highlighted in the study - the *declining rate* of resource productivity and environmental improvement in PVC production. From our analysis, the rate of environmental performance change in the 1980-95 period will generally not be repeated in the following 15 years, although there are exceptions.

Another illustration of this is expressed in Figure 2 which shows rates of change across key parameters of environmental performance for paper production in Scandinavia for the 1980-95 and 1995-2010 periods. In these models we have used 'base case' assumptions about technology. Only electricity consumption shows a more rapid improvement in the later period. Interestingly, the much higher rates of recycling assumed for the 1995-2010 period do not produce a 'step jump' in fibre use (timber), as might have been expected. Rather, these apparently radical changes (increased paper recovery and de-inking) driven by environmental pressures have merely enabled historical rates of improvement of environmental performance to be maintained.

Another complicating factor is the significance of product innovation. Technologically, producers are facing two options. They can search for ways of saving costs, but they can also search for ways of meeting the market demand for higher quality products. Growing competition, related technological developments (such as those in printing) and changes in market demand have forced producers to become more focused on product quality *and* product innovation. These product innovations can be at the expense of resource productivity, as clearly illustrated in the paper sector.

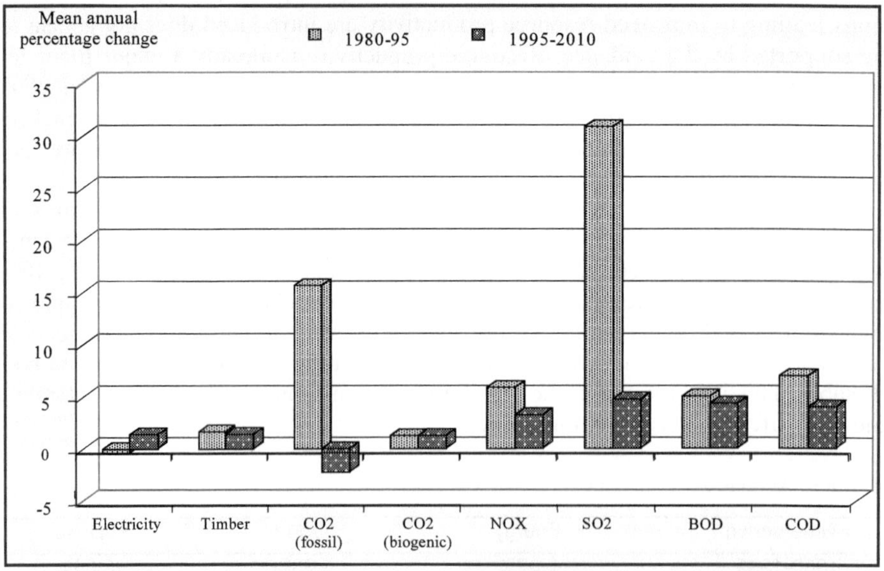

Fig. 2. Environmental Performance Change (Pulp and Paper): Scandinavia 1980-95 and 1995-2010

The results of the study therefore do not support the contention that more rapid improvements in resource productivity are available to firms through the introduction of novel resource-saving technologies. In process industries operating within rather inflexible technological trajectories, and where resource productivity and competitiveness are already strongly aligned, the scope for introducing radically novel resource-saving techniques may be limited. On the other hand, where novel processes have been introduced, the stimulus has frequently been 'environmental' (de-inking and chlorine-free bleaching in paper, and new electrolysis processes in

PVC). Product innovation has become a growing focus for these industries, and may have countervailing effects on resource productivity.[7]

The 'clean technology paradigm' is not borne out in this study: we can identify four 'stylised facts' in the clean technology literature: a) The notion that novel, discreet and identifiable industrial processes – 'cleaner technologies' – will bring step-jump improvements in environmental performance of industries; b) The idea that there will be a *transition* from innovations based on abatement towards innovations based on process change in the search for better environmental performance; c) The assumption that the introduction of novel techniques have a more marked impact on environmental performance than incremental change over time; and d) The argument that accelerated adoption of these novel clean technologies will bring competitive advantage to early movers.

Each of these stylised facts has been considered in the analysis and found to be only partially in accord with the empirical findings. First, on the question of the *identity* of clean technology, there are examples of novel technologies adopted under environmental pressures and with a marked impact on specific dimensions of environmental performance. But these are rare exceptions in a process of technological change that is multifaceted and often incremental. By placing too much emphasis on only one form of technical change, analysts and policymakers risk ignoring other powerful drivers of environmental performance change in industry. Second, we have been unable to find evidence of a *transition* away from abatement and towards novel process changes as a way of achieving environmental performance improvement. While process change is a significant driver of environmental performance change in industry, the reasons for this are not 'environmental' but connected to resource, cost-saving and product innovations. Abatement continues to be an important way for process industries to achieve improved environmental performance *at the same time*. The often-cited dichotomy between 'end of pipe' and process change therefore appears to be false.

Third, there is little evidence that process change really generates more rapid improvements in environmental performance than abatement. Indeed, we find that within dominant technological trajectories (as is the case in the pulp and paper and PVC sectors) process change tends to be rather slow and predictable, and that this will be mirrored in the rates of improvement in resource productivity and pollution burdens. By contrast, the development of new abatement technology can dramatically and rapidly alter certain aspects of the environmental signature of an industry. Lastly, while it is clear that resource productivity is a key factor in the competitiveness of mature process industries, it appears that opportunities to seek competitive advantage through the adoption of novel process technologies leading to environmental performance improvements may be limited. Fundamental uncertainties about the economic benefits of adopting more novel techniques will tend to constrain their diffusion.

Distinguishing between the environmental performance of alternative future technological pathways is difficult: a basic objective of the study was to seek to

---

7 Stronger, more opaque fibres required for lighter grades of paper require more refining of wood fibres in mechanical pulping processes, raising energy use over time.

identify whether alternative technological choices applied to a production system would lead to distinct and quantifiably different environmental outcomes. By conducting a scenarios exercise for the 1995-2010 period, we hoped to be able to distinguish the environmental effects of different trajectories of technological change, represented by alternative bundles of technological options. We posed the question: is it possible to identify a set of technical changes that would bring about a preferred set of environmental outcomes?

The major modelling effort did not yield conclusive results. While there are differences in the shapes of the environmental profiles for each of the scenarios, the main conclusion is that the three alternative technology scenarios generate environmental profiles that are *not* clearly distinct from the base case, or from each other. Within each technology scenario trade-offs are implicitly made between different dimensions of environmental performance. This picture holds for both the pulp and paper and the PVC case studies. Figure 3 illustrates the mix of outcomes across different parameters of environmental performance for the base case and three 'beyond base case' scenarios for the pulp and paper model.

A key result is that no single policy objective translated into specific technology-forcing measures (a focus on encouraging recycling, for instance) is likely to produce generic improvements in environmental performance. Gains achieved across some dimensions of performance may be reflected by losses in others. These trade-offs may to some extent be ameliorated using a mix of policy objectives - modifying losses in performance due to one set of technical changes with a compensating improvement by adopting others. The main conclusion to be drawn is that moving beyond base case performance (defined by best available technology (BAT) standards for 2010) is possible using available technologies, but that there is considerable uncertainty about defining a radically distinct trajectory as measured by environmental performance. Defining a sustainable 'end state' is not easy.

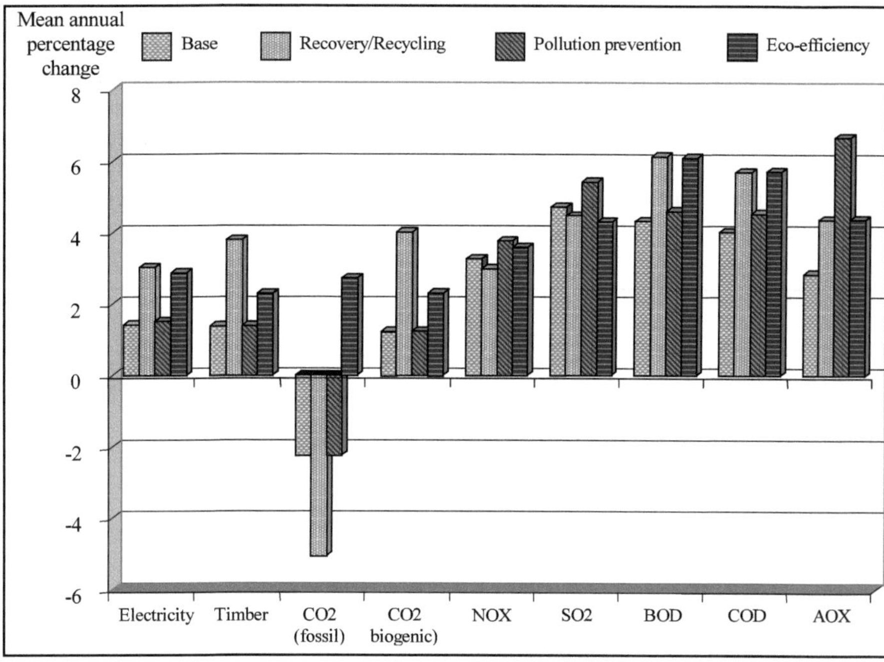

Fig. 3. Environmental Performance Changes for Technology Scenarios: Pulp and Paper, 1995-2010

5. Reconceptualising the Relationship Between Innovation Systems, Technological Regimes and Environmental Performance

We have discussed the validity of claims for two forms of technological transition induced by environmental pressures. The first we may define as a 'micro transition' between abatement technologies and 'clean' technologies. The second we may define as a 'macro transition' from one sectoral technological trajectory to another. We argue that there is little theoretical support and doubtful empirical evidence for the micro transition, and we show that the uncertainties that exist about the environmental outcomes of a macro-transition suggest a reconsideration of a purely objective-driven environmental innovation strategy. Nevertheless, we have also shown that technological change is strongly related to environmental performance at the sectoral or meso-level, pointing to a need for a conceptual model that will allow us to explain how technology and environmental dynamics are coupled. This model should also attempt to place these dynamics in a wider institutional context, linked to innovation and regulatory systems.

One way of conceiving of the link between technical change and environmental performance is in terms of *an innovation triangle* that links changes in abatement technology, process changes, product changes. This dynamic and interactive set of adjustments and adaptations within a technological regime is linked to and co-evolves with broadly autonomous changes in background infrastructures (see Figure 4). Rather than arguing for a progressive shift in the focus of innovation from one form to another (the notion of a *transition* from abatement to process change), this model shows that all technological regimes (and the clusters of firms and other actors that constitute from) face a range of market and social pressures for which a variety of technological responses will be appropriate.

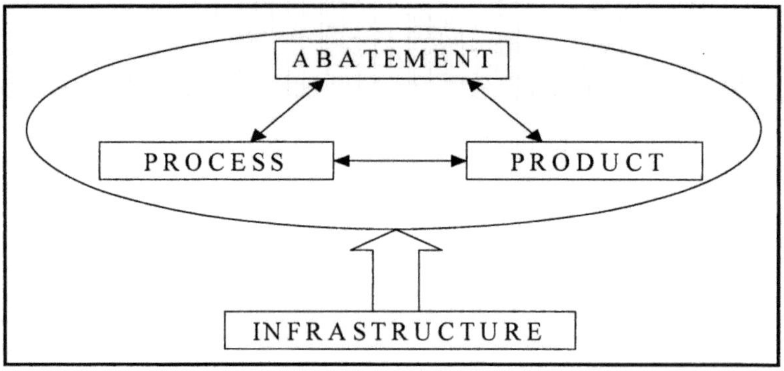

Fig. 4. The Innovation Triangle

Most industrial sectors face environmental pressures at all three corners of the innovation triangle, leading to a number of general conclusions:

- The source of pressures to innovate at each of corner differs: pressure on abatement has tended to come from regulators and neighbours; pressures on process change have tended to come from competitors and customers; whereas pressures on products have come from consumers and pressure groups;
- The innovative response at each corner differs: the technological resources necessary, the source of new technology, the rate of change and so on are all contingent on the technological problem which is being solved; in general abatement technologies are bought in from specialist suppliers, process technologies are developed through partnerships between capital goods suppliers and leading producers, and product changes are managed in-house as a critical source of competitive advantage;
- Changes in one corner of the triangle affect changes in both of the other corners: innovation is dynamically inter-linked and includes incremental and step-like adjustments and changes; this technological inter-relatedness modulates and tempers the opportunities for change that exist throughout the technological system.

We may also be able to generalise these conclusions to argue that the *pattern of pressures and opportunities across the three corners of the innovation triangle differs between sectors*. These differences are determined by economic, as well as social and political factors. In some sectors pressures and opportunities for abatement will be strongest, in others process and product changes will be relatively more important. But these distinctions will be a question of balance. All three types of change will be present, and each will have implications for environmental performance, as well as for competitiveness. Very crudely we may posit that in mature process sectors the focus of environmentally-significant innovation will be along the abatement-process change axis. In consumer goods sectors the focus will be along the product and process change axis, whereas in intermediate sectors within supply chains, the emphasis may be on the abatement-product axis.

6. Conclusion: Path Dependency and Transitions in Technological Regimes

The results of the SCOTCH study and the discussion in this paper lead to a number of observations that are germane to the current debate about system innovations and transitions for sustainability. First, here is little empirical evidence for what we have termed micro transitions between abatement and clean technologies in the paper and PVC sectors. There is a significant process of incremental technological change and some examples of modular reconfigurations at the sub-regime level (both frequently stimulated in response to environmental pressures – market and regulatory). This supports the lock in/path dependency school account of change in technological regimes.

Second, reflecting on the two examples of paper and PVC, the problem of systems innovation appears to be less one of a smooth reorientation of prevailing trajectories, and more a problem of reversal or extrication - with one regime being replaced by another which is morphologically and institutionally separate and distinct. In the choice between reorientation and substitution, the history of technology appears to favour substitution. This means that the problem of technological transitions should not begin with describing the inter-linkages between micro-, meso- and macro- innovations within the context of an incumbent technological regime (transition management). Truly revolutionary innovations are likely to start small, and they will come to define through co-evolutionary processes a new regime *for themselves*. In doing this, they will need to overcome 'barriers' (technological, institutional, economic, political) which stand in the way.

Third, there are a number of policy implications: the need to encourage new incipient regimes; the need to facilitate competition 'early on' (by reducing switching costs, by reducing barriers); and the need to intervene in processes of regime extrication and extinction (negative incentives to incumbent technologies – by imposing full environmental costs on them, for instance), so creating the conditions for their substitution.

Fourth, having argued this, we must also be constantly aware of the paradox of entrenchment – innovation and adoption of radical and risky new technological regimes is not possible without commitments (overcoming barriers and creating an economic and institutional context for adoption and a new process of 'locking in'). Adoption leads to channelling and the formation of new 'trajectories', but each time this occurs there is a new risk that what may with hindsight be seen as risky or costly technologies gain dominance. We have shown through our scenarios thought-experiment how difficult it is *a priori* to identify which set of technological alternatives is likely to yield the best results – just giving something a green label is clearly not enough. Scenario analysis identified that there are many examples of poor choices in history (supersonic air transport, nuclear power). Scenario analysis identified multiple uncertainties in defining alternative future trajectories even within the prevailing technological regime. This problem will be more serious with more novel and unknown technologies (Freeman 1982).

Fifth, not only is the definition of a preferred trajectory difficult, there is also the question of whether the path that a trajectory follows is 'governable' (and governable by whom – firms or government?), or whether to some extent it follows an autonomous or 'emergent' path. Those strategy authors (Mintzberg 1987) who have stressed uncertainty are also those who have emphasised the limits and dangers of following a 'rationalist' approach to strategy. This they view as strategy-making separated from implementation (leading to risk of a failure to learn), and strategy-making that encourages a new form of lock-in which may fail as circumstances change. They argue that it will not be possible to know the outcome of an experiment until the experiment has been completed and take a more 'adaptationist' view of strategy as something that unfolds as a result of the pursuit of 'routines' by business organisations employing heuristics. Strategy is not formulated, but formed with unforeseeable outcomes. This approach also suggests that the governability of thechnological regimes and trajectories remains open to question, and that there will be technological, economic and other developments that induce changes to any defined strategy. Rational behaviour under these conditions would be to maintain a diverse range of options (to mitigate multiple uncertainties). Here a delicate balance must be struck since it may be that the maintenance of options, and the preservation of the option of reversibility (or retreat) undermines the establishment of a new trajectory (investment, learning, standardisation, network externalities and so on). It also may entail quite substantial cost and manipulation of markets by government.

Can Poland's Success in Environmental Policy Reforms Translate into Technological Innovation for Environment?

Halina Szejnwald Brown

1. Introduction[1]

At the end of the first decade of transition to democracy and market economy, Poland has reversed the pattern of disregard for the environment, and achieved significant improvements in environmental quality, as measured by such indicators as air and water pollution and the pollution-intensity of GDP. These improvements in the environment have been achieved alongside of strong industrial growth and enhanced international competitiveness in many sectors of industry[2]. The rate of growth of Poland's GDP during the 1990s has been among the highest in Europe (U.S. Department of Commerce 1999). Poland is now on the brink of accession to the European Union. How have these changes been achieved? Can the environmental improvements continue into the future? What is Poland's capacity for major technological changes towards environmental sustainability? I address these questions from two entry points: institutional and cultural drivers of the improvement in environmental protection during the past decade; the status of the national system of innovation in Poland. The analysis draws on two sources: our four-year study of Poland's environmental transition during the 1990s; and recent data on the status of research and development sector and the knowledge base of the country's economy. The study we conducted included five detailed case studies of recently privatized firms, extensive policy and data analysis, interviews with key policy leaders, entrepreneurs and government officials, and a survey of over one hundred privately owned firms (Brown and Angel 2000; Brown et al. 2000). I conclude that the future direction of Poland's environmental performance and technological innovation for sustainability are uncertain. While the institutional resiliency and a strive to compete in the global economy favor further progress,

1 This research was supported by a grant from the National Science Foundation, and by the National Council for Soviet and East European Research. We gratefully acknowledge the assistance provided by the Central Institute for Labor Protection in Warsaw, Poland, in particular Danuta Koradecka and Roman Broszkiewicz. We are also grateful to Jerzy Jendroska, Environment and Law Institute in Wroclaw, Poland, for his contribution to this work.

2 Economic restructuring has been less successful among very large state-owned enterprises in heavy industries such as steel and energy production, and coal mining. But even within these industries environmental performance has improved.

the limited R&D capacity and the low political profile of environmental sustainability are of concern. I identify two approaches that may achieve progress in the near future: strengthening the links between industry and academia as well greater internationalization of Polish universities; and greater reliance on policy instruments that promote investments into innovation and clean technologies. The case of Poland should be of interest to other developing economies.

2. A Decade of Progress in Environmental Protection

An environmental regulatory system is typically judged successful if it advances the goals of improved environmental quality without imposing unreasonable social and economic costs, and if it does so in ways that enhance rather than undermine the pursuit of other societal goals, such as improvements in socioeconomic welfare and protection of the rights of individuals. Two categories of indicators suggest that significant progress have been made in Poland towards establishing such a system. The first set of indicators relates to environmental quality in Poland. The second set relates to the direct operation of the regulatory system itself, and is based on extensive empirical data we have collected.

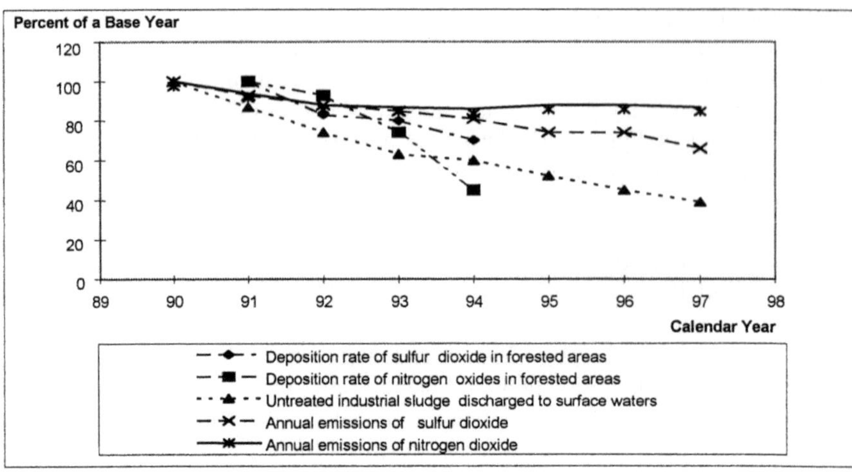

Fig. 1. Trends in Environmental Quality in Poland Relative to a Base Year, GUS 1996, 1998, 1998a

Figure 1 shows that environmental quality in Poland has been improving, as evidenced by trends in air and water pollution. Over the period 1990-97, emissions of environmental pollutants have declined across a range of indicators, including

SO_2, NO_2, and untreated discharges of industrial waste into surface water[3]. Similarly, air emissions of heavy metals in Poland have fallen over the period 1990-96 (not shown here). Significantly, pollution has continued to decline even as industrial output and overall GDP have increased. Poland's GDP recovered quickly from the initial dislocation associated with privatization and economic restructuring, and has been growing at more than 5% per annum for most of the 1990s. Industrial production, which represents approximately 40% of Poland's GDP, grew at more than 10% in 1997, and the vibrant private sector is already responsible for 70% of GDP.

Figure 2 shows the trend in pollution intensity relative to GDP in Poland for the period 1990-97. In the case of SO2, NOX, and total suspended particulates, emissions have declined markedly relative to GDP. There has also been a more modest relative decline in emissions of various classes of volatile organic compounds.

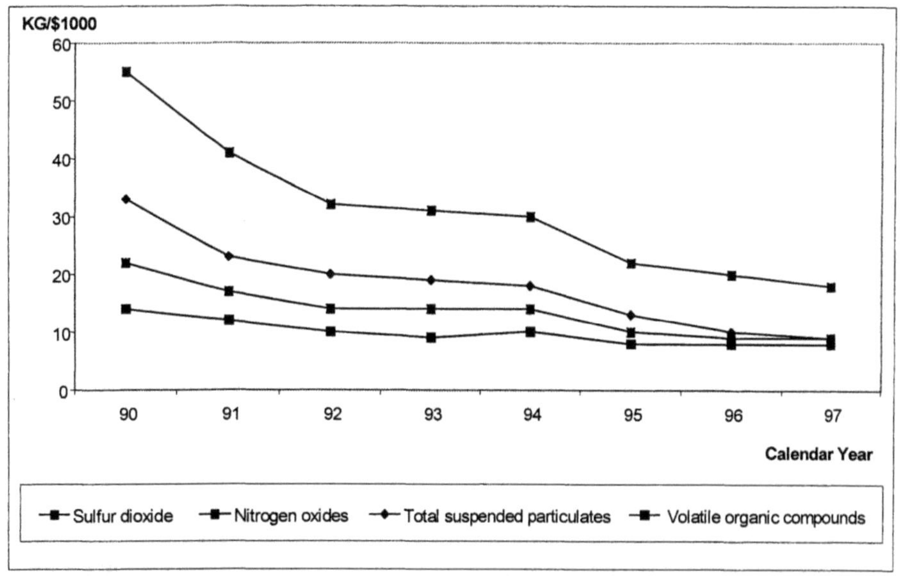

Fig. 2. Trends in Pollution Intensity in Poland (Emissions per Unit of Gross Domestic Product), Stodulski 1999

Declines in emissions relative to GDP, in the context of an expanding economy in Poland, reflect both a restructuring of the economy away from pollution intensive industries, and reduced pollution at the plant-level for a significant number of firms.The latter is has been particularly the case with regard to the energy sector where major advances have been made in switching to higher-grade fuel and installation of pollution control technology.

3 In the case of NO_x, total emissions have declined over the period 1990-96 despite an increase in emissions from the transportation sector.

The downward trends in pollution are encouraging. But the aggregate statistics need to be treated with caution. The closure of a small number of very large point-source producers, or the shift out of certain sectors of heavy industry can have a major impact upon overall trends in pollution, even as little progress is made with a majority of firms. Moreover, short-term gains in environmental quality do not guarantee long term performance, which depends, among other things, on the effectiveness of the permitting and compliance process. For this reason, we also considered the actual operation of the regulatory system in Poland. Here we drew upon three sources: government statistics regarding implementation and compliance, detailed case studies of five recently privatized manufacturing firms, and the results of a mailed questionnaire survey of a random sample of privatized industrial firms.

The case studies revealed a high degree of compliance with the system of facility permitting for air emissions, wastewater and solid and hazardous waste. All of the regulators required the submission of environmental permit applications, collected environmental fees, and imposed surcharges on firms that were operating without permits.

In one case the authorities decided to close a facility for repeated failure to improve its environmental performance, and after exhausting all interim attempts to solve the pollution problem – even though this would mean the loss of several hundred jobs in a city with 20 percent unemployment. The attitudes of the case study firms were consistent with those of regulators. Without exception, the managers of the five firms accept the current system of environmental fees even though in some cases these represented substantial financial burden (for example, 0.7% and 5% of total sales, respectively, in two of the firms). Company managers also regarded the process of setting emission limits as fundamentally fair. When asked whether the ambient air quality standards were too stringent and whether it was unfair to consider existing background levels in issuing air permits, they generally defended the process, citing the potential impact on public health.

The results of the survey confirm these findings. All of the surveyed firms that had air emissions subject to regulation had submitted the required technical assessment of their air emissions and permit applications. Of 109 firms in the survey, only four reported that they had operated without a legal air permit during the previous five years. Completion of the permitting process does not imply that firms are meeting all applicable environmental standards. Indeed, approximately 28% of firms in our survey reported that operation of their facilities would result in emissions that would exceed one or more limits set in their air permit. But in contrast to the widespread practice prior to 1989, the issuance of permits and the identification of firms out of compliance are now the basis for subsequent regulatory enforcement and intervention. The survey also documented the broad support for the regulatory system among the survey sample of private firms. 84% of the firms surveyed reported that permitting regulations and procedures were about right or too lenient while just 16% considered them too tough.

The case studies revealed a very high level of familiarity by the regulatory authorities with the firms under their jurisdiction, and demonstrated a capability to use this knowledge to push for environmental improvement. By way of illustration, the most financially stable and technologically advanced of the five firms was forced by the regulatory authorities to dramatically upgrade its technology to meet very strict standards for hydrochloride emissions. In contrast, another firm was allowed to operate without a permit (by paying modest surcharges) instead of being forced to pay fines for noncompliance with a permit. The authorities did this because the firm's activities were relatively benign, and because the facility could not afford the technological changes required to meet the national ambient air quality standards. The survey data are consistent with the findings of the case studies. Thus 36.4% of surveyed firms had been visited at least once in the previous year by the enforcement branch of the environmental protection ministry. Among larger firms with more than 250 employees, 54.3% had been inspected in the previous year.

Government statistics generally support these empirical findings. Over the period 1993 to 1997 the number of inspections increased 5%, and the number of fines imposed increased 32% (GUS 1998). The total value of environmental fines imposed increased five-fold between 1991 and 1995; principally reflecting substantial increases in the level of fines. The collection rates for all environmental fees have been generally above 70%, and above 80% for air pollution fees.

In short, building upon existing policies and institutions, and after a decade of reforms, Poland now has an operational system of environmental regulation of industry. In particular, Poland has overcome a recent history of widespread disregard for the regulatory process on the part of firms and weak enforcement by the regulatory authorities, and now has what Bell (Bell 2000) refers to as the "culture of compliance". This has been achieved in Poland despite only modest NGO involvement at the national level, and little public engagement in policy formulation (Andersson 1998; Glinski 1998).

This is not to say problems and challenges do not exist. The environmental protection system is intensive in human resources, and staffing shortages are a chronic problem. Many firms do not meet the emissions limits set in their operating permits. This is particularly the case with the highly polluting coal mining industry, which has not been confronted for fear of causing major social eruption (Andersson 1998). More generally, improvements in environmental performance have taken place in Poland during a period of robust economic growth, and may suffer setbacks should the economy falter. In addition, Poland has not leapfrogged, as some had hoped, to a more ambitious national environmental strategy aimed at a sustainable future. The decade of reform focused primarily on reducing pollution and the energy-intensity of economic activities, and on greater use of end-of-pipe pollution control, while progress in ameliorating the environmental effects of continuing economic growth and consumerism has been modest (Stodulski 1999; Mundl et al. 1999).

These outstanding issues have created divisions within an environmental policy elite in Poland, once united by a shared goal of moving environment into the forefront of the national political agenda and eliminating the legacy on non-compliance. As described by Andersson (1999), by the end of the first decade of societal transition three coalitions have emerged, with their individual problem definitions and favored policy tools. The mainstream view, reflected in the reforms implemented during the 1990s, calls for a pragmatic approach to regulation and enforcement, and favors flexible case-sensitive policy instruments (see for example, Jendroska 1996; Bell 2000; Bell and Bromm 1997). This position is challenged by the green advocacy coalition, which would like to see environmental goals dominate the national agenda, and which favors green taxes to control the environmental effects of growing consumerism in Poland. The proponents of this perspective are much less satisfied with the direction of the reform program than the mainstream coalition, but its political voice is relatively weak (Karaczun 1997; Stodulski 1999; Mundl et al. 1999). The second group of dissenting voices calls for greater emphasis on economic analysis of policy options (Zylicz 1992, 1998). By this account, the current emphasis on flexibility, case-specific implementation, and balancing of competing objectives in individual decisions is not efficient economically, and may be ineffective in the long run.

That said, Poland has done in a single decade what some feared it could not, namely, reverse a long history of non-compliance with environmental regulation, and put in place a vigorous functional system of environmental protection capable of controlling pollution from the emerging private sector.

3. Deconstructing the Elements of Poland's Progress

In recent years various researchers have identified the elements of an effective system of environmental regulation of industry (see, for example, Davies and Mazurek 1998, for a recent version of such a policy template). In previous publications we have identified six *structural* characteristics that, in our view, are most predictive of success in efforts to realize environmental goals without excessive social and economic costs. First, a regulatory system must be able to translate national environmental policy goals into clear and consistent performance expectations for industry. Uncertainty about environmental performance expectations, and lack of predictability in enforcement, lead to sub-optimal corporate decisions regarding technology and management choices (Judge 1998; Steinzor 1998). The availability of appropriate policy instruments is another key structural feature of a successful regulatory system. So is the ability to learn and profit from experience. A fourth structural feature of an effective regulatory system is broadly shared ownership. By this we mean that regulators, and other interested parties are participants in the assessment and reform of the regulatory system, and that the system's failure is inimical to their own interests. Two other structural features predictive of success are a capacity for making case-specific implementation decisions, and availability of information on which to make such case-specific decisions.

In general, the regulatory system that emerged in Poland during the 1990s matches up well against such a template. Enhanced clarity of performance expectations derives from codified procedures for obtaining permits and executing appeals, the existence of standard fees and fines, and reliance on national ambient quality standards for defining permits. Even in the case of negotiated compliance, the flexibility allowed refers to timing and technology choice, not to the ultimate performance requirement.

During the 1990s, Poland has experimented with a wide range of policy instruments, both coercive and market-based, including pollution trading schemes, technology-based regulations, experiments with negotiated permits and integrated multimedia pollution permits, loans from environmental investment funds, and fees and fines. Many of these experiments are ongoing, and it is premature to judge how appropriate the policy mix is. And while opinions on the effectiveness of individual instruments vary, largely along the lines of the three policy coalitions discussed earlier, there is little doubt that the overall effectiveness of the available instruments has improved. There are also glaring gaps in the menu of policy instruments. One of those is the absence of any formal process to publicly disclose disaggregated information about firms' performance, such as licensing conditions, emissions, and others. Another is the lack of effective instruments for including local and regional authorities, and the public, in the policy process. The proposals to include such measures in the amended Environmental Protection and Development Act have been debated for years, with little progress to date.

During the past decade, the environmental protection system in Poland has shown a remarkable capacity for learning and profiting from change. This capacity is grounded in a vigorous environmental policy coalition, which formed after the 1970 Stockholm conference, and which by the 1980s included academics, bureaucrats, members of the ruling party apparatus, and environmental advocates. This coalition was instrumental in sustaining a lively policy debate and implementing the present reform program (Cole 1997; Hicks 1996; Mundl et al. 1999). Notably, absent from the debate are organized labor, regional and local administrators, local governments, and most of the industrial sector (the energy sector is the exception). It is therefore particularly notable that these neglected yet crucial participants in facility-level decisions have responded so well to the recent reforms, an issue that we return to below.

Two other intertwined features have been crucial to environmental progress in Poland during 1990s: case-specific decision-making, and the availability of information on which to make such case-specific decisions. Our case studies uncovered a wide spectrum of case-sensitive decisions on a local and regional level, from imposing harsh permit requirements or fines, forcing technological change or even plant closure, to adopting negotiated enforcement schedules, or delaying permit decisions for plants that are clearly unable to meet them. Some of these decisions followed formal guidelines, other were informal.

Case-specific decision-making has been identified as a tool for improving the efficiency and reducing the social costs of environmental regulation (Bell and Bromm 1997; Scholz 1991). Others have voiced concern that such context specific decision-making can lead to lax enforcement and co-option of the regulatory personnel (May and Winter 1999). Part of the issue here is distinguishing between enforcement styles (e.g. more or less conciliatory) and enforcement strategy, such as the degree of discretion available to enforcement personnel (May and Burby 1998). A growing body of research suggests that case-specific decision-making and other aspects of negotiated compliance are most effective when backed by a system of deterrence (Ayers and Braithwaite 1992; Caldart and Ashford 1999; Gunningham 1999). The willingness of regulators to impose fines and to issue cease-and-desist orders is an important dimension of the environmental regulatory strategy in Poland. Finally, it appears that regulatory decisions are not made in a normative vacuum but are driven by three implicit principles: to push firms to the highest level of pollution control that is technologically and financially feasible, to prevent acute threats to public health and the environment regardless of the cost to the firms or the economic loss to the community, to reward firms demonstrating a commitment to environmental improvements with more flexible treatment.

That is not to say that confrontations never occur or that firms meekly acquiesce to administrative decisions. Our research indicates otherwise. However, we interpret these data as evidence of the systems' capacity to accommodate conflict and to provide mechanisms for its resolution. They also suggest that in Poland the preference for non-confrontational approaches are strategic choices by participants rather than a structural necessity.

4. Roots of the System's Effectiveness

To what do we attribute Poland's ability to develop an environmental regulatory system, which possesses numerous structural strengths? The answer seems to lie in three fundamental factors having to do with both the trajectory of the system's recent evolution as well as its relationship to the broader social context. These are: a high degree of pre- and post-1989 continuity in institutions, policies, and modes of conducting societal transactions; wide sharing of certain values and attitudes among the key societal actors; a broad support for the rule of law and due process.

Institutional continuity in Poland, rather than leading to inertia or rigidity, has allowed regulators to draw upon considerable accumulated technical, administrative and human capital in implementing reforms. In the absence of the communist-era external constraints and disincentives, policies and practices that were ineffective prior to 1989 have emerged as important resources within the privatized market economy. Beyond specific institutions and policy instruments, there has also been a broader level of social continuity. This is well illustrated by a decisive preference for negotiation over confrontation in the conduct of social transactions by firms and regulators alike. Other elements include close interactions among firms, regulatory authorities, and independent technical experts, and the value placed upon building dependable and predictable working relationships over time and maintaining reputations and trustworthiness in relation to other parties.

A second theme explaining Poland's success is that the three groups of actors most active in environmental policy making and implementation – industrial managers, regional and local government officials, and national policy makers – appear to share with each other two fundamental values and attitudes towards environmental regulation: that public policies aimed at protection of the environment are necessary and legitimate, and that regulatory decisions require balancing multiple objectives for the common good. Our finding of pro-environment attitudes is not surprising, since these attitudes have a long and rich history in the Polish society (Cole 1997; Jendroska 1996). Just how stable such attitudes towards environmental protection will be likely depends substantially on continued broad-based improvements in socio-economic welfare. While some are critical of segments of the emerging private market economy that tend toward a kind of 'cowboy capitalism' (Horst 1997), there are also very strong external drivers toward the continued balancing of environmental protection with economic development. Foremost among these are the conditions associated with entry to the European Union, viewed by the business community and other groups as critical to the economic future of the country. The details of the European Union acquis require alignment of Poland's environmental protection system with the standards, rules, and procedures of the European Union.

The third theme emerging from the analysis deals with attitudes towards the rule of law and due process. The reform program in Poland, which focused heavily on institutional and policy changes, was premised on a plausible but still untested assumption that all the key actors – the regulatory authorities and industrial managers and executives – would respond to legal and policy changes by changing their behavior. Our study generally provides support for this assumption. While Poland is not a highly legalistic culture, we do not find widespread cynicism about environmental policies and social obligations. Rather, there is considerable support for their legitimacy in the free market economy emerging in Poland.

Additional support for our conclusion that the effectiveness of the environmental reforms in Poland has deep roots in prevailing societal norms and values comes from the recent study comparing the behavior of the state-owned firms with privately-owned firms, based on a survey identical to the one described earlier (Broszkiewicz and Brown, yet unpublished). The survey results show that despite differences various characteristics, such as financial circumstances (state-owned firms are often less profitable), size (state-owned firms are larger), pollution (state-owned firms are 'dirtier'), and the relationship to the state, we found a similar record of compliance with, and enforcement of, environmental standards for both groups of firms.

5. From Institutional Reform to Technological Innovation

The main challenge of the first post-communist decade was to create workable pollution control policies and a culture of compliance. It required that reforms target policies and policy instruments, human and institutional attitudes and behaviors, strategic investments in pollution controls, and imaginative financing schemes for the environment. The challenge for the future requires addressing the consequences of continuing economic growth and prosperity. Meeting this challenge will depend, among others, on reducing environmental intensity of production and consumption. Technological innovation and system innovation will be essential to that success.

The concept of 'national innovation system' is helpful in evaluating the capacity for innovation from a country perspective. The national innovation system is the country's capacity to support and improve the efficient functioning of procedures for creation, accumulation, transmission and application of knowledge. This capacity is grounded as much in complex interactions among institutional actors (for example, as measured by formation of joint ventures or inter-institutional joint research) as it is grounded in direct investments into innovation research (represented, for example, by industry investments into R&D, or status of higher education), or in direct governmental policies regarding foreign industrial investments, trade barriers, and others. Several authors have highlighted the importance of the networking among the key actors, such as companies, research centers, government innovation bodies, universities (Freeman 1987; Porter 1990; Lundvall 1992; Nelson 1993). Elsewhere in this volume Saviotti expounds on this concept by describing the co-evolution of institutions and technologies.

Partly in an attempt to capture the complexity of the factors influencing the overall national capacity to innovate technologically, OECD has developed a set of indicators for the national innovation capacity. The indicators cover both the main drivers of knowledge-based economy, and measures of innovation outputs. Below, I reflect on the status of the national innovation system in Poland, using the OECD metrics as well as description of some key institutional and political issues.

Institutions and political context: The impediments to progress are severe. First, with some notable exceptions, the political will to address the issue of environmental sustainability is generally low (the energy sector, which has been part of the negotiations of the EU accession, is one such exception). In general, environmental issues have been steadily dropping on the political agenda since the high point in 1989 when they were included in the roundtable negotiations between the outgoing communist regime and the Solidarity coalition. Partly, this is because of the weakness of the organized environmental movements, and partly because of the relatively low level of public interest. Therefore, and notwithstanding the inclusion of sustainability in the language of the new environmental protection legislation, it is unlikely that the fractured and constantly shifting political coalitions in the democratic Poland will make the environmental sustainability its high priority. This is particularly the case in the current climate of a cooling economy: in 2000 the growth rate in GDP was 4%, after a decade of 7-8% annual growth rate, and the forecast for 2001 is less than 2%. Environment was not even mentioned in the recent (fall 2001) expose by Poland's Prime Minister on the national priorities.

Second, the industrial research and development capability in Poland has suffered a major setback during the past decade, as did the basic research. This is to a large extent the result of the post-1989 social changes and economic pressures on the government. Prior to 1989, the applied R&D, basic research, university teaching and industrial production were carried out separately. The primary function of the universities was teaching, while basic research was conducted in various institutes of the Polish Academy of Sciences. Similarly, the R&D functions for the industry were carried out in various sectoral institutes established for that purpose and supported with public funds. During the past decade, these research institutes – both for basic and applied research – experienced major budget cuts and loss of talent, as the market economy created new professional opportunities. The void thus created has not been filled by the newly emergent private sector. Neither have the foreign firms investing in Poland been interested in the transfer of their R&D. The collective result is a rather low level of competitiveness characteristic of the Polish industry (EU 2000).

Standardized metrics. Recent statistics from OECD allow a closer examination of Poland's climate for technological innovation as well as a comparison with other OECD economies, including Poland's post-soviet peer economies (OECD 2001). The OECD analysis uses four classes of indicators to assess the vibrancy of the 'knowledge-based industries' among its members: indicators of the intensity of the knowledge base of the economy; indicators of the diffusion of new information technology into commerce and society at large; indicators of integration of economic activities into the global economy; and indicators of industrial growth and productivity (knowledge-based industries include: manufacturing in aerospace, medical equipment, pharmaceuticals and general machinery, and services in post, telecommunications, finance, insurance, and general business, such as accounting). The indicators measure a wide range of conditions, such as: investments in R&D, in education, in venture capital, in basic science, and in ICT hardware and software; trends in new patents in high knowledge industry; diffusion of knowledge within and across economies (as measured by international authorship of scientific publications, intensity of foreign direct investments in the high-knowledge industry, and international ownership of new patents) and others.

The results for Poland are mixed. On the low side, the expenditures into R&D by industry are low, and the indicators of R&D intensity in information and communication technologies (ICT patents relative to all patents, and business R&D expenditures as percentage of GDP) place Poland among the worst two performers in 1999 (figures 3 and 5). Poland's decline in industrial R&D come at the time when business' contribution to the total share of R&D expenditures in OECD countries have risen. For example, during the 1990s it increased from 57 to 67% in the United States and from 52% to 55% in European Union. Not surprisingly, the indicator of research intensity in business in Poland is also relatively low: in 1999 Poland reported 33 researchers per 10,000 workers (a 1 % decline since 1994), as compared to 75 in the U.S. and 55 in the E.U. This number is, however, comparable to Poland's European peer group: 36, 31 and 26 in Slovakia, Hungary and Czech Republic, respectively (Hungary, in fact, has experienced a 7% decline since 1994).

Poland also scores low on the diffusion of information technology into commerce and society. For example, in 1999 Poland had in 35 telecommunication paths per 100 inhabitants and in 2000 it has 10 internet connections per 1000 inhabitants. In comparison, Hungary and Czech Republic had, respectively, 60 and 60 telecommunication connections (per 100), and 20 and 15 internet connections (per 1000).

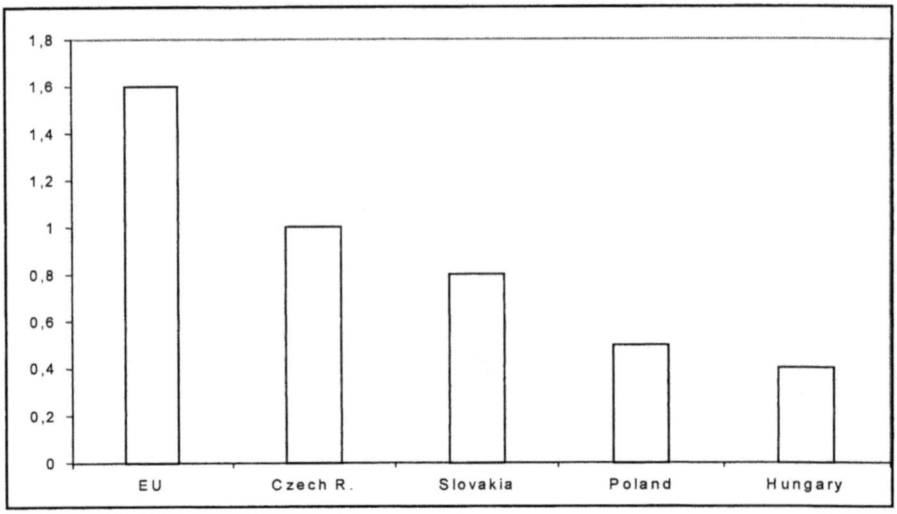

Fig. 3. R&D Expenditures by Industry as Percentage of Domestic Product of Industry (1999), OECD 2001

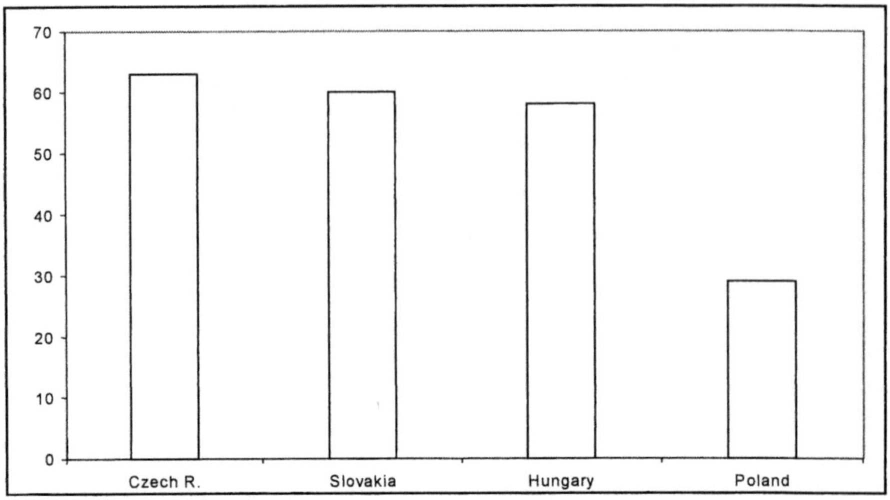

Fig. 4. Average Imports and Exports as Share of GDP (1999), OECD 2001

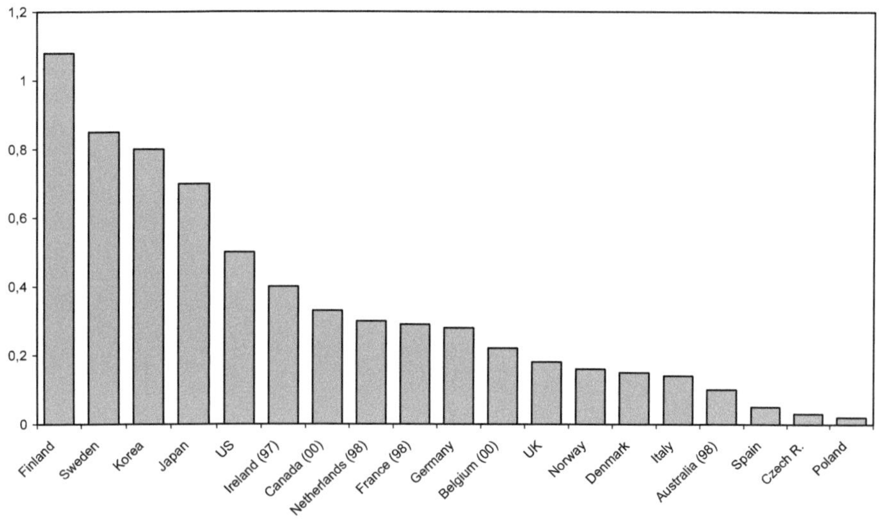

Fig. 5. Business R&D Expenditures by Selected ICT Manufacturing Industries as Percentage of GDP (1999 or latest available year)

On another front, Poland has shown disappointing results in the metrics of openness to international competition (as measured by the ratio of foreign trade to GDP) and in attracting foreign investments (measured by input flows of FDI as percentage of GDP). In 1999 these indicators were 29% for the former and 2% for the latter. Figures 4 and 6 show that Poland trails behind its peer group in terms of integration into the global economy.

On the other hand, the public sector in Poland has been investing in R&D and basic research during the 1990s, despite the initial shock waves experienced by the national research infrastructure. For example, Poland's total national expenditures into R&D (0.5% of GDP in 1999) and into basic research (0.2% of GDP in 1998) are comparable to those in the E.U. and among its post-soviet European peers. Also, the investments into R&D by business, though relatively low (figure 3) has nonetheless shown healthy annual growth rate of 5% during the 1990s. For comparison, the annual growth rate in the EU and Hungary were 4 % and 1 %, respectively (the data for the Czech Republic and Slovakia are not appropriate because of the split of Czechoslovakia during that period).

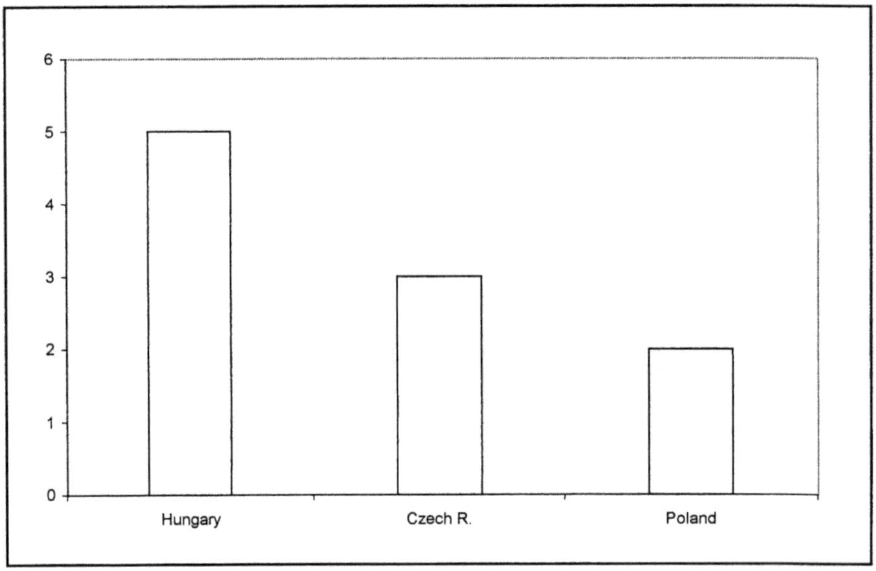

Fig. 6. Inward Flow of Direct Foreign Investments as Share of GDP (average for 1990-1998), OECD 2001

Other considerations. There are other signs of positive changes. For one thing, the government has recognized the problem in the recently *inter alia* adopted document "Assumptions of innovation policy until the year 2002". Second, changes are emerging within the traditionally very conservative academic community. For example, the well-regarded Warsaw Technical University has just established Environmental Protection Department, and newly established business schools are thriving in the major cities. Polish research community is also well connected internationally. Thus, 45% of scientific publications in Poland have foreign co-authors (OECD 2001). This number most likely underestimates the extent of international research collaborations in Poland because it does not count the more-coveted publications in the English language scientific journals. Notably, this integration appears to be unidirectional. At less than 1 %, Poland reports the lowest proportion of foreign students at its universities among the OECD member states. For comparison, US, Hungary, Czech Republic and Slovakia have 4%, 3%, 2% and 2% of foreign students, respectively (OECD 2001).

There are also signs of previously non-existent linkages emerging between the academic sector, research institutions, and policy makers. This has been vividly demonstrated in the course of the deliberations leading to the development of the national air quality strategy in Poland. As described by Botcheva (2001), during the 1990s several major economic analyses were performed to inform the policy making process sponsored by European Union, World Bank, Polish energy sector, and others. These analyses were conducted through close collaboration of the policy makers, universities, the research community, and the power-generating sector. The results have considerably advanced the policy making process by creating a strong base of political support and high regard for the technical credibility of the analysis.

Third, the global opening of trade and investment, and the upcoming accession to the European Union should provide incentives for improved environmental performance. The Polish Chamber of Chemical Industry (PCCI) – the sector which in Poland represents approximately 10% of industrial employment and 11% of industrial exports (mostly to OECD) – actively seeks international markets, and considers environmental performance to be among the key elements in improving its international competitiveness. Adoption of ISO 14000 standards is actively promoted among the PCCI members. Further evidence of a positive impact of the EU membership comes from a study performed by Botcheva in 1998 of over 2000 highly polluting firms in Poland and five other CEE countries (Botcheva 2001a). The study found that export oriented firms consistently performed environmentally better than those with mostly domestic markets. The author attributes this performance difference to reputational and normative pressures, and does not exclude the better access to clean technology as another factor. While these trends are encouraging, they may be, however, self-limiting over time: after the initial improvements in eliminating the worst sources of pollution in these "dirty" industries, the firms may plateau in terms of performance and technological change.

Fourth, the Polish society has demonstrated great enterpreneurship during the past decade, as well as the willingness to innovate and experiment, while public institutions and policy leadership have shown resiliency and learning capacity.

In short, the future direction of Poland's environmental performance and technological innovation for sustainability are at crossroads. The low political profile of environmental sustainability, and the emphasis on pollution abatement and end-of-pipe regulations, suggest that domestic pressures will not play a significant role in future progress towards environmental sustainability. These impediments are magnified by an underdeveloped knowledge-base of the economy, which is necessary for creating the climate of technological innovation. The latter manifests itself in weak industrial R&D, limited information and communication infrastructure, low direct foreign investments, especially in the R&D area, and disappointing record of trading at foreign markets by the knowledge-based industries. On the other hand, the overall research enterprise in Poland shows signs of vitality, entrepreneurship, and integration with the international research community, despite the setbacks it has experienced during the 1990s. The strong desire to become integrated into the European Union is also a key driver of progress.

At this point, the most promising near term changes may include: (1) increasing the reliance on policy instruments that emphasize investments in clean processes and products. This approach will build on Poland's culture of compliance, case-sensitive policy implementation, and the well-established institutional openness to experiment with various policy instruments; (2) Encourage stronger links between academia and industry, and between Polish and international students. The existence of lively interactions between Polish academics and their Western counterparts should facilitate the transfer of this societal model.

The case of Poland should be of interest to other developing economies confronted with an underdeveloped environmental regulatory systems for controlling industrial pollution. It suggests that, while it is not necessary to reenact the evolution that has taken place among the developed economies during the past three decades, neither can they expect to 'leapfrog' into the transition towards sustainable economy. It also shows that the success in the first phase is not a predictor of a success in the second phase. The types of institutions and national capabilities for each are different.

This is particularly apparent when we compare Poland with its European peer economies: Hungary, Slovakia and Czech Republic. Poland has been a leader in this group in environmental policy innovation and reforms during the first post-soviet decade. However, with regard to development of elements of national innovation systems that may be crucial for continuing environmental progress it appears that at this point in time Poland is lagging behind its neighbours.

Sustainable Development and the Regional Dimension of the Innovation System

Gerd Schienstock

1. Introduction

During the last ten years, the 'systems of innovation' approach has developed into a useful tool for studying, explaining and, to some extent, even for influencing innovation activities and technological change. This approach shares with other theories the assumption that innovation is an important factor that stimulates economic growth but that has developed into a more complex model. A system of innovation brings together all the major factors that affect technological progress (Edquist 1997) and tries to form a systemic model of these factors. Such an exclusive focus on technological change and economic growth, however, is not enough to tackle the current socio-economic problems, as innovation does not represent a positive sum game (Boden and Miles 2001).

While the system approach has mainly dealt with determinants of knowledge creation and knowledge diffusion, it has hardly paid attention to socio-economic consequences resulting from the diffusion of innovation. As a focus on the creation of a knowledge stock and its diffusion is too narrow and not sustainable, we have to study feedback on radical product and process innovations more carefully (Lundvall and Archibugi 2001). Taking into account that the innovation system is part of the economy and the wider society and as such impacts on other natural, social and technical systems, such as the labour market or the ecological system, feedback becomes an important aspect of innovation processes. It might be the case that an increasing innovation capability of a territory only creates short-term advantages but actually undermines its capability to produce long-term economic growth. This is the case when the costs of dealing with social and ecological problems as unintended consequences of technical development become bigger than the benefits resulting from innovation and change. Moreover, it is through feedback that the innovation system shapes the framework conditions for its own functioning (Cooke and Schienstock 2000).

The fact that costs of innovation can more than compensate possible gains makes it necessary to introduce sustainable development and sustainable competitiveness as an important criterion of innovation processes and the innovation system as a whole. Here we apply a wider interpretation of the term sustainability; it integrates the ecological, social and economic dimension into a holistic approach. Thus sustainable development and sustainable competitiveness imply simultaneously an environmentally, socially and economically compatible development, as is reflected in the so-called 'three-column model'. In the following, we will direct our attention primarily to environmental sustainability.[1]

2. From Path Dependency to Path Creation

The system approach considers innovation processes to be evolutionary and has focused mainly on path dependency. The strength of this concept is in that it does not separate innovation from past developments, but assumes some kind of continuity in the process of technical change. Innovation lines up with earlier changes, which means that it has historical antecedents of novelty. "(...) the dynamic process itself takes on an essentially historical character" (David 1985). The term "path" means that a specific technology has been chosen among a variety of different alternatives, which leads to large profits for the innovator and the first users but does not necessarily result in macro-economic advantages in the long run. Path dependency always carries the risk of turning into a so-called 'lock-in' (Grabher 1993; Johnson 1992; Schienstock 1997). Traditional technology can lock the market into an inferior development option and may result in a loss of competitiveness and retarding economic growth in the long run.

Does environmentally beneficial technology constitute a new path of economic development? There are quite a few scholars arguing that the sustainable development imperative is exerting a force to change the dominant technological paradigm and to shift the existing technological trajectories, although it will probably not constitute a new upswing in the long cycles (Fukasaku 1999). Nevertheless, integrating the aspect of environmental sustainability as an aimed output of the innovation system means that we have to change our analytical perspective.

Instead of focusing on the path dependency of technological innovation, we have to analyse processes of unlocking and path creation. The traditional economic model is based on a number of technological development paths, which have been successful due to framework conditions of an insufficient economization of natural resources (Hübner and Nill 2001). Sustainable development implies that the economic and social value system underpinning the current development path is changing. Traditional technologies have created a 'lock-in' since they tend to undermine natural capital. Environmental sustainability, therefore, requires a change in the mode of operation of the economy. There is a need to create a new development path oriented more towards saving resources and complying with both environmental sustainability and economic competitiveness.

---

1 Environmental sustainability, however, is often closely linked with social sustainability. Modern societies, as some scholars argue (Bergmann et al. 1969), are characterised by new forms of social inequality which do not result from class structures. Instead, they represent the accumulation of several disadvantages, in which the environmental factor plays an important role. Highly polluted areas, for example, are unlikely to attract new industries which create highly skilled jobs; instead, these areas may only provide low-skilled and poorly paid jobs and, in addition, cause major health problems.

It is very difficult to get out of path dependency, however. Perez (1983) has pointed out that fundamental technological novelties can only become transformative together with organisational and institutional changes. It is likely that the social and institutional framework, hospitable to one set of technologies, will not be suitable for a radically new technology. Whereas incremental innovations can be easily accommodated, it may not be the case with radical innovations, which by definition involve an element of creative destruction. A new fundamental and path-creating innovation, we can conclude, requires the development and co-ordination of a vast array of complementary tangible and intangible elements: new management techniques, new organisation forms, new kind of workforce skills, and new habits of mind. But many other types of institutional changes such as standards, patents, new services, new infrastructure, government policies and public organisations are also called for (David 2000; Freeman 1997).

However, being locked in a specific technological development path is not irreversible, as some scholars seem to argue. Garud and Karnoe stress the importance of entrepreneurship in the processes of path creation. Path creation, as they argue, "provides a way of understanding how entrepreneurs escape 'lock-in'" (Garud and Karnoe 2000). Path creation, according to the authors, is a process of mindful deviation; it implies de-embedding from the structures that embed economic actors.

There is no doubt that individual entrepreneurs and entrepreneurship have a key role to play in the process of creating a new techno-economic development path. But the creation of such a new path is first of all a collective process. Freeman (1997) argues that the deflection from an existing development path and the diffusion of a new techno-economic paradigm is a process of trial and error, involving a great institutional and organisational variety. Sabel (1995; see also Schienstock 1996) has characterised path creation processes as 'bootstrapping reforms', also stressing the importance of trial and error. He argues that stable and lasting processes of path creation and diffusion can only develop when all actors are marching in steps, monitoring each other's change processes and adapting to them. But it is beyond the capacity of social actors to come to terms with the future techno-economic structure, simply because it is unknown. What they can do, however, is continuously reflected on the previous change processes and in the light of their experiences and based on the diversity of knowledge to make corrections and change directions, if needed. To be able to create and stabilise a new development path, continuous exchange of information and knowledge in dialogues and multilogues is needed. Interactive learning is a precondition for the establishment of a new development path. Interactive learning means that learning is co-dependent on the communication between people or organisations with different types of required knowledge (Meeus and Oerlemans 1999).

The greening of the innovation system and the development of a sustainable development path are special insofar as they extend the idea-innovation chain to the households as end-users. Their consumption habits and practices can support but also hinder the greening of the economy to a great extent. So far, systems of innovation, if we take a narrow definition, identifies the R&D departments of firms, universities, research institutes, technology transfer institutes, and government agents involved in technology and innovation policy as social actors. Only those organisations are included that are directly related to the process of search for new knowledge. Not even a wider definition of the innovation system[2] includes end-users as actors in the innovation process. This is due to the fact that the focus of research lies mainly on knowledge creation and effective distribution of an innovation, while the aspect of use and possible consequences of use patterns are more or less unrecognised.

3. Innovation as a Strategy to Achieve Environmental Sustainability

The view that the greening of the economy is associated with a new technological development path entails that economic growth and environmental degradation can be de-coupled by the creation and diffusion of new technologies. There are, however, different views on whether economic growth and sustainable development are compatible with each other. Many scholars are sceptical in this respect; they see economic growth and sustainable environmental development as two contradictory aims (Altvater and Mahnkopf 1999; OECD 1992). The argument is that economic growth depends on the use of additional natural resources, and because these are limited, further growth is not possible without neglecting the principle of sustainable development. This implies that if we want to achieve environmental sustainability, we have to preserve our natural resources through reducing economic growth. In this scenario, the consumer has to play the most important role. Therefore, environmental issues that have received attention were less associated with production processes and more with consumption and post-consumption (Howes et al. 1997)

To use natural resources sparingly will probably have positive environmental effects. But climate change, waste reduction and sustainable transport technology are all problems that concern consumers as well as producers (Fukasaku 1999). One can doubt whether it is possible to achieve environmental sustainability through strategies of changing consumption patterns only, particularly in a period of stiffer global competition, in which the capability to continuously produce new products becomes the most important competitive edge (Schienstock 1999). It seems that strategies of sustainable development without product and process innovations have little economic and political chances to become successful; on the contrary, technology is critical in securing sustainable development goals. We can therefore characterise the new development path as 'innovation-oriented development model of environmental sustainability'.

Notions such as eco-efficiency and zero emission indicate the growing perception that a strategy to achieve ecological, social and economic sustainability needs to be based on environmentally beneficial technological innovations (Petschow et

---

2 A wider definition also including 'higher-level organisations', whose objects are to facilitate learning processes and that can provide additional input into the innovation process (Teubal 1998).

al. 1998). "(...) companies can improve resource productivity by producing existing products more efficiently or by making products that are more valuable to customers – products customers are willing to pay for" (Porter and van der Linde 1995).

In de-linking economic growth from environmental degradation and unsustainable resource use the development of environmental beneficial technology becomes critical (OECD 1999). However, only if technological advancement is accompanied by organisational, institutional and behavioural changes, will it be possible to achieve sustainable development goals like reduction in energy consumption, pollution emission, and waste production. These issues call for technical as well as organisational innovations. In addition, systemic innovations characterising processes of fundamental transformation also have to include institutional changes; they are indispensable in devising sustainable transport systems, for example.

This suggests that the greening of the economy depends upon the business sector as a carrier of innovation to a great extent. The advantage of an innovation-based strategy in achieving environmental sustainability is that it not only introduces novelties into the economic analysis, but also triggers a process of creative destruction (Schumpeter 1939) and therefore supports the transition from one development path to another. But what are sustainable technologies?

4. The Concept of Sustainable Technologies

Innovation is generally defined from the perspective of the creator of new products and processes. Innovation is then conceptualised as something fixed, as a well-defined 'objective' artefact. However, technology by itself has no value; its value comes from beneficial use (Dearing 1999). The locus of innovation is social practice; we can speak of an innovation only when a technology is in use (Tuomi 2001). Particularly the focus on sustainability makes it necessary to define innovation primarily from a user perspective. A technical artefact itself has no environmental impact; only if it is used in concrete production and in consumption processes does it become ecologically relevant. Here we will concentrate on production processes.

Besides a technical dimension, social practices also have an organisational and a cultural dimension; innovation therefore implies the concurrent emergence of technical, organisational and cultural changes. The use of new technology is associated with new forms of division of labour and co-operation as well as with new meanings. If we apply such a user perspective, then the concept of environmental beneficial technology means more than reducing emissions technically, for example. To speak about sustainable technology implies that a new technology needs to be embedded in new sustainable practices and it has to be given a new environmentally beneficial meaning. Progress towards sustainable development requires not only new ways of doing business, but also far-reaching shifts in corporate attitudes. Environment has to be internalised in corporate culture (Schmidheiny 1992).

If we look at concrete company strategies, we can identify a focus on technologies of waste disposal and recycling; they are mainly end-of-pipe technologies. These technologies are supplemented to the original production process without the introduction of major changes into the technical system. They are additive in the sense that the existing technology is supplemented by a new component with the aim of avoiding or reducing the damage to the environment, caused by the traditional technical system. Typical are filtration and purification plants, deposition methods and recycling technologies.

'End-of-pipe technologies' do not contribute to creative destruction; their aim actually is to continue production without changing the existing technical system. End-of-pipe technologies stabilise the existing technological system by repairing possible environmental damages (Diekmann and Preisendörfer 2001); they do not trigger the development of a new technological development path. They only produce incremental improvements along established pathways and may in the end lead to a situation of 'lock-in' (OECD 1999).

Applying the above user perspective, we may actually doubt whether we can characterise 'end-of-pipe technologies' as environmentally sustainable technologies. We have argued that innovation incorporates technical, behavioural and organisational changes as well as changes in the meaning of technology. However, 'end-of-pipe technologies' do not introduce significant changes in existing social practices. Employees involved in social practices do not have to change their work behaviour to a great extent; they are not forced to learn. Moreover, the meaning given to the existing technological system does not change; its major aim is to produce products or services in the most efficient way and not to ameliorate the ecological environmental.

From our viewpoint, only 'integrated environmental technologies' can be characterised as sustainable innovations because they introduce changes in social practices. They represent technical solutions that do not produce or at least directly reduce environmental damages (Diekmann and Preisendörfer 2001). As they aim at preventive avoidance of environmental damage either in the form of clean processes or clean products, the technology is also given a new meaning. While 'end-of-pipe technologies' contribute to the stabilisation of the existing development path, 'integrated environmental technologies' contribute to creative destruction and thereby help creating a new development path.

5. Coping with Uncertainty as a Rationale of Companies' Investment in Clean Innovation

There is, however, as some scholars have observed, a change in corporate environmental strategies from a defensive, reactive attitude to a pro-active and positive one. While the traditional 'resistant adaptation' resulted in the use of 'end-of-pipe technologies', there is now a shift to the development and use of cleaner products and processes (Fischer and Schot 1993). How can we explain this shift?

Cleaner technology is not a criterion for practical technology choice. Companies' engagement in clean technology cannot be understood as being motivated by a moral change in values. Companies do not diverge from the economic logic for the benefit of an environmental conviction. They do not invest in clean technology if they do not expect to be rewarded by the market. "(...) corporate decision-making", as Fukasaku argues, "is rarely based on purely environmental considerations, or on the selection of cleaner technologies for their own sake (...)" (1999). On the other hand, companies' environmental activities are seldom motivated by short-term profit expectations or by the aim to cope with actual damage (Dresel and Blättel-Mink 1997).

It is argued that companies increasingly associate clean technologies with a win-win situation (Porter and van der Linde 1995). The authors assume that clean technologies can generate environmental benefits and are at the same time cost-saving from companies' point of view. This argumentation seems to be too simplistic because, on the one hand, an investment in clean technology is often very costly, while, on the other hand, economic returns are highly uncertain, because investment in cleaner technologies will probably pay off only after a longer period of time. Therefore, it is only seldom the case that companies consciously use the concept of clean technology to shape their strategies (Dresel and Blättel-Mink 1997).

Rather, investment in clean technology represents a more general change in corporate strategy development. In an increasingly turbulent environment companies try to get control over the areas of uncertainty which could have an impact on their long-term strategies and revenues. Innovation can then be seen as a corporate activity aiming at getting control over situations of uncertainty by reacting to anticipated events and changes. "Clean technology" as Fukasaku argues, "is not a criterion for practical technology choice, but rather an element of broader corporate strategy, which can refocus it at a higher level in such a way as to build environmental criteria into decision making and the technology development process" (1999). Investing in clean technology therefore entails the same business logic as quality management, human resource management, or improved customer relationships. All these activities represent an attempt to give companies more control over situations of uncertainty in order for them to be able to secure the achievement of long-term economic goals.

6. New Insights into the Innovation Process and the Regional Dimension

The so-called linear model of innovation was the dominant approach in innovation research for quite a while. It mainly deals with explicit knowledge developed in research processes. In this model, basic research is placed at the beginning of a causal chain that ends in productivity growth mediated by innovation and diffusion. Each level in the linear model produces outputs that are transferred to the next level as inputs. The flow of knowledge is unidirectional, which means that later stages do not provide inputs for earlier stages (Kline and Rosenberg 1986). The main assumption of the linear model is that new knowledge will always find its way into marketable products without major transformation problems. But as Freeman argues, there is now increasing evidence that the linear model of innovation represents an exception rather than a rule (1987).

The 'circular model' suggests that, instead of interpreting innovation as a linear process, we have to understand the creation of novelty as a recursive process (Schienstock 1999). This means that we have to take into account complicated feedback mechanisms and interactive relationships involving science, technology, learning, production, and demand (Edquist 1997). While explicit knowledge is in focus in the linear model, the circular or recursive model emphasises tacit and codified but sticky knowledge. It conceives of innovation as an interactive process of a social nature, emphasising co-operation, not competition (Lundvall 1999). Much more than with the linear model, the focus is on the connection among company-internal, company-external, and technological factors (OECD 1992). As there is no clear development logic, an efficient innovation and knowledge management within and increasingly among firms becomes crucially important. Networks among firms and with knowledge producers are seen as the most efficient way of organising innovation processes. The main argument is that networks allow companies to specialise because they can expect to receive complementary knowledge from their network partners. And networks support inter-organisational learning, which is crucial for the necessary trial-and-error approach in innovation processes.

Particularly for the successful development of radical innovations, including sustainable technologies, spatial proximity and efficient knowledge management becomes crucial, as communication and knowledge exchange is increasingly difficult because codes, developed to communicate a constant, or a gradually changing technology, become inadequate. On the one hand, producers who have followed a given technological trajectory will have difficulties in evaluating the potentials of the new paradigm. Users, on the other hand, will have difficulties in decoding the communications coming from producers developing new products built according to the new paradigm. "The lack of standardised criteria for sorting out what is the best paradigm implies that 'subjective elements' in the user-producer relationships – like mutual trust and even personal friendship will become important. These subjective elements are not easily shared across regional borders" (Meeus and Oerlemans 1999)

We can assume a close relationship between the level of tacitness and stickiness of knowledge and the importance of spatial proximity. The fact that a larger territorial space may contain more diversity will not necessarily lead to innovation as long as there is not enough proximity to support intensive communication. The above argument suggests that, due to their radical character, codified knowledge is less relevant for environmental beneficial products and processes. Instead, the development of these technologies depends to a great extent on the exchange of tacit or sticky knowledge on the basis of trust and social capital. Also, the fact that concerned people and households have to be involved in the creation of a new development path points to the great importance of spatial proximity.

It is because environmentally beneficial innovations involve a great degree of tacitness and stickiness of knowledge that the regional innovation system is put into the spotlight. Efficient management of such knowledge is more easily achieved at the regional than at the national level. Some scholars have recently pointed to the key role of social capital in innovation processes (Lundvall 1998). Transformative innovations, it is argued, depend on trust-based relationships and a high amount of accumulated social capital as the production of these novelties often involves intensive exchange of confidential information and tacit knowledge. Dense and frequent links between people and organisations are more likely to develop in regions than in large countries, which explains their relevance for the greening of innovation systems. As regional institutions provide the basis for the development of trust and the accumulation of social capital, we will probably see intensive co-operation among involved organisations and open knowledge exchange supporting interactive learning and collaborative innovation processes.

7. Instruments of Sustainability-Oriented National Innovation Policy

So far, an innovation policy for sustainability is still far from being developed. Traditionally innovation policy was conceptualised as a dualistic model. According to Braun (1994), innovation policy has two functions. It should support, enhance and accelerate the development and use of technology, with the ultimate goal of strengthening the economy. But, in addition, it should also regulate the use and development of technology in such a way as to minimise risks posed by technology to health safety, the social fabric and the natural environment.

There is no doubt that risk management through environmental regulations has led industry to develop and adopt various pollution-control techniques and equipment, for example. In addition, as Porter and van Linde argue (1995), companies that are forced to adapt to high environmental standards may benefit from the 'early-mover advantage'. However, such a command and control approach (Fukasaku 1999) has seldom stimulated radical technical change. In many cases, forms of environmental regulation have been a predictable stimulus to small, incremental improvements along established pathways, often in the form of 'end-of-pipe technologies' (OECD Working Group on Technology and Sustainable Development 1999).

It is very unlikely that the regulatory approach solves the problem of under-investment in sustainable technologies.[3] To overcome under-investment in sustainable technologies, the state may be legitimated to intervene in the market process. A more dynamic environmental policy is needed, which focuses on the development and diffusion of clean technologies and integrated approaches aiming at promoting prevention rather then abatement (OECD 1999). Schienstock (1994) differentiates between two broad categories of innovation policies, namely direct technology policy which offers financial incentives to companies for their innovation programmes and innovation-enabling policy which focuses on public supply for infrastructures with the aim to attract companies to set up research activities and innovative production. The aim of direct technology policy is to help environmentally beneficial technologies over the initial barriers to acceptance by giving them a selective advantage via the tax system or direct subsidies. The fact that environmentally beneficial technologies compete against other technologies which have been developed under selection criteria where ecological aspects have widely been ignored, legitimates public intervention through the use of economic instruments.

The key problem with direct technology policy is that the government needs to be able to pick up winners. But more radical innovations are characterised by high technological and market uncertainty. Why should state bureaucracies be able to deal with these uncertainties in a more effective way than corporate management and select the most promising technologies? Because it is becoming increasingly difficult to anticipate technological, economic and social aspects, public agencies do not often have a solid basis for directing the change process and defining clear strategies of change, although technological foresight may help them to identify useful areas for technological development and to decide where to put the money.

It is certainly the case that the innovation-enabling policy type has increased, particularly due to the fact that economic competitive advantages can be constructed deliberately. In a competitive global environment, governments have to upgrade their institutional infrastructure to attract technology-intensive activities. Particularly in the case of developing a new sustainability-oriented techno-economic path, the countries that can provide the needed support for systemic innovation will have a competitive edge. The greening of the whole economy is not possible without institutions that provide new scientific knowledge, new skills and competencies, new legal regulations, needed financial resources, and a proper communication infrastructure. But while national governments can set up new supporting institutions, it is more difficult for them to make the system working

---

3 Two aspects have been mentioned to explain under-investment in sustainable technologies. First, the traditional market failure argument (Arrow 1962), which holds true for innovation in general, is particularly applicable for sustainable innovations (Hübner and Nill 2001: 73). Second, sustainable technology has to be developed outside the existing development path and therefore requires particular effort.

because here proximity is often decisive. In this respect, regional governments may be more efficient.

8. Regional Policy for Sustainable Development

As environmental problems are often of a local character, it is obvious that regional authorities have a monitoring function concerning environmental pollution, for example. However, regional governments can play a much more active role in the greening of systems of innovation as the above new insights into the nature of innovation processes suggest. The specific instruments of innovation policy need to be adapted within a broader framework that stresses the importance of policy coherence and of inter-linkages within innovation systems.

The development of a systemic vision, or a new *Leitbild* of techno-economic and social development, can be seen as a key element of the network-enabling innovation policy. A *Leitbild* can be defined as a symbolic scheme for creating reality; it includes general ideas about the future structures of the economy and society (Berger and Luckmann 1966). But a *Leitbild* also has a normative dimension, as which it becomes the basis of practical restructuring processes. A major advantage of a *Leitbild* is that it makes communication among social actors possible, even if they have different interests and preferences.

In order to foster the greening of the economy, propagating 'sustainable development' as the new *Leitbild* of economic development can be seen as an important element of modern innovation policy (Renn 1997). But too general a *Leitbild* hardly releases concrete restructuring activities, as it becomes very difficult to deduce strategies for solving existing problems. A Leitbild developed on the regional level is probably closer to concrete problems than the one developed at the national level and is therefore more likely to become the basis of practical restructuring processes. This is the more the case, the more interests are represented in the process of creating the new Leitbild. Again, regional systems have an advantage because they represent genuine communities of economic interest and can take advantage of true linkages and synergies among economic actors (Ohmae 1993).

9. Supporting Innovation Networks as a Core Element of Regional Policy of Sustainable Development

Policies that promote research collaboration, facilitate firm networking and clustering, encourage institutional ties, and involve people concerned are taking on new significance (OECD Working Group on Technology and Sustainable Development 1999). A new type of innovation policy that can be characterised as network-facilitating policy emerges (Schienstock and Hämäläinen 2001)[4]. In the en-

[4] The following part relies heavily on this publication.

vironmental realm, partnerships and networks are valuable because processes of specialisation can accelerate the development and diffusion of clean technologies and reduce obstacles (OECD Working Group on Technology and Sustainable Development 1999). As networking becomes the core of a new sustainability-oriented innovation policy, regions assume a much more important role in this field.

The specialisation, dynamism and social embeddedness of networks makes network-facilitation a demanding challenge for policy-makers. Sophisticated interventions require deep knowledge about the major problems to be dealt with, the relative efficiency of different organisational alternatives, as well as the specific strengths and weaknesses of potential support institutions and partner firms. Regional governments often have an information and knowledge advantage over national agencies in this respect. Moreover, since the feasibility of carrying out complex inter-organisational co-operation declines with geographical distance and increasing knowledge diversity, the preconditions for successful networking are also best at the regional level (Scott and Storper 1992). It is similarly important that involving households and people concerned in sustainable innovation processes can best be practised at the regional level.

From the viewpoint of industrial innovation and sustainability, regions have an additional advantage. Due to their proximity and flexibility, regional networks provide an ideal platform for carrying out social innovation experiments which are often very complex and involve a great number of actors, needing close interaction between various kinds of firms, consumers and government agents. Meyer-Kramer (2001) mentions the change from product-ownership consumption to use-oriented consumption as an interesting social experiment in which regions can provide an appropriate basis for the implementation of social experiments, and the experimenting with various options. The aim here is to encourage households to change their patterns of behaviour significantly, to transform products into services, and to stimulate new technical concepts.

The practical problems of networking change in different phases of the networking process. The following analysis of such problems follows the phases of a typical networking process (Schienstock and Hämäläinen 2001): (a) firms' awareness of networking opportunities, (b) search for partners, (c) building trust and a shared knowledge base, (d) organising the network, (e) adding complementary resources, (f) stimulating demand, (g) involving concerned persons, and (h) stabilisation of co-operation.

The nature and potential benefits of network co-operation are not always very well known and internalised, particularly not among small firms. This information problem may slow down organisational adjustments among firms that could benefit from active network co-operation. Regional governments can promote firms' awareness about networking, for instance, by arranging seminars, by distributing information and by trying to get the media to cover successful examples of networking. It is important to form a 'critical mass' of firms and other knowledge creating partners for the formation of innovation networks.

Finding appropriate partners for co-operation involves another problem. Many surveys have shown that the most important reason for not participating in cooperative networks is that there are no suitable partners available or that they are difficult to find. Particularly SMEs have difficulties in finding partners within universities or other research institutes that can provide the knowledge needed to develop clean products or eco-efficient processes. Governments can support firms' own search for network partners with information, brokerage and matching services (Lundvall and Borrás 1997).

Finding potential networks and partners is not easy. It requires deep knowledge about firms' specific strengths and weaknesses and about how they could complement each other. This is particular the case when the aim of network formation is fostering the development and diffusion of sustainable technology, as this implies the co-operation of firms from different industries. With respect to sustainable technology, the diffusion aspect is of particular importance. As knowledge-intensive business service (KIBS) firms play a crucial role in the process of knowledge diffusion, it is important to involve them in the process of network formation from the beginning. However, it is often the case that KIBS firms specialised in problems of sustainable development and sustainable technologies do not exist; public network policy therefore also needs to pay attention to the development of such KIBS firms.

Experiences suggest that the search for potential network partners should take place very close to firms, an aspect, which again favours regional solutions. Furthermore, it suggests that network policies should not aim to create new networks from scratch: network promotion could be focused on emerging but fragile networks, which require further encouragement and support. To solve the problem of finding adequate partners, public/private partnerships to conduct applied research in the field of sustainable technologies can also be set up.

Before networks become more stable and co-operation functions efficiently, many mental barriers must be overcome. In fact, the mental rigidities and old behavioural routines of entrepreneurs are often seen as the biggest hurdle to effective networking. Potential partners need to learn more about each other's worldviews, beliefs and attitudes, values, business strategies, and operating methods. This can only be done through an intensive and open discussion in which the participants gradually build trust and a shared knowledge base. Being a neutral and trusted 'third party', regional governments can often reduce the suspicions and reservations that firms have toward closer inter-firm co-operation and co-opera-tion with research institutes, particularly when some partners come from traditional smokestack industries.

Building shared understandings and trust takes time. As a result, regional governments should favour policies which provide firms with adequate incentives to continue participating in the networking process long enough to build the necessary shared knowledge base and social capital. Setting up long-term network facilitation programmes and building inter-firm meeting arenas may be more productive than trying to more directly match potential partners who have not had enough time to learn to know each other well or to build shared understanding and trust. Once firms understand and trust each other, they can start to build a shared vision, strategy, structure and behavioural rules for the network. A shared vision of the future is an important co-ordinating mechanism, particularly in sustainability-oriented innovation networks.

New, emerging networks do not often have all key resources and capabilities required for competitive success. For example, a key technology or other input may not be available from the existing network partners. In addition, network-facilitating innovation policy also needs to focus on the development of learning organisations and competence building within networks (Lundvall and Archibugi 2001), as organisational and other social innovations are of particular relevance for the greening of the economy. SMEs, being left alone, will hardly undertake a fundamental transformation of their business structures to be able to continuously develop their ecological competence and improve their environment-related knowledge. To overcome these difficulties, governments can, for example, focus on workforce training and encourage managerial and organisational changes among firms to improve their ability to assess and adopt sustainable technologies (OECD Working Group on Technology and Sustainable Development 1999).

It is often not missing knowledge but a lack of demand that limits the technological progress of sustainable technology. Companies invest only if the market rewards the production of green products and services and the application of eco-efficient process-technologies and consumers demand such changes. This implies that consumers themselves have to change their consumption behaviour, everyday buying patterns and life styles, and to refrain from the current resource intensive habits and practices. However, while consumers have become more aware of environmental issues, it has not yet translated into far-reaching changes in actual life. Regional governments can take initiatives to shift demand towards products that are more supportive for environments. They can, for example, encourage reporting by enterprises on emissions and the environmental implications of their activities. And they can overcome information deficits by increasing consumer knowledge of the ecological impacts of their consumption pattern and product choice through launching their own campaign to foster demand for sustainable products (OECD Working Group on Technology and Sustainable Development 1999).

Technology-related discourses involving various stakeholders can be seen as important co-ordination mechanisms of transformation management. Broader societal participation must be guaranteed and households should be involved in such technology-related discourses, not only to influence their consumption practices. Their involvement is also crucial to get their backing for more concrete steps towards the greening of the economy and to avoid public resistance and serious conflicts in later stages of the creation and development of the new development path. There are major trust implications for the acceptance of specific technological paths and for reaching an agreement on how to manage technological risks. Informing the public is not enough; instead, it is important to establish a discursive confrontation between the persons and organisations who gain from the renewal of the economy and who may suffer from the technical and social innovations (Renn 1997). A technology-related discourse can be viewed as a platform to jointly create and exchange information among social actors. Discursive co-ordination is not intended to create consensus among the participants in the first place, but it aims at initiating learning processes.

To improve long-term perspectives of business partnerships, regional governments can set up specific institutions that provide the needed services for the stabilisation and further growth of co-operative networks in close co-operation with national agents. Here we can mention centres of expertise in environment technology, eco-industrial parks, or regional environmental cluster programs. Network policy, however, can also produce failures. Business networks may become dependent on state support, which may actually hinder necessary change processes. Lundvall and Borrás (1997) mention the integration/flexibility dilemma. The advantage of networks is seen in their flexibility and openness; however, in later stages and due to invested interests, they can become mechanisms which prevent network partners from adapting to new conditions. Particularly in a period of rapid technological change, specific networks may become inefficient and block the environmentally beneficial renewal of regional economies.

While there is general agreement that establishing co-operation networks, including technology-related discourses, becomes an important instrument in the creation of a sustainable development path and gives regional governments an important role in technology policy, there is little knowledge about how such a policy can be conducted. What exactly is the new role of regional policy-makers in the greening process of the economy? How should they intervene in the transformation process?

10. The Role of Government and Policy Learning

Particularly the region-state, as we have argued above, has an important role to play in the process of path creation and transformation management. However, it becomes quite clear that in a period of a changing development path, the role of the state must be reconsidered. Governments can no longer assume the role of a sovereign economic actor steering the innovation process through bureaucratic forms of control. Creating a new development path implies a lot of uncertainty. Therefore, in a transformation period, the significance of technical, macroeconomic management may decrease but the role of the state as a facilitator and orchestrator of different interests of various social actors remains strong (Hirst and Thompson 1992).

The role of the state in a transition period towards a green development path can be described as a catalyst for innovation processes, a supporter of ongoing research and innovation activities, a facilitator of co-operation in R&D, an organiser of a dialogue between various social actors about future development, and as an initiator of critical questions and new tasks. Sabel (1995), as we have mentioned earlier, characterises the role of the state in transformation periods as an initiator of bootstrapping reforms; his main task is to get actors moving in the same direction because it might be more risky to stay put than to move in the wrong direction.

In the context of a major transformation, we can characterise innovation policy as a process of policy learning. Such an interpretation is quite different from traditional conceptualisation of innovation policy, which assumes that a decision-making process consists of three clear-cut stages: setting goals, developing programmes, and implementing projects. The learning approach, on the other hand, provides a fluid perspective of a policy process in continuous transformation and evolution where no such stages can be discerned (Lundvall and Borrás 1997). Policy learning relies on intelligent benchmarking, policy evaluation, technological foresight, and assessment studies. The main aim of these instruments can be seen in promoting a dialogue among users, producers, other social groups concerned as well as policy-makers.

11. Conclusion

In this paper we have argued that the traditional focus of systems of innovation on knowledge creation and knowledge diffusion might be too narrow. We have stressed the need of feedback analysis pointing to possible unintended consequences of an accelerating innovation dynamic such as environmental damage and social segmentation and exclusion. This implies that sustainability needs to become a key aim of systems of innovation.

We agree with Fukasaku, who argues that the sustainable development imperative implies a change in the dominant technological paradigm to shift the existing technological trajectories (1999). Environmental issues have to be associated with production technology, we can no longer focus on consumption and post-consumption; instead, sustainability concerns business. Talking about sustainable technology, however, means more than ‘end-of-pipe technologies’; they actually do not trigger a fundamental transformation process of the innovation system. Integrated environmental technologies, on the other hand, contribute to creative destruction and thereby bring about the basis for a new development path, demanding complementary behavioural, organisational, and institutional changes at the same time.

New insights into the nature of innovation, we have further argued, have led to the adoption of new instruments of innovation policy, which focus on the inter-linkages within innovation systems. Policies that promote research collaboration, facilitate firm networking and clustering, encourage institutional ties, and involve people concerned are taking on a new significance. At the same time, regional governments become key players in the innovation system. Developing business networks, including the establishment of technology-related discourses involving a variety of different stakeholders, we have stressed, are in the centre of regional innovation policy. In this paper we have discussed the various stages and strategies of such a network policy. We have also argued, however, that such a policy needs to be applied with care because, particularly in a period of rapid technological change, specific networks may become inefficient and block the renewal of regional economies.

While it is important to better understand the new role of regional governments in innovation processes aiming at sustainable development, it is also crucial to link regional activities with processes at the national and trans-national level. Many problems are too complex for only national or even concerted international activities to resolve. Innovation policy that aims at sustainable development therefore needs a much broader approach combining various instruments and integrating various policy levels. We are far from the application of such an integrative approach and we have definitely not enough knowledge of how to design and implement it.

Green Innovation in Nordic Energy Industry: Systemic Contexts and Dynamic Trajectories

Atle Midttun and Anne Louise Koefoed

1. Introduction

Greening of industry is becoming part of government policy in most developed countries, partly as a reaction to the global climate debate, but also as reaction to more specific national environmental needs. In the electricity-sector one of the major greening initiatives has come through innovative development and use of new renewables. The Nordic countries have been among the forerunners in this development, and have achieved impressive results with development of new and more efficient windmills and incorporating them in the electricity system; with development and improvement of combined heat and power (CHP)-technology; and with trying out new types of biofuel-feedstock.

A striking feature of these innovation and diffusion successes has been the diversity and complexity in institutional and commercial settings and the variation in trajectories under which they have taken place. This indicates that we have to deal with innovation systems rather than innovation within well-defined boundaries of firms or political entities.

The article summarises research into the institutional framing and commercial development of six cases of successful green innovation of Nordic energy systems, including wind energy and biofuels in Denmark, combined heat and power and biofuels in Finland and Sweden. Taking a systems of innovation perspective it discusses the interplay between processes within a broad set of institutional contexts, ranging from politics to markets and their importance for successful innovation.

The article also focuses on the dynamics of innovation systems. The cases exhibit interesting dynamic variations where the innovation has passed through a sequence of institutional contexts. Not only the types of institutions, but also their sequencing provides important clues to the innovation and diffusion success.

The five case studies presented in this report are selected because they represent impressive success-stories in commercial and technological innovation. Firstly, they have taken large market shares of electricity generation and heat supply. In addition, most of them have impressive technological and industrial spin-offs. These spin-offs also generally generate large export income.

2. Theoretical Framework

Our case studies of advanced green technological innovation in the Nordic energy systems show that major re-orientation of these systems is a complex process, typically involving: change in production processes; change in consumption patterns; institutional change, and change in social and political practices. This complexity has to do with the fact that energy industries, besides being major industrial sectors in most countries, traditionally have had focal attention in public planning and public debate.

The study of complex processes involved in green energy-industrial growth has therefore required a parallel richness in theory. We find theoretical basis for such a broad analysis in the theory of innovation systems, which focuses on the systemic character and complex organisation of innovation. The Innovation systems concept was introduced in the late 1980's and established in the innovation literature as a result of the collaboration between Freeman, Nelson and Lundvall (Dosi 1988). Studies pursued by Freeman and his colleagues at SPRU in the beginning of the seventies (Rothwell and Zegweld 1981) gave strong support to the idea that success in innovation has to do with long-term relationships and close interaction with agents external to the firm.

As pointed out by Lundvall (1998) and Edquist and Mc Kelvey (2000), the most fundamental reason for innovation scholars to begin to think in terms of systems has to do with the fact that it was increasingly realised that innovation is an interactive process. This interactive process is seen as taking place within national institutional contexts and suggests that country-specific technological trajectories are shaped by the systemic and structural components of society and diffusion of knowledge required for industrial innovation (Bartholomew 1997). There is, in other words, a co-evolution between institutions and technological development taking place.

Closely related to this conceptual approach is also Michael Porter's influential work on industrial clusters (Porter 1980, 1990, 1998), which may be seen as an important precedent to the innovation systems perspective. Industrial clusters, as Porter sees it, represent a kind of new spatial organisation of technical-economic processes in between arm's-length markets on the one hand and hierarchies, or vertical integration, on the other. According to Porter, the clustering process is critical to industrial dynamics by increasing the productivity of companies based in the area; by driving the direction and pace of innovation; and by stimulating the formation of new businesses.

A central factor in both the innovation systems and the cluster theories is the concept of positive externalities. Both perspectives recognize that knowledge typically transcends the boundaries of even large firms. The systems of innovation and cluster perspectives, therefore, explore ways to unleash these externalities within environments rich in complementarities.

Besides highlighting the complex interplay between the firm and its commercial, institutional and political environment the innovation systems and industrial cluster- perspectives also lead to the recognition of the importance of national contexts for innovation. The innovation systems perspective, particularly in its national systems of innovation version, here shares many features with a more general national style or path dependency perspective, which focuses on national/regional differences in how firms and markets are structured and operate in general. This general argument is developed under several labels: Business systems (Whitley 1992), social systems of generation (Campbell et al. 1991) and modes of capitalist organisation (Orru 1994).

Implicitly, and sometimes also explicitly, the national-styles literatures links up to a path dependency argument which claims that industrial systems cannot develop independently of previous events and that local positive loops serve to propagate traditional patterns into future strategic decisions (David 1993). Thus, the path dependency and national-styles literatures emphasises that national differences in strategic orientations, including innovation, are likely to be maintained even under international competitive conditions. Innovation systems are, in other words, likely to specialise and to utilise specific competencies and resources within their respective industrial and national milieus. This is indeed illustrated by the technological diversity in the greening of electricity in our case-examples.

In the case of greening of electricity industry we are analysing innovation in a sector, which has traditionally had strong ties to the political system. This implies that the system boundaries in this case should be defined so as to include political elements. We have chosen to pull such an extended politico-economic analysis together by using the concept green energy-industrial clusters, drawing on Michael Porter's analysis of business clusters as a core element in the innovation systems. However, our analysis expands beyond the business strategy context to encompass also the political and societal processes by which government policy is legitimised. Porter's attempt to break out of a limited market analysis into a broader strategic focus is, in other words, taken one step further, where the political and societal processes and institutions are studied more thoroughly. In this wider interpretation, the cluster and innovation system concepts have allowed us to integrate the different elements of a fairly rich framework for understanding innovation departing from a focus on a given technology and its commercial and political-institutional context. We shall briefly expand on three elements: the commercial core, the societal basis and the political-administrative system.

2.1 The Commercial Core

According to Porter's cluster-perspective, a basic part of economic analysis should focus on geographic concentrations of interconnected companies and institutions in a particular field. Given the centrality of both market-diffusion and product-innovation processes, it must include both production side and market side theory.

In our case, the cluster-perspective may be applied at two levels. Firstly, the green energy-industries constitute their own value chains, with their own commercial dynamics, and their success may presumably be analysed in terms of standard "Porterian" business strategic analysis, as interaction between agents belonging to a green innovation cluster.

Secondly, emerging green energy-industrial clusters may also be analysed as new entrants in the value chain of incumbent energy-industries. In this case they are seen in relation to a larger system, where they may function as rivalling or innovative elements. While our focus is primarily the first, we shall also visit the second focus as part of our analysis.

2.2 Societal Basis

The commercial core, both in terms of productivity, market penetration, innovation and formation of new business, is obviously embedded in a wider societal arena, which becomes highly important when we focus on a highly politically sensitive area such as energy & environment. This embeddedness creates what could be termed a "bottom up" basis for technical and commercial development, as it provides societal predispositions that may be utilised both for production and marketing purposes.

In the words of Aldrich (1999) we may - in addition to the commercial logic, talk of a socio-political legitimacy, referring to: "the acceptance by key stakeholders, the general public, key opinion leaders, (and government officials)[1] of a new venture as appropriate and right".

2.3 The Political-Administrative System

Given the high level of public attention to energy and its environmental impacts, the political-administrative system easily also becomes a central part of the innovation system. Some of the primary functions of the political-administrative system vis a vis desirable technical-commercial activity is to provide an institutional framework, to mobilise resource flows, and, if necessary, to provide legitimation. It is likewise the prerogative of the political-administrative system to meet undesirable commercial activities with institutional obstacles, demobilised resources and to generally undermine their legitimacy. How green innovations are favoured or discouraged by political processes and administrative procedures may, therefore, be of critical importance to its success.

2.4 The Extended Cluster Concept: A Summary

To sum up, the tree major elements that are included in our extended innovation-cluster- concept and some core relations between them are presented in figure 1.

---

[1] Sections of the quote in parenthesis refer to the next section "political administrative system. For the sake of brevity the entire quote is placed in this setion.

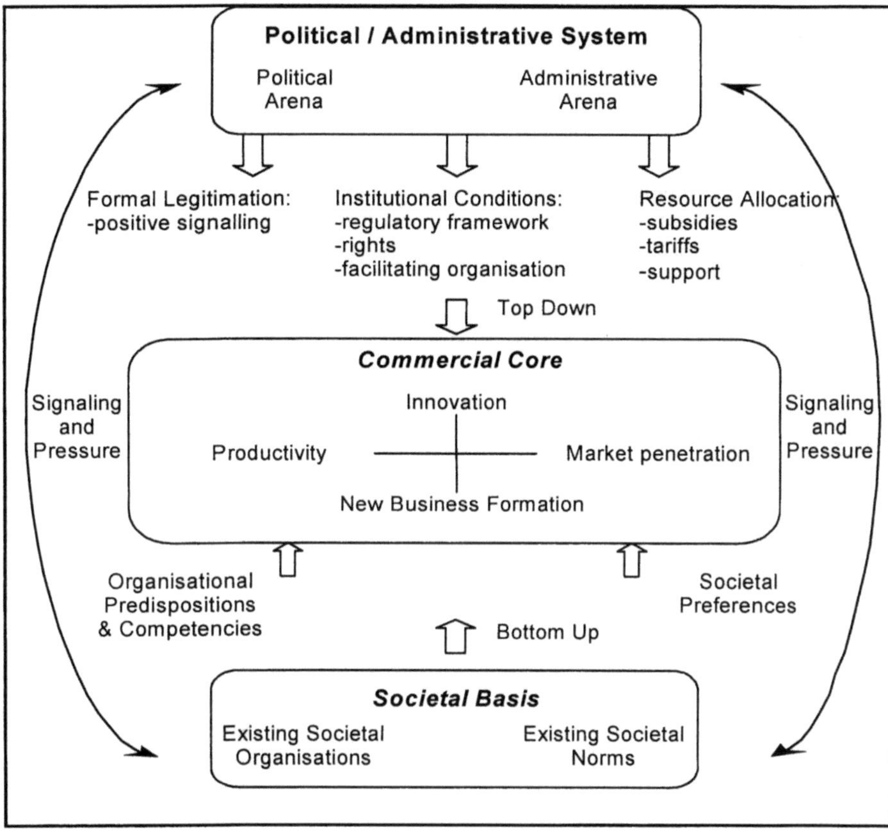

Fig. 1. Core Elements of the Model

The commercial core, understood in terms of Porter's clustering is posited in the middle, including both supply side and demand side aspects of innovation and new business formation. This commercial core is then related both to political-administrative (figure 1) and societal/prepolitical forces (figure 2). The former operates through formal legislation, institutional conditions and resource allocation. The latter operates through organisational predispositions and competencies and societal preferences.

2.5 Dynamics of Innovation Systems

In addition to the complex interplay between a number of institutional elements and societal spheres, the cases also exhibit interesting dynamic variations where the innovation and diffusion has passed through a sequence of institutional contexts. Not only the types of institutions, but also their sequencing provides important clues to the innovation success. Arguably, all cases involve a trajectory from experimentation through selection and to institutionalisation/ retention of the new green technologies. However, these functions are served by fairly different institutional contexts, in each of our cases.

As a framework for the dynamic analysis we have chosen the Vernon – supply side – and Kotler – demand side-models interpreted within an evolutionary framework (figure 2).

The so-called product-cycle model (Vernon 1981, 1985) spells out core functional stages in innovation. Here industrial transformation is seen as going through distinct stages 1) innovation and development, 2) production, and 3) mature production.

A complementary model of product-adoption dynamics, associated with Philip Kotler (1991) sheds light on dynamic processes seen from a marketing point of view. According to this model, consumers are generally classified as innovators (the first group of customers to buy a product) early adopters, members of the early and late majorities and laggards. In addition come non-adopters, who do not buy the new product at all (Kotler 1991).

Both the Vernon and Kotler models may be interpreted within an evolutionary perspective where variation is generated in an early phase, followed by later selection and retention and further diffusion.

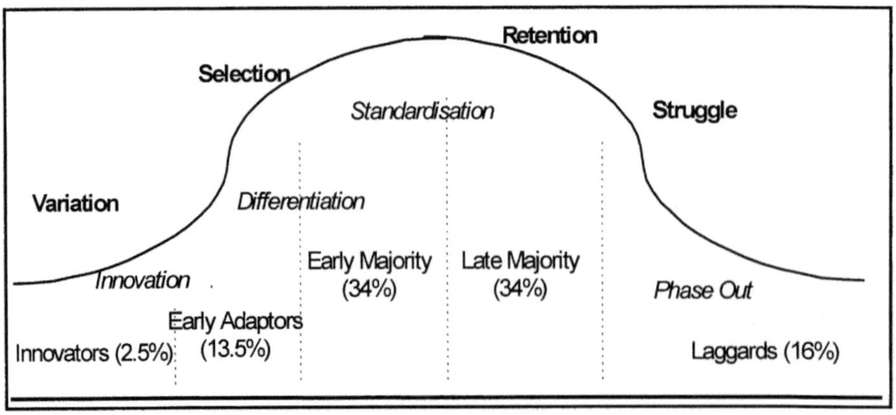

Fig. 2. The Vernon-Kotler Models in an Evolutionary Framework

The extended product-cycle-evolutionary model allows us to move from a discussion of the composition of elements of the green energy innovation and diffusion systems to how these elements are sequenced over time. This focus brings out dynamic aspects of innovation-systems that have been under-explored in the literature.

In the following sections, we shall first present a general analysis of case studies of successful greening of Nordic industry with a focus first on the technical economic core of the innovation system and then on societal and political-administrative aspects. Thereafter, the paper pursues a dynamic analysis of two of the cases, the Finnish Chp and Danish wind power.

2.6 Success Cases

The five case studies presented in this report are, as already mentioned, selected because they represent impressive success-stories in commercial and technological innovation. Firstly, as indicated in table 1, they have taken large market shares of electricity generation and heat supply.

Table 1. Market Shares for New Energy Clusters

CHP % of EL consumption (1997)		**Bio fuel** % of energy consump. (1997)		**Wind** % of EL consump. (1997)	
Denmark	39%	Finland	25%	Denmark	6 %
Finland	34%	Sweden	19%	Netherlands	0.5%
Netherlands	30%	Denmark	10%	Germany	0.4%
Austria	23%	Switzerland	5.6%	Spain	0.2%
Czech Republic	18%	Norway	5%	Sweden	0.16%
Germany	14%	France	4%	England	0.16%
Portugal	13%	Canada	3.8%	Ireland	0.06%
Hungary	12%	Austria	3%		
Italy	11%	USA	3%		
Poland	10%	Germany	1%		

Inside Energy 1997, IEA/OECD statistics-Electricity Information 1997, Energistyrelsen 1997 statistics

In addition, most of them have impressive technological and industrial spin-offs. These spin-offs also generally generate large export income. To briefly recapitulate:

Danish wind power now holds more than 7% of the total electricity supply in Denmark; Danish windmill industry has supplied 50% of the worlds windmill capacity; Danish wind industry has 10.000 employees, and is Denmark's 3rd largest export-industry. Total sales amounted to 5 billion DKK in 1997 and 12.5 billion DKK in 1999.

The Finnish bio-energy industry supplies 25% of all energy in the country, and thereby has the industrialised world's highest use of bio-energy. Together with Sweden, Finland is in a world-leading position in combustion, harvesting and logistics/transport techniques related to bio-energy industry. Together with CHP-related technology, bio-energy production and consumption technology provided Finland with 12 billion FIM in export value in 1998.

Swedish bio-energy industry had a market share of 19% of total Swedish energy consumption in 1997. The industry consumes more than half of the 87 TWh of bio-energy used in energy production. 50% of district heating in Sweden is based on bio-energy, equivalent to 25 TWh. Since CHP is still not extensively applied in Sweden, only a small share of the electricity production is based on bio-energy. Like Finland, Sweden is a major exporter of various biofuel-handling technologies.

Danish bio-industry supplies 6% of total energy supply (1997); if incineration of waste is included, we are talking about 10%. Sales from bio-related industry and bio-related product categories are the second largest renewable export area after wind power. The value of bio-related technology exports was 750-1000 million DKr in the early 1990s. When direct energy- and heat producing companies are excluded, some 160 Danish firms have been indicated to be somewhat linked to the energy sector's use of bio-fuels.

Finnish CHP ranks as the primary technology choice for electricity production, accounting for 1/3 of total Finnish electricity production. District heating related CHP furthermore provides 1/3 of the Finnish heating market. Developments in the Finnish CHP-sector have also appended a broad supply industry from valves, boilers and piping to advanced combustion technologies. Technology-export of energy related products, where CHP-related products constitute the major part, provided Finland with 12 billion FIM in 1998.

3. Analysis of the Technical-Economic Core

Generally speaking, the case studies support Porter's cluster theorem, that successful industrial development is related to the interplay between national factor endowments, advanced supply-industry and demanding consumers in addition to rivalry between competing firms. They also support the system of innovation perspective that innovation takes place in rich knowledge environments with learning across firm- and institutional boundaries. However, the case studies also illustrate how green innovation in energy industry may take on a large variety of forms, depending on industrial structure, traditions etc.

3.1 Finnish CHP and Bio-Energy Industries

As indicated in table 2, the Finnish CHP and bio-energy industry cluster successes both rely heavily on their links to the Finnish forestry-based industry. With its world-leader position, the forestry-based industry has had the volume necessary for creating a significant internal market for CHP, as well as for supplying CHP-production with wood waste products from its own primary production, at an impressive scale. The Finnish CHP and bio-energy cluster success can therefore to some extent be explained as a spin-off from the Finnish forestry industry, which created an early home market both for fuels and energy technologies.

One of the success-factors of the Finnish cluster, is the modification of boilers to let coal be replaced by local industrial wood waste and by-products from the forestry-based industries. The competency was later transferred from the forestry industry to the municipal district heat producer in Helsinki, and thereafter to other municipalities, responsible for their local heating systems.

The fact that the "green energy systems" have been adopted as attractive options for municipal heat and electricity companies implies that they have found a wider market outside the forest industry, and hence even larger volumes on which to base their product innovation.

The success of Finnish CHP and bio-energy industry in terms of traded volumes clearly relates to their dual market-structure: on the one hand, they serve the need of the Finnish forest industry and provides process heat for electricity production as a by-product. On the other hand, they supply an expanding Finnish district heating system that began its development during the 1950's and 1960's. These heating systems were largely transformed to CHP-technology in the 1970's and 1980's, after the heat-load was first established.

However, there are also notable differences between Finnish CHP and bio-energy developments regarding cause and effect relationships. Whereas the greening of the energy sector may be viewed as a *consequence* of enhanced penetration of industrial CHP to serve largely industrial needs, the greening of the energy sector through bio-energy developments was strongly politically *motivated*. The different degrees of politicisation also probably, to some extent reflect timeframes within which CHP and biofuel developments emanated.

Table 2. Main Features of the Industrial Economy of Finnish Biofuel and CHP Innovation

Finnish bio-fuel	BioEnergy: • Early stages (prior to 1970) use of by-products (black liquors and wood waste) in the forest industry for internal generation of process heat and industry autoproduction in Chp-boilers • From mid 1970s: increased use of peat in industry electricity production (mostly in conventional boilers, not Chp-boilers) • A massive growth of municipal district heating using peat as a fuel, greater involvement from energy companies • From 1990: refocus from peat to wood waste both in industry and municipal energy company Chp-boilers Late 1990s: increased focus on gasification technology for the use in electricity production at CCGT-plants, involving the largest Finnish energy companies.
Finnish Chp	CHP: Early stages (1920-30s): strong anchoring in Finnish manufacturing forest industry, in which CHP was used for internal heat and power production. The adoption and diffusion of CHP within industries owes to a number of factors: • Availability of cheap local fuels (black liquors and wood wastes) • Poorly developed rural electricity networks • Guaranteed internal markets (process heat and power) • Technological competence and 'know-how' • Abundant process heat allowed for rural electricity supply From 1950-60s: anchoring in district heating • Access to existing grid • New guaranteed market From 1970s: involvement of municipal, industrial and public energy companies • Transformation from separate heat and power networks to CHP networks • Anchoring in public policies aiming to maintain a versatile and flexible energy sector

3.2 Swedish Biofuel Industry

As indicated in table 3, the Swedish bio-energy industry resembles the Finnish in so far as it has large factor-endowments in waste from forestry-based industries. Like in Finland, the Swedish bio-energy industry developed in close interplay with these industries. In an early phase, bio-energy was developed for use in internal markets within the wood/paper & pulp industries, and technology-development took place within this framework. The market for bio-energy in Sweden gradually expanded to district heating systems in the municipalities surrounding the forestry-based industries. Because of high transport costs, the more or less unrefined by-products had to be used only for local heat production.

Main features of the industrial economy of Swedish biofuel innovation[2]:

- Early stages: Strong anchoring in paper & pulp industry, which provided black liquor and wood waste as fuel in autoproduction
- From mid-1970s: municipal district heat creates a new market for wood waste in forest-rich areas
- From early 1990s: refined and dried wood-based energy (pellets) produced with greater involvement from energy companies. Pellets produced for national and international markets, long-distance district heat production and a new market niche in direct heating of detached houses. Involvement of large energy companies and oil companies
- Late 1990s: increased focus on gasification technology for the use in electricity production at CCGT plants, this involving the largest Swedish electricity companies

However, when dried and compressed bio-energy in the form of pellets was developed in local energy companies, the Swedish bio-energy industry expanded earlier local markets to supply distant district heat production companies and boilers in detached houses. Easy and cheap transportation of pellets later led to production capacity being established in other Nordic and Baltic countries, implying international competition for Swedish producers, and to a temporary surplus production capacity occurring.

With the development of national markets, energy companies, both at the local and national/international levels, major energy-companies, such as Sydkraft and Vattenfall have taken more active part in the bio-energy industry, with the aim to develop next generation biomass gasification-technology for electricity generation in CCGP-plants. Innovation in production technology has, in other words, produced biofuel products that are more easily transportable, which again has made it attractive to larger market players.

---

[2] The table summarises findings in Eikeland (1999).

3.3 Danish Wind Power and Biofuel Industries

Compared to the Finnish CHP- and biofuel cases the Danish wind power success has had less direct industrial, and more articulated political-administrative ties (table 4). However, given local and national demand for wind energy systems, Danish agro-mechanical industry with traditional ties to Danish agriculture, proved instrumental in developing state of the art wind technology. Later on the national technology centre at Risø stimulated further improvement of the windmill-industry by certifying technical solutions and stimulating further technological developments.

Table 3. Main Features of the Industrial Economy of Danish Biofuel and Wind Innovation

Danish biofuel	• Largely straw-based • Deliveries from Danish agriculture • Large Danish plant- and boiler industries are using biofuel technologies as part of a strategy to become comprehensive turnkey technology providers, this incorporating renewable energy and environmentally efficient power plant technologies • Denmark has become a leader in strawfired-technology for CHP and DH production • Industrial movement and electricity sector movement towards a 'multi-fuel' power plant concept
Danish wind	• Productive interplay between small-scale initiatives and industrial competencies with roots in supplies to farming and mechanical industry fishery. • Homemarket stimulated with demand from private co-operative owners • Strong cost-reduction • Diffusion to larger national and international markets.

With the home market as a basis, Danish wind-mill producers have built up a world leading export industry with up to 60% of the world market. A largely politically orchestrated domestic energy-system conversion has thus fostered high level industrial competency and supplier industry, which has also strengthened the industrial base for Danish mechanical industry. Throughout its fairly short modern history, Danish wind energy has achieved remarkable cost-reductions. From 60 øre / kWh in 1984, the long term marginal cost has decreased to 25-32 øre / kWh in 1997.

Compared to Danish wind-industry, Danish bio-industry, as indicated in table 2, has been more directly linked to the Danish agrarian economy. As contrasted with the dominantly wood-based Finnish and Swedish bio-fuel industry, the Danish bio-fuel priorities have focused on straw and manure from live feedstock production, this reflecting the different resource-endowments of the agrarian Denmark as compared to the extensive forestry opportunities of its Nordic neighbours.

The more or less exclusive use of straw in biofuelled Danish CHP production has allowed conversion from food to fuel production as Danish farmers have been obliged to reduce arable land for food crops as part of the EU programme for reduction of agrarian over-production. The public policy dictated strategy for straw-based biofuel supply thus presents Danish farmers with an alternative market for a by-product that has little other commercial options. However, similar to the wind industry, the Danish biofuel conversion has led to new advances in supply industry, hence, Denmark has developed advanced competency in boiler technology with multifuel applications.

Referring back to our discussion of the systemic character of innovation and diffusion we have argued that innovation will be shaped or influenced by a broad set of institutions and processes with the segment of the economy that are the targets of our analysis. Given the strong political attention attracted by the energy-system, Porter's cluster concept and technology-based systems of innovation perspectives seem too narrow to capture major innovation-drivers and needs to be complemented by a broader focus on political and societal institutions.

4. Societal Basis and Political-Administrative Anchoring

Our summary of key elements from the case-studies in tables 5-7 indicate that, similar to the industrial anchoring, both the political and societal anchoring of green clusters may take a variety of forms:

4.1 Finnish Bio-Energy Production and CHP Development

As indicated in the table, in the case of Finnish bio-energy production and CHP development, cluster-expansions have been closely integrated with district policy, and have, as such enjoyed great socio-political legitimacy. As electricity supply based on endogenous resources ranks high on the political agenda, both at the local and national level, extended local self-supply is all the more positively viewed, as it alleviates the load on fairly weak local electricity grids.

Although mainly supported at the local and industrial level, the Finnish State has also been central in bio-energy development, not least in the case of peat fuels. The state owned company VAPO Oy is by far the largest producer of peat fuels, and has been instrumental in cost-reducing technology development, bringing peat to the cost-effective fuel it is today. This political promotion of peat as an energy-fuel, came as a response to the oil crisis in the early 1970's, and was largely targeted at introducing domestically produced feedstock in district heating.

Compared to district policy concerns, environmental policy has carried considerably less weight in legitimising bio-energy and CHP technology. Only during the 1990s has the climate change issue become an important issue that led to a refocus from peat fuels to further wood-based energy technology development. The basic economic strength gained from the link to the forestry industry entails that massive public subsidy from the Finnish State has not been necessary. This holds for the introduction of by-products from the forestry-industry as fuel, and as well as for the introduction of CHP-technology in industry.

In the absence of substantial governmental subsidies, municipalities wishing to build DH plants were forced to obtain loans on commercial grounds. The risks were moderate, however, as the municipalities usually would pass the bill on to the consumers by raising electricity prices. However, investors in DH networks were in general granted loans with low interest rates. The rural policy element has continued in the form of investment support. The production of bio-energy has been particularly subsidised in regions with high rates of unemployment.

Table 4. Political and Societal Anchoring of "Green" Clusters in Finnish Biofuels and Chp[3]

Finnish bio-fuel	BioEnergy: • Peat-production has traditionally been linked to a strong Finnish rural policy • State support of peat production in early years. Later, peat production basically commercially viable without government support. The state company VAPO Oy central in peat production and peat harvesting technology development • Municipal DH systems represent major markets for peat • In the 90's public support to commercialise the production of wood fuels and development of wood-based technologies, similar to the policies supporting peat fuel development at an earlier stage. This time, policy is legitimised as part of climate changes mitigation policy Development support for bio-energy and CHP is part of a more comprehensive industrial policy
Finnish Chp	CHP: • CHP was attractive to municipalities as a means to self-sufficiency and independence • CHP played a key role in post-war modernisation and industrialisation processes • Formal and informal networks with bonds to the forest industries have provided legitimisation for both CHP and biofuel developments Alliances and partnerships between municipal, private and state-owned energy companies secured beneficiary institutional conditions for CHP

4.2 Swedish Biofuel/CHP

The similarity in Swedish and Finnish bio-fuel/CHP-clusters/innovation systems, as far as the industrial dynamics is concerned, is not paralleled in their political and societal anchoring. Compared to the Finnish case, Swedish policy has been more environmentally oriented, as Swedish bio-fuel industry is part of a broad movement, with a strong public opinion behind it, to substitute nuclear power with a CO2-neutral energy (table 6). Nevertheless, at the local level, bio-energy is part of local industrial development with strong local support as it provides alternative employment, following rationalisation and reduced employment in the traditional Swedish forestry industries. Bio-energy has become part of the broader Swedish

3 The table summarises findings from Charistiansen & Tangen (1999) and Eilekeland (1999).

industrial policy, with extensive resources allocated to development of new fuel handling and production technologies, gasification and combustion technologies, etc.

Part of this combined environmental/industrial focus has led to establishment of incentives for bio-energy production and consumption without seriously damaging the Swedish electricity-intensive export industry. Another part has been massive government R&D funding for technology development aimed at further commercialisation of new technologies in the production, transportation and combustion of bio-energy. A range of investment support programmes aimed at supporting the development direction in heat and electricity production has been given. In addition comes various development and demonstration projects funded through the Energy Technology Fund. The fuel choice in the increasing number of district heating plants has therefore changed substantially. The substitution of fuels in district heating in the late 1980's and in the 1990's has mainly been from coal to bio-energy

Political and Societal Anchoring of "Green" Cluster/ Innovation System in Swedish bio-fuels[4]:

- Bio-energy development is part of larger Swedish government plan and popular movement for the substitution of nuclear energy and CO2-emitting energy sources
- Bio-energy industry is part of local industrial development with strong local support. An important legitimacy base is that it provides alternative jobs to the traditional forestry-based industries
- State support for introduction of CHP in municipal district heating plants.·
- State support of bio-energy combustion technology
- Large-scale industrial support programme for bio-energy-related technology development
- Tax reforms have favoured the use of bio-energy in heat production

Other changes in the Swedish tax regime have also had noticeable effects on fuel composition in Swedish heat and CHP production. In 1983, VAT was removed on indigenous fuels and taxes on other fuels increased significantly. From 1990 onwards, fuel taxation has continued to increase, without any taxes levied on forest based fuels.

4.3 Danish Biofuel and Wind Industries

Danish biofuel industry and biofuel priorities are supported both by key domestic economic sector interests (agriculture, forestry) and through the priority put on bio-energy by a broad parliamentary majority as a result of the tradition for integrated energy and environmental planning (table 7).

---

4 The table summarises findings from Eikeland (1999).

The support from sector-interests is rooted in the fact that non-environmental benefits (indigenous fuel use, local job creation, income generation, and use of surplus agricultural land, nutrient recycling and waste control) accompany bio-energy developments. Use of bio-energy particularly enjoys wide support from Danish agriculture, which through the use of bio-fuels in the energy sector has an alternative market for by-products (slurry and straw) from their primary food - or livestock production.

The political support is rooted in a consensus on 'no future coal developments' and the conversions of existing coal-fired plants to bio fuels. Besides Parliamentary support, this consensus is anchored in an extensive organisational network, including the environmental movement, which also favours de-central solutions. The national consensus over environmental policy therefore also enjoys a strong local/ municipal anchoring, and reflects a policy-style with strong elements of participant democracy.

Table 5. Political and Societal Anchoring of "Green" Clusters/Innovation Systems in Danish Biofuels and Wind[5]

Danish biofuel	• Broad political and parliamentary consensus on biofuel priorities and agreements • The Danish energy- planning tradition creates a favourable context for bio-energy • Supportive lobbying from Danish farm and forest interest associations • Heavy public investment and R&D support for biofuels • Feed-in obligation on Danish electricity sector from de-central units • State support to technology development • Mandated agreements on biofuelled capacity instalments between electric utilities and national government
Danish wind	• Broad political and parliamentary consensus on wind energy plans and political agreements • The Danish energy- planning tradition creates a favourable context for wind energy • Supportive lobbying from Danish environmental movement and local wind owners - and industry associations • Initial investment subsidies to wind investors functioned as a support policy in start up and development phase, terminated in 1989. • Feed-in obligation for Danish electricity industry legislated upon in 1992 after failure of voluntary agreements between utility sector and the manufacturers- and owners associations. • Regulated terms of payment for wind generated electricity • State support to technology development, R&D efforts • State support of public information activities • Mandated agreements on wind power capacity instalments between electric utilities and national government.

Even more than the Danish biofuel industry, the *Danish wind industry* has been able to rely on strong local co-operative organisation that have combined the role of demand for windmills with mobilising support in local and national decision-

[5] The table summarises findings from Koefoed (1999).

making. This social mobilisation was initially strongly motivated by the debate over introduction of nuclear power in the 1970s where wind energy was presented as an alternative energy source. The emergence of strong turbine-producers'- and wind power producers' interest associations with active information to the public has also served to strengthen the position and priority on wind energy.

The relationship between wind energy producers and the electricity supply system has been rather ambiguous. In an early phase, the electricity sector voluntarily negotiated favourable feed-in conditions for wind self-generators. In a later phase, however, the emerging wind energy system gradually met a more restrictive attitude from electricity industry. In this phase, wind energy came to rely on legislatively sanctioned feed-in obligations imposed on the electricity industry. The feed in rights were synchronised with ambitious plans for wind power in national indicative planning. Finally, public support has also been extended to R&D efforts and technological certification via the national research centre at Risø.

5. Dynamic Issues and Evolution

Taking the previously described Vernon-Kotler models and the evolutionary framework as a point of departure, two selected cases of successful green innovation in electricity industry: Finnish CHP and Danish wind, can be shown to illustrate two widely different institutional trajectories. The Finnish Chp case illustrates an innovation and diffusion process conceived in heavy industry and gradually adopted in the public electricity supply, and with spin off effects in international technology markets. The Danish wind case illustrates and innovation and diffusion process, in the first phase driven by local communities then spreading out to the public energy system and subsequently fostering a large windmill export industry.

5.1 Finnish CHP

The dynamics of the Finnish CHP case is characterised by an innovation and diffusion process over three arenas: an industrial arena, a municipal arena and an international arena for technology export (figure 3).

Historically, the use of CHP in Finland was conceived in the paper and pulp industries during the 1920s and 30s, in response to increasing heat and power demands. As already mentioned, in the previous section, CHP proved useful as it allowed the forest industries to utilise cheap local fuels for stream production, and at the same time relieved them of waste. Finnish industries' use of wood fuelling also implied development of new burner-technology, as burners were traditionally designed for coal. This provided advances in multi-fuel boiler technology and in particular the development of fluidised bed combustion technologies. Other important CHP related innovations developed by Finnish companies are technologies for exploiting peat resources and modular district heating networks.

Since the innovation in CHP technology took place under the ownership of the potential user of the same technology, the self-supplying internal industrial market for CHP served to radically diminished demand side risk exposure. Industrial adoption was facilitated by the stakes that industrial actors had in the innovation process and by the specific tailoring of the innovation to the industrial needs. The double economy of waste management and energy-production served to make CHP and attractive option in this specific industrial context.

Furthermore, the close ties to paper and pulp industry also provided a valuable knowledge base for technology development. The internal industrial context allowed careful tuning of the CHP innovation to industrial needs and selection of best practice in internal iterative solutions.

Adaptation of CHP technology in public heating and electricity generation produced a second arena for commercial expansion. District heating networks, constructed and expanded by municipal companies in most cities during the 1950s and 1960s provided rapidly increasing municipal heat load that allowed transformation to municipal CHP. The already existing adaptation of CHP technology by industry implied that subsidies were not necessary to promote the technology in public electricity and heating "markets".

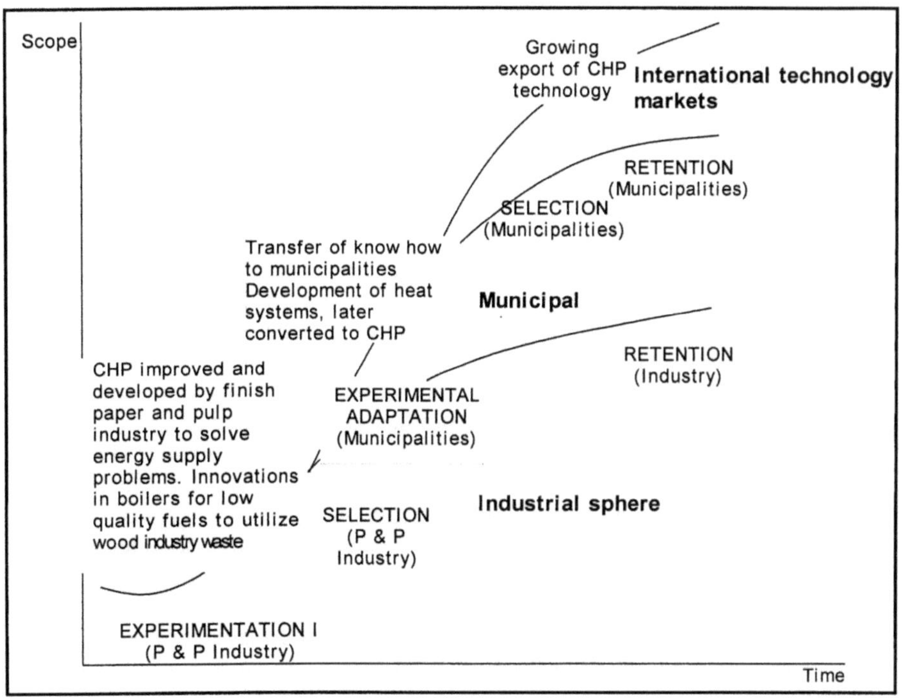

Fig. 3. Dynamic Trajectories of Finnish Development

The new adaptive innovation cycle for the municipal sector could start with a well functioning industrial technology, and utilise the knowledge base already created within industry. The challenge was here to undertake a general transposition from industrial to municipal use.

Like in the industrial context, the municipal market could also start with a guaranteed customer-base and with financing supplied on favourable terms and with security provided by local supply monopolies. Standardised application of the municipally adapted technology could take place within a large municipal market, where diffusion could take place within internal communication channels within the municipal system.

A spin-off from both the industrial and municipal markets has been an extensive industrial infrastructure supply industry. Apart from steam and gas turbines, all required parts in CHP are manufactured in Finland and has developed into an important export industry, indicated in the third cycle in figure 3. The experimentation and product development within the domestic industrial and municipal arenas supplied state of the art technology with an international market-potential. In this arena Finnish exporters could meet a more open competitive market with development costs covered in the two domestic markets. This obviously served to make the risks of open market-competition easier to carry.

5.2 Danish Wind

The dynamics of the Danish wind-case differs extensively from the Finnish CHP particularly with respect to the role of politics. As indicated in figure 4, we may therefore speak of an innovation and diffusion process over three arenas: a local cooperative arena, a public planned economy arena, and an international market arena for technology export. The first two arenas were largely driven by direct political motivation, though translated into domestic industrial competencies. The third arena driven by commercial forces and the impressive standing of Danish windmill technology on the world market.

In the local cooperative phase, the Danish windmill initiatives originated with individual investors organised in local turbine guilds, establishing small-scale community owned wind power systems. Technology was supplied by local machine tool industry with a connection to agriculture. These initiatives were strongly politically motivated, and part of a public reaction against a large-scale nuclear energy strategy and represented a softer small scale ecologically founded alternative. The political mobilisation around small-scale wind energy projects also served to establish generous framework conditions, notably financial support schemes that made local windmill projects a lucrative commercial option.

The experimentation with small-scale co-operative windmill projects was thus much more politically and socially driven than in the Finnish CHP case. The grass root mobilisation behind the Danish windmill initiatives was of a prepolitical social character that only later fed into regular politics.

The establishment of extensive subsidies provided the commercial basis for retention of the small-scale windmill development program. The collective movement and the many local windmill associations provided a market for windmill development and also secured sites for windmills on their own land.

A second phase in diffusion, and further innovation came with the broad adoption of windmills as a major cornerstone of the Danish energy policy to fulfil the country's very ambitious climate policy obligations. This brought with it a systematic involvement of the Danish electricity system both in the form of guaranteed feed-in rights as well as research and investment support. This large-scale strategy has taken the windmills out of the local farmers cooperatives and moved them into large offshore platforms with energy companies as the major investors. In parallel the institutional framework has been changed so as to allow subsidies, feed in rights and consumer obligations under new commercial market regime, to continue support of the market expansion.

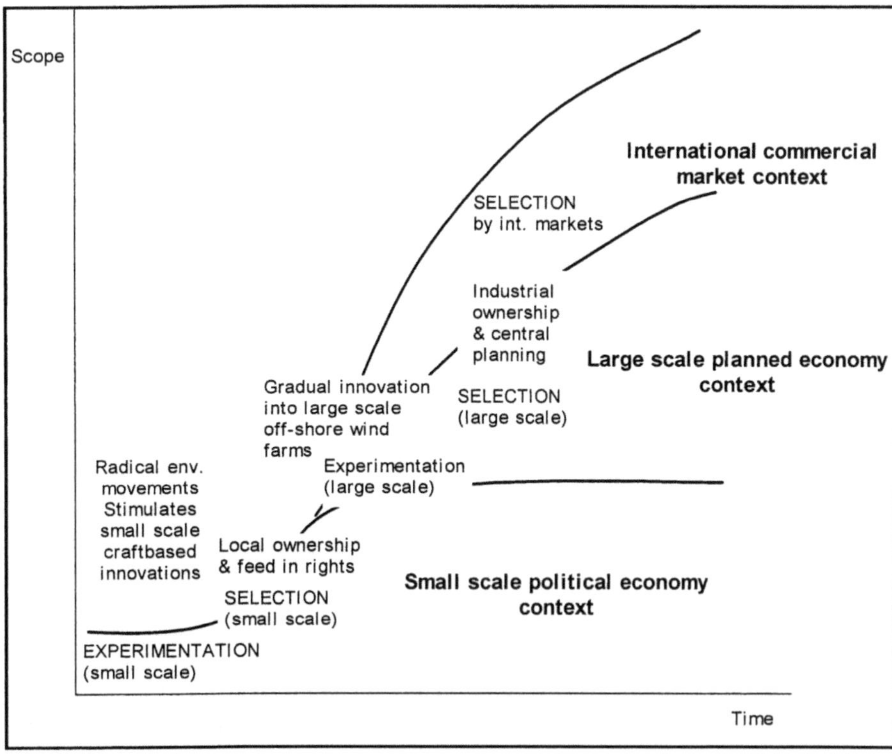

Fig. 4. Dynamic Trajectories of Danish Wind Mill Development

The ongoing subsidisation of windmill technology continued to supply a stable financial framework for further technical development also under its large-scale expansion phase. The highly ambitious Danish plans for CO_2 reduction created a demand-pull for continued retention of wind technology.

In parallel to the domestic small- and large scale expansions, Danish windmill technology has enjoyed extensive success on the international market, turning this into one of Denmark's largest export industries. As already mentioned, Danish windmill producers have had up to 60% of the work market, and have generated one of the largest Danish export industries.

While targeting different market-segments, the three arenas for Danish wind power innovation and diffusion are clearly mutually dependent. The large-scale expansion of offshore windmill farms could clearly profit from political and institutional mobilisation as well as technological breakthroughs provided by the earlier wave of small-scale windmill development. Again, the ability to handle challenges on the open market-arena for Danish windmill export has been buffered by the stability of the domestic market with guaranteed demand and investment subsidies.

6. Normative and Theoretical Iimplications

The five case studies have left us both some theoretical and normative insights:

Our case studies highlight the multilevel interplay between local, national and international arenas in greening of energy industry, both in terms of markets, technology and commercial dynamics, as well as political decision-making and societal anchoring. This indicates that innovation policy is well advised to consider coordinating local, national and international approaches and to develop policy tools that reach across local-regional and national levels.

The comparative observations from the two cases that were explored dynamically are that the local context may mobilise complementary resources and focus in a flexible way, which is difficult to match in an early phase on a larger national arena. However, at a certain point, the local arenas become to limited for further expansion and technology-development, and spill over to larger national and/or international market arenas become important.

In the case of the Danish windmill industry, we have noted the crucial importance of the local initiatives, linked up to Danish agro-mechanical industry, which seems to have played a central role in mobilising focus, competencies and resources around local wind energy projects. The later transposition of this process to central planning for the Danish system as a whole, took the early locally developed initiatives into a larger "market" and technology context, where the offshore windmill platforms represented a further innovative development clearly transcending the local context. By serving as reference-markets for state of the art wind technology, Danish firms also expanded into the international windmill markets where they took impressive market shares. The policy prominence of the windmill programme supplied it with financial resources through a generous subsidy scheme, originally adapted to local conditions, but then gradually opening up for broader large scale national energy actors.

In the Finnish forest industry case, the technology-development within an industrial setting also provided a locus for the innovation project within a restricted context capable of active resource-mobilisation in a critical innovation stage.

The attractive simultaneous solution to a combined waste and energy problem for Finnish Paper and pulp industry attracted industrial resources for development of new CHP technologies. Industry simultaneously provided a first protected market for CHP in an early differentiating stage. The spill over to the municipal heat and electricity arena after industrial standardisation of CHP technology boosted further technical development.

The success-cases of greening of energy-industry also display a high variety of trajectories, involving both political and commercial dynamics. This indicates that there does not seem to be one "optimal" sequence, but rather that supportive interplay between several arenas seems to be essential. More specifically, the comparative observations from the two cases that were explored dynamically is that the role of public vs. private initiatives and institutions in various functional contexts of the innovation-process varied extensively. This variation may have to do with the types of externalities attached to the innovation in its early non-commercial phase. If the externalities can be captured within the context of the firm, then the firm or a group of firms with similar challenges may constitute the context for innovation/experimentation. If the externalities are primarily policy/societally-oriented, then the political/societal arena is more appropriate.

The sequencing of the innovation and distribution in both cases, through several institutional contexts indicates that the success of the innovative technology arises as the function of a sequential interplay between several arenas, where products generated in one arena (such as industrial CHP) is further developed and modified for another arena (municipal Chp) and entails externalities (technology-development and supply-industry) which generates value in a third arena (international markets for advanced systems-systems).

Successful innovation, in this perspective, is conditioned on a sequence of public and private initiatives where complementary societal needs and markets work together to extract the full value of the innovation initiative. The sequencing of these institutional contexts may depend both on the externality-characteristics, the learning capabilities and the financial abilities of each institutional setting.

Analytically, the case studies seem to validate the necessity of a broad political-economic perspective in order to capture the complex interplay between factors in greening of energy-industry. The boundaries of the systems of innovation or the industrial clustering in these cases must obviously be broadly defined. The case studies indicate the need to transcend technical-economic structures and processes and also study innovation-drivers in political-administrative institutions, both in the early initiation- and later selection phases of the innovation-process.

Furthermore, our case studies indicate strong technological and institutional path dependency, which makes direct transfer of models of greening of energy-industry difficult. The highly differentiated set of "green" energy technologies found in the Nordic countries, clearly illustrate the variety of resource endowments, industrial use, institutional structures and policy preferences in innovation. This provides a path dependency in the ways problems are perceived, and in approaches that characterise national and sect oral initiatives, and thereby in the variation of commercial and political "greening- projects" that are produced. The forest industry basis in Finland and Sweden thus naturally set these countries on a different track of green energy innovation than the agrarian Denmark. Our studies indicate that different national resource- and competency-bases and institutions may make for large national variation in innovation within environmentally orientated energy technology. Learning from these case studies must therefore obviously imply transferring insights from the underlying "logics" rather from the prima facie-events.

Public Policy, Voluntary Initiatives and Water Benign Process Innovations: Empirical Evidence from the West German Chemical Industry during the Mid-1990s[1]

Frank Becker and Frank C. Englmann

1. Introduction

The purpose of this paper is to exam the role of public policy and voluntary initiatives for promoting technological process innovations that reduce water pollution, taking the chemical industry in West Germany as example. More generally, we will analyse the firms′ reasons for (i) carrying out or (ii) refraining from process innovations to reduce water pollution. Understanding the innovation behaviour of firms and their establishments is a precondition for an effective support of public policy towards an environmentally safer development. In doing so, the 1990s are especially interesting because in this period chemical firms and their establishments increased technological activities to tackle water pollution at source or to recycle valuable material in effluent polluting streams (production-integrated technologies), rather than implementing additional processes that treat waste water after its emergence (end-of-pipe technologies).

We focus on the *chemical industry* because it plays a key role in reducing the environmental pollution of industry as a whole. Chemical products such as polymers and various other chemical substances and materials are intermediate inputs in several industrial sectors such as the electronic, car, and building and construction industry, i.e. chemical products "are the basis of the economy of virtually every industrialised nation" (Anastas and Williamson 1998a).

Furthermore, we concentrate on one specific type of environmental pollution: *water pollution* caused by the production of chemical substances. Water pollution is one of the main sources of industrial pollution in the chemical sector. Waste water is produced by cooling chemical processes, by cleaning equipment and pipes, by the employees of chemical companies, and by joint products in chemical reactions. Waste water resulting from joint products is an idiosyncrasy of chemical industry (Müller-Fürstenberger 1995): The conversion of a starting material and at least one reaction partner (supported by the addition of further substances such as solvents, some auxiliary substances and a catalyst) into the desired chemical sub-

[1] This study was carried out as part of the project 'From Science to Products: A Green Paper for the European Chemical Industry'. Financial support from the European Commission is gratefully acknowledged.

stance is *necessarily* accompanied by producing some undesired joint products which may contaminate water.

The rest of the paper is organised as follows. Section 2 develops a framework for examining the innovation behaviour of chemical firms and their establishments to reduce production-related waste water. Section 3 shortly presents an overview of various types of innovation indicators, section 4 the sample of chemical firms and their establishments which were selected for our postal survey. Section 5 and section 6 deal in more detail with the questionnaire that was used for the postal survey: section 5 concentrates on the questions concerning the innovations, section 6 on the questions concerning reasons why establishments undertake innovations or why they refrain from doing so. Section 7 treats some methodological issues related to the methods used. Section 8 presents the empirical results including econometric estimates. Section 9 concludes.

2. Factors Influencing Water Benign Process Innovations

In general, process innovation activities depend on incentives, like e.g. expected cost reductions that outweigh necessary R&D expenditures or compliance with governmental regulations, and on available resources, like e.g. the existing knowledge stock inside and outside the firm. Incentives, in turn, crucially depend on appropriability conditions. Furthermore, the innovation process can be subdivided into several stages that are mutually interlinked (Grupp 1998; Kline and Rosenberg 1986).

With respect to water benign process innovations, we pre-selected a list of influencing factors that was based on existing literature (see for instance Freeman 1984; Hemmelskamp 1997; Kemp 1997, 1998; Kemp et al. 2000; Yakowitz 1997) and on interviews with representatives of chemical firms. Later, these factors were tested empirically. While some of these factors refer to process innovations in general, other factors are specific to process technologies that modify waste water after it has already emerged (end-of-pipe technologies) and process technologies that reduce waste water at source or re-use valuable material that otherwise pollutes water (production-integrated technologies).

To begin with, there are at least three reasons for firms and their establishments to carry out water protection process innovations (see also figure 1):

- Reaction to environmental regulations by the government and public authorities
  By setting tighter emission standards government forces firms to implement process technologies that reduce water pollution in order to avoid prosecution. By setting high emission charges government makes incumbent production methods costly to use. Thus, firms may adjust technologically to reduce charge payments.
- The opportunity to receive reputation gains vis-à-vis different stakeholders
  By reducing (potential) conflicts with different stakeholders by implemening water benign process technologies firms receive reputation gains which in turn

affect future profits. Because chemical firms and their establishments often belong to the upstream sector local activism and the government may be particularly important stakeholders among a large number of stakeholders including e.g. insurance companies, the media, consumers, environmental pressure groups, and employees (Hoechst 1996).

- The opportunity to reduce production costs by reduction of input material
  The development and optimisation of production processes to reduce production costs has played a major role since the beginning of industrial chemistry. Here, the re-use of joint products has played an important role in reducing production costs[2]. By recovering valuable material in polluting streams firms reduce production costs *and* pollution.

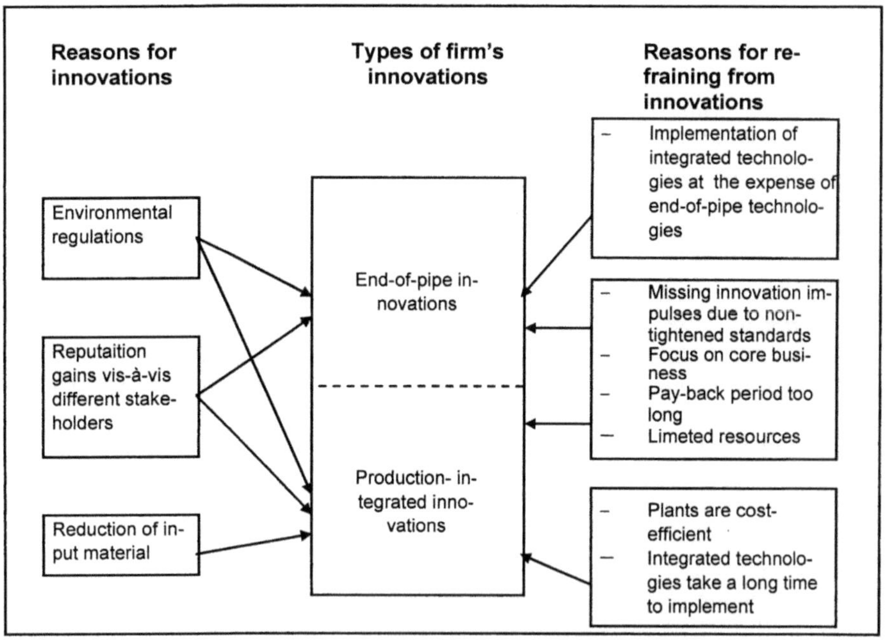

Fig. 1. Factors Influencing Water benign Process Innovations

Firms may refrain from water benign process innovations if incentives of regulations, for instance, are absent. Additionally, there is a host of reasons for firms and their establishments to refrain from (further) end-of-pipe or integrated innovations that reduce water pollution. They will be briefly described in turn.

2 At BASF the extent of integrating intermediate and final chemicals is unique in the chemical industry and the term "Verbund" has virtually developed as a brand name for BASF (Kempf 2000, p.13).

Missing impulses due to non-tightened standards / emission standards are met due to earlier innovations
The model of Downing and White (1986) shows, that the innovation incentives of emission standards will be low if the polluting firm already meets the standards. The intuitive reason for that conclusion is that an additional process innovation such as a waste water treatment plant for instance would cause extra costs but might only provide a small amount of extra gains (the gains of meeting the actual standard at lower costs due to process innovations).

Pay-back period is too long
In chemical industry the stipulated pay-back period is about four to six years. If future cost reductions of water protection process technologies are too low the pay-back period may be too long. Then, "investment" in water protection process innovations will not be undertaken.

Focus on core business
Establishments aim to focus on activities that they can carry out particularly well. A pharmaceutical firm for instance, focuses on product rather than process innovations. Thus, firms and establishments whose core business is product rather than process innovation are also more likely to refrain from carrying out water benign *process* innovations.

Limited knowledge resources
Firms may lack technological competencies to internally develop and implement water benign technologies. Furthermore, lack of co-operation partners and outside suppliers may hamper water protection process innovations.

The following reasons for firms and their establishments to refrain innovations refer to *either* end-of-pipe *or* integrated innovations:

Cost-efficiency of incumbent production process
If an establishment has adjusted to a set of actual prices such that there is no available alternative process which is preferred in terms of production costs, the plant manager will not have a reason to alter production methods. Changing production methods would decrease profits or increase prices or a combination of both.

Integrated technologies take a long time to implement and do not eliminate the need for end-of-pipe technologies
Production-integrated innovations reduce the amount of water pollution. However, production of chemical substances with zero emissions is usually impossible. Hence production-integrated innovations do not eliminate the need for end-of-pipe innovations. If polluting firms urgently have to reduce water pollution in order to meet standards they may decide to implement an end-of-pipe technology at the expense of an integrated technology because implementation of an end-of-pipe technology requires less time than replacement of an incumbent production process. Again, the time horizon matters.

Implementation of integrated technologies at the expense of end-of-pipe technologies
The reduction of water pollution by an ongoing addition of end-of-pipe technologies requires ongoing material input and energy consumption while the incumbent production process remains unchanged. Thus, end-of-pipe innovations tend to reduce productivity and thereby competitiveness in terms of costs. Therefore, a polluting firm will favour to implement production-integrated technologies at the expense of end-of-pipe technologies as emission standards get tighter and tighter for instance.

Between some of the mentioned innovation incentives a relationship *may* be established. For instance, the threat of government to impose a tightened legislative level for emissions into water positively affects firms to commit themselves to implement new or improved processes to reduce water pollution. The increased reputation of firms may avert governmental actions.

Furthermore, there is a circular relationship at least between environmental regulations and water protection process innovations. For instance, tightened emission standards positively affect technological innovations, and the existence of advanced technologies positively affect the tightening of technology-related emission standards (with a time lag).

Similarly, between some of the impediments for process innovation a relationship can be established. For instance, establishments will refrain from innovations if they already meet emission standards because (i) they want to focus on their core business, (ii) their incumbent plants are already cost-efficient, or (iii) investments in innovations do not pay back within the required time. Another example is the relationship between the factors "focus on core business" and "lack of technological competencies". Establishments whose core business consists of *product* rather than process innovations may have limited competencies to develop and implement new or improved water protection *process* technologies.

Additionally, there is a circular relationship at least between the impediment "lack of technological competencies" and innovation itself. Because of limited competencies, establishments refrain from innovating, and absence of innovations in turn limits technological competencies.

3. Innovation Indicators

The quantitative evaluation of corporate water benign process innovations and their influencing factors requires the selection of indicators representing the terms of the conceptual model. We applied two criteria for the selection of appropriate indicators: First, the indicators have to provide data at the level of firms and their establishments. Second, the indicators have to capture data separately for end-of-pipe and production-integrated technologies.

In the literature one can find a host of indicators for process innovations and related activities. Usually, the variety of indicators is categorised into two or three classes. Grupp (1998), for instance, distinguishes (i) input indicators, such as R&D personnel, internal R&D expenditures, and investment in R&D-intensive equipment, materials, compounds, (ii) R&D output indicators, such as scientific publications, and patent applications, and (iii) progress indicators, such as innovation counts by questionnaires, factor productivity, and market shares. See figure 2.

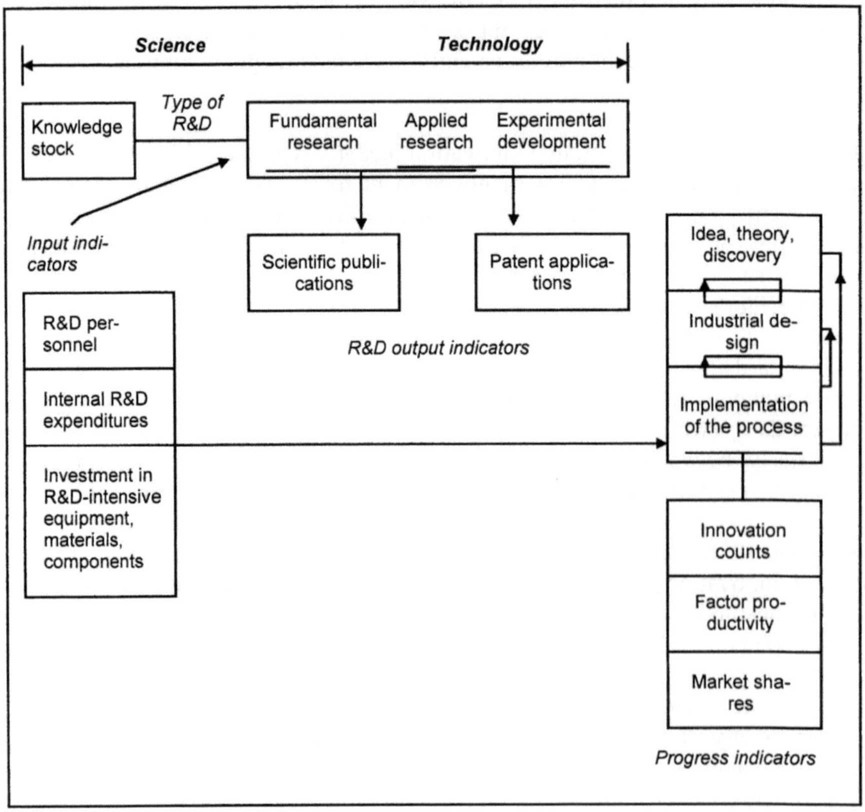

Fig. 2. Indicators of Process Innovation and Innovation Stages, the Figure adapts Grupp 1998

In particular, evaluation of patent applications and innovation counting by questionnaires allow to capture data separately for end-of-pipe and production-integrated technologies at the level of firms and their establishments. Hence, we used both indicators (Becker 2002). Innovation counting allows to measure process innovations, whereas patent applications measure R&D output. As the emphasis of this paper lies on the incentives to undertake water benign innovations we concentrate on the dichotomous process innovation indicator resulting from innovation counting by questionnaires. Thus, we had to draw a sample of firms to whom we sent the questionnaire.

4. Description of the Sample

For a number of reasons we limited our sample to West German chemical firms (FRG) and excluded East German chemical firms (former GDR). First, in the mid-1990's, firms of both regions were working with a different age and composition of capital stock. For instance, production of chemical substances was based on brown coal in former GDR, whereas it was based on oil in West Germany (Wirtschaftspolitische Diskurse Nr. 19 1991). Additionally, West German firms have had to account for many environmental regulations when developing and operating new plants. Therefore, the knowledge base of firms of both regions was different before 1990. The existing knowledge base of firms in turn influences their capability to discover new technologies. Hence, for firms of both regions, the competitiveness and the capability to discover new processes was fundamentally different *even after* German unification. Even eight years after unification both chemical sectors look very differently as depicted by table 1.

Table 1. Economic Performance of West and East German Chemical Firms in 1998

German region	Employees	Employees / firm	Turnover (DM) / employee	Turnover by foreign trade/ turnover (%)
West Germany	465,177	412	475,323	47.1
East Germany	26,923	177	331,555	27.7

Statistisches Bundesamt 1999

Second, the different age and composition of the capital stock have produced a different quality of environmental pollution. In some regions of the former GDR such as Bitterfeld-Buna-Leuna, environmental pollution was extremely high. Thus, the importance of environmental regulations for the reduction of pollution can be expected to be systematically higher for firms of former GDR. In order to control for those differences we only focus on West German firms.

The selection of chemical firms and establishments is based on the company data base of the "Verband der Vereine Creditreform (VVC)". VVC is the largest credit ranking agency in Germany. According to the data base of VVC, the chemical sector in West Germany contained 1698 chemical firms having more than 19 employees in 1998 (see table 2, second column). In contrast, the German Federal Statistical Office (Statistisches Bundesamt) counts only 1127 West German chemical firms with more than 19 employees (Statistisches Bundesamt 1999). The rather large difference may be due to two factors: First, about 10 percent of the chemical firms which we selected, using the VVC data base, do not regard themselves as belonging to the chemical industry (see table 2, third and fourth column). Second, the German Federal Statistical Office classifies firms according to their *main* business activities in order to avoid double counting (Fachserie 4 Reihe 4.1.1 1999). In contrast, by using the VVC data base we account for firms with at least one business activity which belongs to the chemical sector irrespective of its relative importance.

Table 2. Classified Sample of West German Chemical Firms in 1998

Classes according to firm size	Total of West German chemical firms	Selected firms	Adjusted sample	Respondent firms
4999>	20	20	18	9
500-4999	174	20	19	13
50-499	844	20	16	5
20-49	660	20	19	4
Total	1698	80	72	31

Data base of the VVC; own survey

In order to select the individual firms for the empirical investigation we classified the population of chemical firms into four groups according to the number of employees (see table 2, first column). Within each class we chose 20 firms randomly (within the class with more than 4999 employees the firms were selected with a probability of one). Hence the sample contains 80 chemical firms of different firm sizes. The idea behind that method of selection was twofold: First, establishing different classes of firm size ensures that our sample contains a cross section of the German chemical industry. Second, by choosing at least the 20 largest chemical firms we cover a fairly large part of turnover and thus production and water pollution of chemical industry because business concentration within the chemical sector is very high. Thus, the 25 largest firms (with respect to turnover) count for more than 50 percent of turnover of the overall sector (Statistisches Bundesamt 1998). Hence, the largest share of chemical substances is produced by a small number of large firms. This is due to *economies of scope* in innovation and *economies of scale* in production. Economies of scope in innovation arise, for instance, because of positive spillovers between different R&D projects which are run within a large firm. Economies of scale in production arise because the equipment of chemical plants primarily consists of pipes and cylinders. Thus capital costs depend on the surface area of pipes whereas capacity depends on their volume (Freeman and Soete 1997). Hence – as a rule of thumb – if capacity increases by one capital costs increase only by about one half.

As already mentioned, 72 out of the 80 selected firms regard themselves as chemical firms. Altogether, 31 out of the 72 firms responded to our questionnaires. Response was particularly low within the last two classes: about three-fourth or four-fifth did not respond to our questionnaires. These firms refrained to respond because they (1) perceived our questionnaire as too 'thick', or (2) had a bottleneck of personnel, or (3) perceived the water pollution problem as not important for their firm because they produce low levels of effluent process waste and cleaning water.

The 31 respondent firms were asked to refer their replies to only one establishment. Of course, the establishments were smaller than the firm itself. Table 3 (second column) shows the distribution of the 31 establishments according to the number of their employees. Not surprisingly, only four establishments have more than 4999 employees. The largest number of establishments of our sample have less than 5000 but more than 49 employees.

Table 3. Size of Establishments on which the Firm Replies Refer to

Size of establishments (employees)	Respondent establishments	Approximate number of employees within each class
5000 and above	4	81,140
500-4999	13	30,060
50-499	10	2,890
20-49	4	110
Total	31	114,200

The *number of establishments* we selected was small. However, when looking on the *number of employees* who work for the firms′ establishments of our sample the result is different. The selected establishments count for nearly one-fourth of the total of employees of the West German chemical sector (see table 3). Thus, the empirical results based on our sample are meaningful. Furthermore, concentrating on a small number of rather important firms had two distinctive advantages. First, the small sample enabled us to carefully select the most suitable respondent in each firm. That increased the quality of answers and the unit response rate. Second, we were able to carry out follow-up telephone interviews with each respondent to clarify surprising answers or to reduce item non-response. Thus, the empirical results based on our sample are meaningful even if the empirical base is a bit weak. In any case, more research is necessary to improve the empirical base in the field of 'green' innovations in the chemical and other industries.

5. The Questionnaire: Innovation Counting

As pointed out above, we used a postal survey to obtain direct information from chemical establishments separately for end-of-pipe and production-integrated innovations and data on factors influencing these innovations.

The dichotomous process innovation indicator provides us with data on process innovations by directly asking firms. In the questionnaire we asked:

"Did your company introduce new or improved end-of-pipe technologies that treat waste water between 1996 and 1998?"

"Did your company introduce new or improved production-integrated technologies that avoid or reduce waste water between 1996 and 1998?"

The terms of 'end-of-pipe' and 'production-integrated' technologies may be interpreted differently by the firms of our sample. In order to reduce differences in interpretation we defined both classes of technologies:

"End-of-pipe technologies that treat waste water: technologies that are added to unchanged chemical processes and that treat waste water."

"Production-integrated technologies that reduce waste water: technologies that alter or substitute current chemical processes and aim at reducing or avoiding production-related waste water and thus reduce or avoid sewage sludge."

Furthermore we gave some examples for each class in the questionnaire to reduce the variance of interpretations:

End-of-pipe innovations:

"waste water treatment plants that treat a larger number of pollutants than incumbent treatment plants."

Production-integrated innovations:

- new syntheses,
- use of new catalysts,
- use of new solvents,
- change in operational settings,
- change in input material.

So far, the dichotomous process innovation indicator is a crude measure of process innovation, in particular for the large variety of production-integrated technologies. In order to obtain more detailed information on production-integrated innovations we asked the firms to specify the production-integrated efforts according to their technical characteristics (new syntheses, catalyst, solvent, change in operational settings, alternative input material) and environmental effects (reduction of COD, AOX,[3] other substances, and the amount of waste water).

6. The Questionnaire: Applied Indicators of Explanatory Factors

Following the proposal of the Oslo-Manual (1997), we measured the reasons for water protection process innovations via the objectives of firms (separately for end-of-pipe and production-integrated technologies):

"What importance did the subsequent objectives have for all innovation activities between 1996 and 1998 in the field of waste water treating end-of-pipe technologies (waste water reducing production-integrated technologies)?" (with 1 denoting no importance at all and 5 very large importance)

Because the term "innovation" may be interpreted differently by firms of our sample we defined it in our questionnaire:

"We define innovation as the introduction of technologies that are new or improved from the viewpoint of your firm. These technologies can be either acquired externally or be developed by internal resources."

According to our conceptual model, we proposed the following objectives:[4]

- compliance with emission standards of appendix 22[5]
- compliance with emission standards of other appendices
- compliance with emission standards of the Bundesland

3 COD denotes chemical oxygen demand, AOX absorbable organic halogens.

4 The respondents could add other objectives.

5 The appendices refer to the appendices of the Waste Water Directive (Abwasserverordnung (AbwV)).

- compliance with emission standards of the municipality
- anticipation of stricter emission standards in the future
- reduction of waste water charges
- reduction of sewage sludge to reduce disposal costs
- reduction of input material
- environmental policy of your company: reduction of possible conflicts with the neighbourhood, non-governmental organisations (NGOs), and employees
- compliance with voluntary self-commitments

Finally, the reasons that hamper water protection process innovations were measured by the following question:

"What importance did the subsequent reasons have for either not carrying out innovations at all or for not carrying out additional innovations between 1996 and 1998 in the field of waste water treating end-of-pipe technologies (waste water reducing production-integrated technologies)?" (with 1 denoting no importance at all and 5 very large importance)

Again, the proposed items were selected according to our conceptual model. The impediments included:[6]

- Pay-back period too long
- no need for innovations because emission standards are already met
- focus on core business
- lack of technology suppliers
- lack of opportunities for co-operation
- lack of qualified personnel
- technology does not belong to the technological competence of the firm
- implementation of integrated technologies at the expense of end-of-pipe technologies
- integrated technologies take a long time to implement and do not eliminate the need for end-of-pipe technologies
- no need for innovations because the plants are already cost-efficient

7. Methodological Issues

7.1 Validity of the Dichotomous Process Innovation Indicator

The most important property of indicators is their validity. Validity is concerned with whether an indicator really corresponds to a respective variable.

By asking whether an establishment uses a novel or improved process technology within the production fabric or not we obtain information about the process innovation itself. Information about the actual application of the discoveries of

6 The respondents could add other impediments.

R&D is important because only the use of discoveries results in economic and environmental benefits.

In the literature, particularly two issues are emphasised with regard to the interpretation of this dichotomous process innovation indicator. First, the evaluation of the company efforts are based on corporate self-estimates or *self evaluation* (Grupp 1999). Some firms may denote an effort as process innovations while others may not. And second, the dichotomous process innovation indicator is very crude. It does not provide information about the extent of process innovations and their economic and environmental benefits.

In order to reduce the subjectivity of the data in the questionnaire we provided a definition of the term "innovation" and gave examples of end-of-pipe and production-integrated innovations that reduce industrial waste water. Thus, our most important objectives by using the dichotomous process innovation indicator will be attained: To provide basic information about both end-of-pipe and production-integrated innovations and to explore the relative importance of the establishments' reasons to carry out either end-of-pipe or integrated innovations. In order to obtain information about the *extent* of process innovations and their economic and environmental benefits we included a follow-up question in the questionnaire asking for the environmental benefits of the most important water protection process innovation the establishment carried out in the second half of the 1990s.

7.2 Validity of the Indicators of Explanatory Factors

We used a questionnaire for a postal survey in order to obtain information about the importance of different factors influencing process innovations. Kleinknecht et al. (Kleinknecht 1993) investigated the validity of answers in questionnaires on that kind of information in order to develop a common questionnaire which resulted in a well-known document called "Proposed Guidelines for Collecting and Interpreting Technological Innovation Data" (OECD 1997).

Among others they analysed whether the respondents provided valid answers to factors promoting and hampering innovations. First, they received evidence for an *interpersonal bias* (person A and person B of the same firm are asked independently and give different answers) and concluded that "(...) we cannot circumvent the problem that we probably do not measure the objectives (...) important to *the* firm, but rather the specific view of whoever happens to be the respondent" (Kleinknecht 1993). And second, they mentioned the problem of *"average" answer*. This problem will arise if the firm has several innovation projects running simultaneously where different objectives and bottlenecks may be important. "Hence, when asked about the bottlenecks experienced by *the* firm, the respondents are forced to give some "average" answer across the various projects" (Kleinknecht 1993).

In our study, we did not systematically test for interpersonal bias. However, in follow-up interviews with some respondents it turned out that in small and medium-sized firms (firm with less than 500 employees) the questionnaire was actually filled in by the person we contacted by telephone before sending the questionnaire - the managing director or the plant manager. Because these persons decide whether to introduce a new or improved water benign process technology they know best the factors influencing water benign process innovations. In larger enterprises we tried to choose (by telephone) the most suitable respondent or responsible person for co-ordinating the replies if several people were involved – often an executive of the corporate environmental technology department. He may have the best overview about what has happened in environmental technologies and why it has happened.

The problem of "average" answer is reduced by three steps in our study. First, the answers only refer to one establishment rather than to the whole firm. Second, the answers only refer to the last three years. And third, the answers refer to two very special fields of innovation projects – namely projects that aim to reduce industrial waste water either by end-of-pipe or production-integrated technologies.

7.3 Response to Questionnaires in Postal Surveys

Face-to-face interviews require skilled personnel carrying out the interviews. In addition, travel costs will be high if the respondent firms are spatially distributed. Therefore, we decided for a postal survey which is comparatively less expensive. However, the comparatively lower unit response rate is an important issue in postal surveys (OECD 1997). If the unit response is too low the empirical results loose in meaningfulness. We tried to increase the unit response rate by (i) getting into telephone contact, (ii) sending a cover letter (which states the rationale for the study) and (iii) promising to send the respondents the main findings of the current postal survey. By doing so, we tried to secure interest and consent to participate in the postal survey. Table 2 shows that we had a fairly high unit respondent rate among the larger firms.

8. Empirical Results

8.1 Water Benign Process Innovations

Between 1996 and 1998 there were 12, respectively 15 establishments that implemented end-of-pipe, respectively production-integrated technologies (see table 4). The sub-sample of innovating establishments contains a number of large establishments. Hence, this sub-sample represents quite a large part of chemical industry in terms of employees, production and pollution.

In order to get a better idea about the technical nature of production-integrated innovations the innovative establishments were asked to specify their most important production-integrated innovation during the last three years (1996-1998). Altogether, 13 innovative establishments responded. In most establishments the production-integrated innovation consisted of a change in operational settings or was linked to it. A change in solvents or input material rarely occurred (see table 5).

Table 4. Innovating Establishments (1996-1998)

	Innovative establishments	
Size (in employees)	End-of-pipe technologies	Integrated technologies
5000 and above	3	2
500-4999	6	9
50-499	2	4
20-499	1	-
Total	12	15

Table 5. Examples of Production-integrated Technologies Implemented by Chemical Establishments (1996-1998)

Cases	Number of cases
Change of operational settings *and* change of the chemical reaction, the catalyst etc.	5
Change of operational settings	6
Change of solvent	1
Change of input material	1

On average, the reduction of the value of the AOX-parameter was most difficult to achieve with production-integrated innovations (see table 6, second column). Furthermore, the success of different establishments to reduce water pollution was most different with respect to the value of the AOX-parameter (see table 6, third column). Highest reductions were achieved with regard to the amount of waste water. Here, the success of the individual establishments was most similar.

Table 6. Examples of Environmental Benefits of Production-integrated Technologies Implemented by Chemical Establishments (1996-1998)

Environmental parameter and amount of waste water	Mean of reduction in percent (approximate estimation)	Standard deviation divided by the mean (in percent)
COD	30	120
AOX	5	160
Amount of waste water	31	80

8.2 Incentives for Water Benign Process Innovations and the Role of Public Policy

In order to determine the relative importance of the three classes of innovation incentives we proceeded as follows: First, we counted all establishments that attributed at least one type of regulation the highest rank among all incentives (on an ordinal scale ranging from 1 to 5). Establishments that attributed the same highest rank to several innovation incentives were added if the combination of equally important incentives included at least one type of regulation. The sum was then related to the number of innovating establishments. This procedure was also carried out for the remaining two classes of reasons for innovation: reputation gains and reduction of material input. Hence, we calculated the proportions of establishments for which environmental regulations (or reputation gains, or reductions of inputs) were the most important innovation incentive or belonged to the most important innovation incentives[7].

There is suggestive evidence that reactions to environmental regulations were by far the most important reason for carrying out both end-of-pipe and production-integrated innovations (see table 7). For about 80 percent of the innovating establishments environmental regulations were the most important incentive or belonged to the most important incentives for end-of-pipe and integrated innovations.

Table 7. Reasons for Carrying out Innovations (1996-1998)

	Innovations	
Reasons for carrying out innovations	End-of-pipe (n=12)	Integrated (n=15)
Reaction to environmental regulations	83%	80%
Reputation gains vis-à-vis different stakeholders	50%	47%
Reduction of material costs	-	27%

For nearly every second establishment reputation gains vis-à-vis different stakeholders were the most important incentive or belonged to the most important incentives for end-of-pipe and integrated innovations. Reputation gains were obtained by reducing potential conflicts with neighbours, NGOs and employees (information obtained via questionnaires) and regulators (information obtained via follow-up telephone interviews). Thus, the importance of reputation gains *may* be positively affected by environmental regulations.

The opportunity to cut production costs by reducing material input was less important for the establishments′ innovation behaviour.

8.3 Impediments for Water Benign Process Innovations and the Role of Public Policy

There is suggestive evidence that successful compliance with emission standards is the most important reason for refraining from (further) end-of-pipe innovations (see table 8). For about 55 percent of all establishments missing impulses due to

7 Often raw data are analysed be calculating the proportion of firms that attribute a large or very large importance (rank '4 or '5') to a single item. However, if the respondents have different perceptions about the word 'large importance' or 'very large importance' the results become inconsistent. In contrast, the method we use already gives consistent results if the single respondent himself gives consistent answers, i.e. if the respondent knows that 'very large importance' means more important than 'large importance' or 'moderate importance' and chooses the correct rank.

non-tightened standards were the most important impediment or belonged to the most important impediments for end-of-pipe innovations. Here, the importance of missing impulses due to non-tightened emission standards as an impediment for end-of-pipe innovations may be positively affected by other impediments such as the establishments′ focus on their core business. Other important impediments for end-of-pipe innovations were the introduction of integrated technologies at the expense of end-of-pipe technologies and a long pay-back period. The remaining impediments for end-of-pipe innovations such as different types of limited knowledge resources did not play an important role.

Table 8. Reasons for Refraining from Innovations (1996-1998)

	Innovations	
Reasons for refraining from innovations	End-of-pipe (n=29)	Integrated (n=28)
Emission standards are met	55%	46%
Incumbent production plants are already cost-efficient	-	50%
Implementation of integrated technologies at the expense of end-of-pipe technologies	34%	-
Pay-back period too long	24%	25%
Focus on core business	17%	18%

In the case of production-integrated technologies missing impulses due to non-tightened standards lose their outstanding role as reason for refraining from (further) process innovations. For about an equal number of establishments cost-efficiency of incumbent plants was the most important impediment or belonged to the most important impediments for integrated innovations. Finally, the establishments′ focus on their core business, and the long pay-back period were similarly important innovation impediments as in the case of end-of-pipe technologies.

8.4 Impediments for Water Benign Process Innovations: Public Policy Versus Voluntary Self-Commitments

Basically, there are two major approaches to reduce industrial pollution: public policy initiatives on the one side and voluntary private activities on the other.

Our analysis of innovation incentives showed that regulations seem to provide stronger incentives for the corporate innovation behaviour than voluntary private activities promising reputation gains vis-à-vis different stakeholders.

We will now investigate the role of public policy and voluntary self-commitments for greening the industry by means of the structure of innovation impediments. In doing so, we compare the innovation behaviour of establishments that participate in the "Responsible Care Initiative" (which is the best known codified collective commitment in chemical industry) with non-participating establishments. Participation in "Responsible Care" can be regarded as an indicator for corporate environmental awareness and the power of stakeholders.

Firms that participate in the "Responsible Care Programme" voluntarily commit themselves to certain management practices which account for environmental issues. The "Responsible Care Programme" was initiated by the chemical industry in Canada in 1984 to anticipate future regulation of the government due to accidents that occurred in Canada (Love Canal) and abroad (Bhopal in India) (OECD 1999). In 1989 the first formal "Responsible Care Programme" in Europe was launched by the Chemical Industries Association (CIA) in the United Kingdom (UK). In Germany, the German Association of Chemical Industry (Verband der Chemischen Industry e.V. (VCI)) adopted the programme in 1991 and established ten guiding principles. Among others, the principles deal with process innovations: "The chemical industry continuously reduces the dangers and risks involved in the manufacture, (...), processing and disposal of its products in order to protect employees, neighbours, customers and consumers, and the environment" (VCI 1995).

It can be argued that if the power of stakeholders and / or corporate environmental awareness is high firms and their establishments will participate in "Responsible Care" and try to implement the principles of the programme. Thus, participating establishments may especially account for integrated innovations (see the cited principle above) and may require less impulses of governmental regulations than non-participating establishments. Hence, it can be argued that voluntary programmes such as the "Responsible Care"-Initiative can – at least partially – substitute public policy.

In order to explore these issues we determined whether the selected establishments participate in the "Responsible Care"-Initiative or not and tested three null hypotheses. First, participation in "Responsible Care" does not affect the perception of 'missing impulses due to non-tightened emission standards' as impediment for end-of-pipe innovations. Second, participation in "Responsible Care" does not affect the perception of 'missing impulses due to non-tightened emission standards' as impediment for integrated innovations. Third, participation in "Responsible Care" does not affect the perception of 'implementation of integrated technologies at the expense of end-of-pipe technologies' as impediment for end-of-pipe innovations. Tables 9 to 11 depict the corresponding 2 x 2 frequency tables. In order to test the null hypotheses we applied a binary logit model with a qualitative independent variable 'Responsible Care'. The used estimation methods account for heteroscedasticity. Rejection of the null hypotheses means that participation in the voluntary self-commitment "Responsible Care" affects the corporate innovation behaviour.

Table 9. Emission Standards are met (EMISSMET) as Reason to Refrain from End-of-pipe Innovations and Participation in "Responsible Care"

		Participation in the "Responsible Care Initiative"		
		Yes	No	Total
EMISSMET is / belongs to the most important impediment(s)	Yes	8	8	16
	No	9	4	13
	Total	17	12	29

Table 10. Emission Standards are met (EMISSMET) as Reason to Refrain from Integrated Innovations and Participation in "Responsible Care"

		Participation in the "Responsible Care Initiative"		
		Yes	No	Total
EMISSMET is / belongs to the most important impediment(s)	Yes	7	6	13
	No	9	6	15
	Total	16	12	28

Table 11. Introduction of Integrated Technologies at the Expense of End-of-pipe Technologies (INTEXPEOP) and Participation in "Responsible Care"

		Participation in the "Responsible Care Initiative"		
		Yes	No	Total
INTEXPEOP is/belongs to the most important impediment(s)	Yes	8	2	10
	No	9	10	19
	Total	17	12	29

With respect to the first two hypotheses, we expect a negative sign of the coefficients. For establishments that participate in "Responsible Care" successful compliance with actual standards should be of lower importance for refraining from further water benign innovations than for non-participating establishments. For the activities of participating establishments should exceed policy requirements.

With regard to the third hypothesis, we expect a positive sign of the coefficient. Participating establishments should favour integrated innovations in order to support a more sustainable development. Hence, for participating establishments the implementation of integrated technologies at the expense of end-of-pipe technologies should be a more important reason to refrain from further end-of-pipe innovations than for non-participating establishments.

The logit regression results reported in table 12 indicate that the estimated coefficients have the expected sign but the p values[8] are fairly high – especially with respect to the first two hypotheses. Because some readers are risk-lovers while

[8] The p value is defined as the lowest significance value at which the null hypothesis can be rejected.

others are risk-averters we leave it to them to decide whether to reject the null hypotheses at a given p value.

Table 12. Binary Logit Estimations (using quasi-maximum likelihood methods (Huber/White))

Equations	Coefficient	*P*-value
Equation to explain whether missing impulses due to non-tightened standards is / belongs to the most important impediment(s) (end-of-pipe technology)		
Independent variable: Responsible Care	-0.8109	0.2996
Number of observations: 29 Jarque-Bera: 4.0913 McFadden R-squared: 0.0277		0.1293
Equation to explain whether missing impulses due to non-tightened standards is / belongs to the most important impediment(s) (integrated technology)		
Independent variable: Responsible Care	-0.2513	0.7430
Number of observations: 28 Jarque-Bera: 4.5956 McFadden R-squared: 0.0027		0.1005
Equation to explain whether the implementation of integrated technology at the expense of end-of-pipe technology is / belongs to the most important impediment(s) for end-of-pipe innovations		
Independent variable: Responsible Care	+1.4917	0.1028
Number of observations: 29 Jarque-Bera: 3.7044 McFadden R-squared: 0.0814		0.1569

From our point of view, at best with regard to the third hypothesis a relationship between participation in "Responsible Care" and the perception of the innovation impediment can be established. In order to determine the impact of the participation in "Responsible Care" on the probability that the implementation of integrated technologies at the expense of end-of-pipe technologies is / belongs to the most important impediment(s) for end-of-pipe innovations we calculate the marginal effect on the mean of the regressor. If an establishment decides to participate in "Responsible Care" because of increased environmental awareness and / or increased pressure from stakeholders, the probability that the implementation of integrated technologies at the expense of end-of-pipe technologies is / belongs to the most important impediment(s) for end-of-pipe innovations will increase by about 32.67 percent.

9. Conclusions

This paper has sought to investigate the reasons for establishments in the West German chemical industry to carry out and to refrain from process innovations that protect water resources. More specifically, we sought to examine the role of public policy and voluntary initiatives for promoting water benign process innovations in the West German chemical industry.

In order to investigate these issues we randomly selected 80 firms from a classified population of chemical firms. Information about corporate innovation activities and their incentives and impediments were received by a postal survey. The selected firms were asked to refer there answers to one establishment separately for end-of-pipe and integrated innovations for the period from 1996 to 1998. Altogether, 31 firms responded, including 9 out of the largest 20 German chemical firms.

This rather small sample size implies that further empirical studies are necessary where more money can be spent on data collection than was possible for this study. Thus, the present study is more of an exploratory nature. Still, our results suggest that the establishments′ reactions to environmental regulations seem to be by far the most important reason for carrying out both end-of-pipe and production-integrated innovations.

Successful compliance with actual emission standards were the most important reason for the establishments of our sample to refrain from (further) end-of-pipe innovations. In contrast to end-of-pipe technologies, the missing impulses due to non-tightened emission standards lost their outstanding role as innovation impediment in the case of integrated technologies. Here, the cost-efficiency of incumbent plants was as important as the missing impulses due to non-tightened standards as innovation impediment.

The remaining innovation impediments could be divided into two groups. For about 17 to 34 percent of all establishments of our sample a long pay-back period, the focus on their core business, or the implementation of integrated technologies at the expense of end-of-pipe innovations were the most important reasons or belonged to the most important reasons for refraining from (end-of-pipe) innovations. In contrast, limited knowledge resources (lack of qualified personnel, lack of outside suppliers) were almost never perceived as the most important innovation impediment.

In order to analyse the importance of voluntary initiatives for promoting water benign process innovations we divided our sample into establishments that participate in the voluntary initiative "Responsible Care" and establishments that do not.

It can be argued that establishments that participate in "Responsible Care" because of high pressure from various stakeholders, for instance, may require less impulses from governmental regulation than non-participating establishments. Furthermore, it is imaginable that increased awareness about environmental issues is the reason for participation in "Responsible Care". Hence, one may expect that participating establishments have a higher preference for integrated innovations to support a more sustainable technology path than non-participating establishments.

With respect to the perception of missing impulses due to non-tightened standards as innovation impediment, both groups do not seem to differ systematically. From our point of view, that result indicates that voluntary initiatives are not able to substitute governmental regulation in the chemical industry. In an *upstream* sector like the chemical industry the power of stakeholders (environmental pressure groups, consumer etc.) to influence firm behaviour may not be strong enough to substitute public policy.

However, participation in "Responsible Care" seems to influence the direction of technological solutions. If an establishment decides to participate in "Responsible Care" because of increased environmental awareness for instance, the probability that the implementation of integrated technologies at the expense of end-of-pipe technologies is / belongs to the most important impediment(s) for end-of-pipe innovations will increase by about 33 percent.

From our point of view, the central policy implication is that if society prefers a low level of water pollution appropriate governmental interventions are still necessary. Related to this, the new EU Water Framework Directive may be a step into the right direction. First, it responds to an actually increased demand of Europe´s citizen for cleaner water. Second, it stresses both quantity rules and price incentives. Standards secure firm response. This is important in regions with relatively high concentrations of harmful substance. Charges in turn have the advantage of permanently providing innovation incentives.

Government and Environmental Innovation in Europe and North America[1]

Nicholas A. Ashford

1. Abstract

This article challenges certain tenets of the theories of reflexive law and ecological modernization. While far-sighted prevention-oriented and structural changes are needed, some proponents of these theories argue that the very industries and firms that create environmental problems can, through continuous institutional learning; the application of life cycle analysis; dialogue and networks with stakeholders; and implementation of "environmental management systems," be transformed into sustainable industries and firms. While useful, these reforms are insufficient. It is not marginal or incremental changes that are needed for sustainability, but rather major product, process, and system transformations – often beyond the capacity of the dominant industries and firms. This article also questions the alleged failure of regulation to stimulate needed technological changes, and identifies the conditions under which innovation for sustainability can occur. Finally, it discusses differences in needed policies for industrialized and developing countries.

2. Introduction

Both governmental and environmental innovations are sorely needed to move industrial economies towards more sustainable transformations. This recognition has given rise to theories of reflexive law and ecological modernization. While ecological modernization theory has its theoretical origins in continental Europe (Andersen and Massa 2000; Mol and Sonnenfeld 2000a; Mol and Spaargaren 2000), paradoxically some of its effects have been felt perhaps most strongly in the United States (US). There, some of its tenets have arguably been incorporated into the anti-regulatory and anti-government Reagan ideological revolution, and into social and environmentalist responses to that revolution. It can be argued that ecological modernization has not been adopted as a theory of environmental governance *per se* in the US[2]. Various "articles of faith" have arisen there, however, about the best way to achieve improvements in environmental quality. These include the use of economic instruments, exploiting industry's potential to engage in

---

[1] Adapted with permission from an article appearing in a Special Issue "Globalization, Governance and the Environment," by Sonnenfeld and Mol.

[2] This is in contrast to the Netherlands (cf. Spaargaren and Mol 1992).

technological innovation, encouraging more voluntarism and stakeholder participation in governance, and promoting demand-side policies focused on green consumer behavior (cf. Fiorino 1999).

These articles of faith have been endorsed by government, industry, and mainstream environmentalists alike. Each was dissatisfied with the gridlock in environmental policy, and the opportunity to try a different approach was appealing – though different actors were more attracted to some initiatives than to others. After two decades of experience with such new approaches, the need for a strong, directive government in fostering sustainable industrial transformations requires examination.

Different "schools" exist within the broad range of ideas labeled reflexive law (Teubner 1983) and ecological modernization (Mol 1995)[3]. Far-sighted prevention-oriented and structural changes as advocated by numerous scholars in these approaches are certainly needed. The weakest feature in the more neo-liberal ecological modernization formulations, however, is that they implicitly argue that the "problem industries/firms" – the very industries/firms that create environmental, health, and safety problems – can transform into "green or sustainable industries/firms." This can be accomplished, it is argued, through continuous institutional learning; the application of life cycle analysis; dialogue and networks with suppliers, customers, environmentalists, and workers; and the commitment to implement "environmental management systems." In all ecological modernization approaches, efforts can be found to influence governmental regulation through consensus, dialogue-driven processes and to encourage governments to use economic instruments. But it is not clear to what extent the different branches of ecological modernization theory see these policy and regulatory innovations as a complement to, rather than as a substitute for, so-called "command and control" regulation.

Further, rather than fostering more effective regulation through using consensus and dialogue-driven processes as an adjunct to regulation, there is increasing evidence that "cooperative" approaches may often actually impede the needed changes and transformations – especially if governmental processes are unduly influenced, or even captured, by the problem industries (Coglianese 1997; Caldart and Ashford 1999). The remainder of this essay, explains why I believe this to be so, and why dialogue and consensus, while useful, on their own are likely not sufficient to transform the industrial system into a sustainable one. At the core of this analysis is the argument that it is not marginal or incremental changes that are needed (Andersen and Massa 2000), but major product, process, and system trans-

3 Different scholars make different distinctions in ecological modernization schools of thought. See for instance, Christoff (1996) and Dryzek (1997) who distinguish radical and reformist versions of ecological modernization, and Mol and Sonnenfeld (2000a) who suggest that ecological modernization theory has developed in three historical phases, each with its own dominant tradition.

formations – often beyond those dominant industries and firms are capable of developing easily[4].

3. Ecological Modernization and Its Problems

Ecological modernization theory apparently arose in response both to those who argue for a transformation of society/the industrial system according to the anti-development formulations of deep ecology – and alternatively to those who were convinced that while historic regulatory approaches were incapable of adequately addressing remaining or new environmental problems, other options might be successful. What emerged as tenets of present formulations of the still-evolving theory are several lines of thought which begin differently in different disciplinary domains, but have since been melded into the theory:

0. unregulated capitalism is responsible for the present ecological and environmental problems, and this is partly because the prices of goods and services do not adequately represent the social costs of production and consumption;
1. historically, "command and control" regulation has been only partly successful in correcting market failures, because it proved inflexible, it under-utilized economic instruments, and it focused on end-of-pipe approaches, rather than on preventive or precautionary "cleaner technologies"; and
2. under thoughtful "reflexivity"(Teubner 1983), the present and enlightened industrial actors can succeed in advancing the material well-being of citizens, contribute to their nation's competitiveness, and can also contribute to the necessary scientific and technological changes (innovations) in products, processes, and services to adequately meet the environmental challenges – especially if a broad array of stakeholders are involved.

---

[4] This argument is centered on the idea of "the winds of creative destruction" developed by Joseph Schumpeter (1939) in explaining technological advance. The distinction between incremental and radical innovations – be they technological, organizational, institutional, or social – is not simply line drawing along points on a continuum. Incremental innovation generally involves continuous improvements, while radical innovations are discontinuous (Freeman 1992), possibly involving *displacement* of dominant firms and institutions, rather than evolutionary transformations. Christensen (1997) distinguishes the former as "sustaining innovation" and uses the term "disrupting innovation" rather than radical innovation, arguing that both sustaining and disrupting innovations can be either incremental or radical. See the later discussion below. In contrast, Kemp (1994 and 1997) argues that 'technological regime shifts' brought about by 'strategic niche management' can result in radical (i.e., disrupting) innovation through a stepwise evolution in learning and experimentation by dominant firms. See also Kemp and Loorbach (2003) and Rotmans et al. (2001).

My purpose here is not to weigh in on one side or the other of the sustainable development/anti-development debate or to address the subtleties of connecting these trains of thought, but rather to argue that some forms of ecological modernization theory as it is developing have the danger of not offering a solution to the problem. In a too narrow or strict or one-sided application, these basic tenets alone are unlikely to be strong enough to guide the policies needed to more closely approximate a sustainable industrial system.

Even if we take comfort from the fact that the eco- or energy efficiency of products and services have made dramatic improvements over the last decade, the fact of the matter is that the rate at which the best technologies are diffused into the world economy and the rate at which consumption is increasing will not be sufficient to address the environmental problems we now face (Andersen and Massa 2000)[5]. Not only are eco-systems seriously endangered by destruction of the ozone layer, global warming, and the global diffusion of pesticides, but new threats are now suggested related to endocrine disruption compromising the reproductive systems of all species at levels of chemical exposure in the parts-per-trillion, rather than parts-per-million range (Colborn et al. 1996). Evidence is increasing that diseases heretofore unconnected with chemical exposures, such as autoimmune disease, attention deficit hyperactivity disorder, and childhood cancers, are in fact consequences of the chemicals-based industrial production and consumption (Ashford and Miller 1998).

Regarding unregulated capitalism, getting the prices right will help, but it will only address market imperfections, not the fact that for some problems, such as global warming, even a perfectly working market is insufficient to address the problem – because of both the disparate time horizons over which present costs and future benefits are distributed, and because equity concerns are not adequately reflected in market decisions.

Some commentators have simply read the history incorrectly or too narrowly with regard to the limitations of command and control environmental regulation. For neo-liberal economists, for instance, command-and-control regulation is too often a conveniently-constructed straw man; it is alleged that most regulation requires specific technologies to address environmental hazards. In the US, regulation has, in fact, mostly been definitive on targets, but flexible on means. Where it has been stringent enough and designed thoughtfully – which as ecological modernization theorists rightly claim has not always been the case – regulation has spurred technological innovation of the kind desired by the ecological modernization theorists (Strasser 1997). In Europe, regulation reflects a softer, less-confrontational, and hence less technology-forcing situation (Gouldson and Murphy 1998; Wallace 1995). This demonstrates the need for a less generic and more context-specific idea of ecological modernization.

Finally, inviting industry to solve the problems by operating in a more modern and enlightened manner, can be a viable addition to conventional ideas of envi-

5 See McDonough and Braungart (1998), who argue that more than "eco-efficiency" is required in the sense advocated by Schmidheiny (1992). Fundamental redesign is required.

ronmental regulation. But counting only or mainly on existing industries for environmental transformation ignores increasing evidence that it is not just willingness and opportunity/motivation that are required for such change, but that a third crucial condition – the ability or capacity to change – is essential (see below). This is unlikely to be present or within grasp of the dominant technological firms (Ashford 2000). Dominant technologies rarely if ever displace themselves in product markets (Christensen 1997), or in general (Schumpeter 1939). In some situations they may do so because society or market demand sends a strong signal, but not in all or even in most of the cases. This belief has, so far, turned out to be mostly wishful thinking.

If rather modest improvements in the eco- and energy efficiency of products, processes, and services were sufficient to address the problems that we now face, then the more neo-liberal, market-oriented and technocratic forms of ecological modernization in the industrialized nations might be of great interest for fashioning solutions to world-wide sustainability[6]. Unfortunately, this is not the case. On the other hand, in developing countries – which to an increasing extent follow, copy, and adapt technologies from the developed world, rather than develop new ones – a theory of governance which involves the stakeholders and provides the most receptive environment possible for technology diffusion – does have merit, especially where traditions of government intervention and regulation are weak[7]. But, the developing world needs something good to copy or adapt. Ecological modernization, at least in its neo-liberal incarnation, does not produce the technological and social innovations necessary to do the job.

The remainder of this essay addresses the alleged failure of regulation, the evidence that cooperative approaches by themselves offer more promise, the conditions under which innovation needed for sustainability can occur in industrial firms, and approaches to resolving the apparent dilemmas in environmental policy and governance.

4. Has Regulation Failed?

The justification for government intervention in activities of the private sector is cited by neo-classical economists to be based on "market failure," the inability of the unregulated market to internalize the social costs of production. These market failures are legend and are manifested as environmental degradation, resource depletion, and compromises to worker and consumer health and safety. In the 1970s, traditional end-of-pipe regulation significantly improved industrial emissions to air, effluents to water, and waste disposal and treatment. In recent times, progress

6 For a discussion of the industrial firm's motivations for making modest improvements of an evolutionary or "sustaining" nature, see Reinhardt (1999).

7 For a discussion of the transferability of environmental regulatory systems from developed countries to developing countries in the context of the experience in Poland, see Brown (2000).

in environmental improvements appears to have been slower. In addition, industry's objection to so-called "command-and control regulation" has been increasingly vocal, especially in the US. At the same time, in both the US and Europe, concerns with increasing levels of unsustainable production and consumption have lead to the realization that major changes in industrial practices are needed "beyond compliance" with current environmental standards and current levels of energy production and use[8]. Thus, frustration has been voiced by both the regulated industries and environmentalists about traditional regulatory approaches. However, US and European responses to this frustration have been different[9].

In the US, the anti-regulatory climate of the 1980s – and continuing to this day – argued for reducing the burdens on industry who "knew better" than the government bureaucrats how to handle responses to environmental challenges. Those observers who were disillusioned with government's inability to deliver better protection argued that "we have exchanged market failure (the original justification for government intervention) for bureaucratic failure limits in central regulatory capacities)" (Fiorino 1999). A closer reading of regulatory history gives a different interpretation. Markets are inherently unable to internalize (unpriced) social costs without intervention. Bureaucracies, while not perfect, are not inherently flawed. In the US, it was national leadership that failed since the 1980s to streamline regulatory processes[10] and to promulgate the kinds of standards that stimulated technological changes that could lead to significant change and win-win scenarios (Ashford 1993). While the US Pollution Prevention Act was enacted in 1990, no serious effort followed at translating this act into meaningful requirements. Industry "picked the low-hanging fruit" and worked at the margin of undertaking housekeeping changes, rather than implementing serious technological change (Ashford 1993; EPA 1991; Hirschhorn 1995). The truth is that there were plenty of unexploited win-win opportunities for pollution prevention/cleaner production. Industry was not very interested in pursuing them, given other interests, reduced pressure from a reluctant, beleaguered, and underfunded regulatory system, and the fact that serious pollution prevention efforts take time, are disruptive to an industry struggling to meet inventory demands, and risky – even with saving of costs and increasing profits in the longer run.

The US Environmental Protection Agency (EPA) understood the need for bureaucratic reform, much in line with more sophisticated and less neo-liberal ecological modernization ideas. It established the multi-stakeholder National Advi-

8 See Porter and van den Linden (1995a,b) and Reinhardt (1999) for a discussion of the conditions under which industrial firms might be motivated to go beyond compliance under regulatory pressure.

9 See Vogel (2002) for an interesting commentary on how the US and Europe have 'traded places' in their approach to stringent regulation.

10 For example, "innovation waivers," allowing more time for compliance in return for innovative approaches on the part of industry, were permitted in statutory provisions in environmental laws. These might have encouraged better and cheaper environmental technologies, but they had hardly ever been used because of the lack of proper incentives for agency personnel to get involved with complex issues (Ashford et al. 1985).

sory Council on Environmental Policy and Technology (NACEPT) as a sister advisory board to its Science Advisory Board to work on accelerating progress and reducing perverse incentives for the adoption of cleaner production and pollution prevention technologies. The Council subscribed to a win-win philosophy, believing that changes in technology could provide better compliance at lower costs. The author chaired the core committee of that council – the Technology, Innovation, and Economics Committee – which issued a series of reports resulting from multi-stakeholder working groups in the early 1990s, laying out the strategies for streamlining and reducing barriers to environmental technology diffusion and innovation, mostly within the existing regulatory structure (NACEPT 1991-1993). Eventually these ideas were transformed in the US EPA's Technology Innovation Strategy. Unfortunately, because standards promulgated in the decade prior to these efforts were not stringent or demanding, industry was not interested in taking advantage of a streamlined regulatory approach to environmental protection[11].

The strategic question is whether and how we can re-conceptualize the regulatory approach which has as its focus the deliberate stimulation of innovative solutions, rather than the historical defining of environmental problems and "acceptable risk" (Finkel and Golding 1994), or whether we should retreat from the traditional path of regulation and only or mainly trust on market actors and consensual styles of environmental reforms.

"Command-and-control" regulations have been the whipping boy of economists and government critics, often with a misconception of what these regulations actually require. The implication of the term is that both the targets and the means for compliance are specified by regulation. In fact, this is only occasionally the case. Many environmental standards are health-based standards, without reference to the means of compliance. Most so-called technology-based standards require the adherence to pollution levels that reference technologies can achieve, but the standards do not require that these reference technologies actually be used. The use of more innovative or cost-effective technologies are, in fact, possible and there are statutory provisions allowing their adoption (Ashford et al. 1985; Becker and Ashford 1995), although – as discussed above – environmental authorities such as the US EPA could do more to encourage their use.

In a number of studies at the Massachusetts Institute of Technology (MIT) beginning in 1979, it was found that in the US, regulations in the chemical producing and using industries did stimulate significant fundamental changes in product and process technology, which also benefited the industrial innovator, provided the

---

[11] One justified criticism of traditional regulation is that it is fragmented on a media-specific basis. Air, water, and waste regulation evolved as separate systems and uncoordinated regulatory requirements do create disincentives for holistic, prevention-oriented technological change. In the US, coordinated, facility-based permitting, while not commonplace, is increasingly implemented to meet this criticism, as are some of the "regulatory reinvention" strategies discussed below. In Europe, some voluntary agreements are multi-media in nature, but most continue to be single-media (or energy) focused. Thus, the fragmentation of efforts across problem areas continues to plague whatever approach to improving environmental quality is currently in vogue.

regulations were stringent and focused (Ashford et al. 1985). This empirical work was conducted fifteen years earlier than the emergence of the much weaker Porter Hypothesis which argued that firms on the cutting edge of developing and implementing pollution reduction would benefit economically through "innovation offsets" by being first-movers to comply with regulation (Porter and van den Linden 1995a, 1995b)[12]. Analysis of the US situation since the earlier MIT studies reinforces the strategic usefulness of properly designed and implemented regulation, complemented – but not replaced – by economic incentives (Strasser 1997).

Perhaps paradoxically, in Europe where regulation was arguably less stringent and formulated with industry consensus, regulation was not found to stimulate much significant innovation (Kemp 1997). In the Netherlands, for instance, a concern with future sustainability heralded a series of National Environmental Policy Plans which were characterized by mandated clear future targets for environmental performance, coupled with a cooperative partnership involving government, industry, and NGOs to achieve those targets through flexible means (Keijzers 2000). Far-future environmental goals were subjected to "backcasting" to determine what changes needed to be put into practice now to achieve those goals (Vergragt and van Grootveld 1994). The Dutch researcher, Kemp, whose views are informed mainly by European environmental regulation (Kemp 1994, 1997) acknowledges that regulation can be an important tool both to stimulate radical (i.e. disrupting) and environmentally superior technology and to yield economic benefits to innovating firms. However, he also expresses faith in evolutionary, stepwise change within the original 'technology regime' to eventually bring about the needed transformations. In contrast, a comparison of the Dutch and UK regulatory systems (Gouldson and Murphy 1998) concludes that stringent regulation, without yielding to the pressure of the regulated firms common in the UK system, is *essential* to bring about significant technological changes. Kemp argues that for a technology regime to shift – i.e., to transform – there has to be a unique/new niche for a radical alternative (Kemp 1994). What Kemp may not fully appreciate is that regulation may be essential to create that niche and such niches are unlikely to be created by incumbent firms. New entrants, rather than the regulated or "problem industries/firms" will be the responders, and it may be that technological innovation by new entrants is what is needed for sustainable development, consistent with the central theses of both Christensen (1997) and Reinhardt (1999)[13]. In his later work, Kemp has come to accept the view that outsiders are essential (Kemp and Moors 2002). However, writings on 'transition management' continue to emphasize stepwise learning and innovation over dec-

---

12 See Jänicke and Jacob (2002). "Ecological Modernisation and the Creation of Lead Markets" for an excellent discussion of the importance of creating lead markets for environmental friendly technologies in the context of ecological modernisation. Unlike, many others, these authors recognize the need for radical innovations and the limits of incumbents to achieve the needed changes (this volume).

13 See also van de Poel (2002) for an insightful discussion concerning the importance of outsiders.

ades as an essential formula for system innovations leading to sustainability (Rotmans et al. 2001; Kemp and Loorbach 2003).

5. Have Consensus-based Approaches Succeeded?

Consensus-based, cooperative approaches have been promoted both in the US (Susskind and McMahon 1985) and Europe (COWI 1997; EEA 1997). It is argued that alternative dispute resolution (ADR) or negotiation can be a useful tool in the establishment, implementation, and enforcement of environmental and occupational safety and health policy. Negotiation can facilitate a better understanding of issues, concerns, facts, and positions among adversaries. It can also promote the sharing of relevant information, and can provide an opportunity for creative problem-solving.

While there is little doubt that constructive dialogue among the stakeholders can reduce misunderstandings, facilitate an appreciation of common ground, and generate solutions that are mutually-advantageous in the context of a clear set of established performance criteria or environmental goals, the superiority of self-regulation or "voluntary regulation" as a replacement for government protection is another matter (Harrison 1999). Of course, as ecological modernization theorists have argued, negotiation can be a valuable tool used as an adjunct to regulation, e.g., in apportioning financial responsibility among a large number of polluters in toxic waste clean-up operations. Alternative dispute resolution seems to work best applied to the means by which targets and goals are achieved, rather than in establishing the targets themselves (see also Mol et al. 2000). In that sense, negotiations and consensus building can complement direct regulation, although the extent to which very much depends on the prevailing policy styles and culture rooted in historical developments. Negotiation can, however, not be viewed as an overall panacea for all the various difficulties that typically confront the regulatory policy-maker[14].

Both Coglianese (1997) and Caldart and Ashford (1999) have reviewed the record of negotiated rulemaking in the United States and are critical of the outcomes of this approach. Aside from not delivering on the promise of faster and less litigious rulemaking, in general the approach appears to offer less protection and inhibit technological innovation. Capture of regulatory agencies by dominant regulated firms remains a serious limitation of the extent to which major innovation occurs as a result of negotiated rulemaking, implementation, and compliance (Caldart and Ashford 1999). In the international context, it is widely acknowledged that the Montreal Protocol, an international agreement to phase out ozone layer destroying chemicals such as chlorofluorocarbons (CFCs), was ultimately supported by a desire of CFC producers to protect their markets by fashioning an agreement that favored their own substitutes (Reinhardt 1999), that were not as

[14] For an international study that makes this point, see Gouldson and Murphy (1998).

protective of the ozone layer as those that emerged from other firms, later than might have been the case.

It is useful to note that "reinvention" initiatives by the US EPA are generally acknowledged not to be successful, notwithstanding rhetoric to the contrary by the agency. Many of these same concerns as are voiced in the context of negotiated rulemaking are apt when negotiation is used in an extra-statutory sense, as it is now being used in US EPA's Project XL and Common Sense Initiative, in an attempt to change regulatory policy. Where there is no meaningful incentive for industry negotiators to move away from the status quo – that is, where there is no impending "default" standard or requirement that they perceive as onerous – they may well be interested only in those regulatory changes that save them money (Caldart and Ashford 1999)[15].

Industry has also created unilateral voluntary programs, such as Responsible Care, that boasts of modest success, but their effectiveness is unclear. Firms tend to respond at their own pace, in their own way, and mechanisms for trade-association monitoring and sanctioning are weak (King and Lenox 2000; Howard et al. 2000). There is evidence that these programs tend to inure to the advantage of large firms over small firms (Nash and Ehrenfeld 1996), possibly favoring undesirable increased industry concentration.

There are also examples where cooperative approaches that include environmental and labor stakeholders have yielded positive outcomes resulting in the adoption of better, but not development of new, technologies. In the context of the existence of clear mandated government targets, labor union participation can help firms comply with environmental requirements (Kaminski et al. 1996). Since workers are often also the residents of the communities surrounding industrial facilities, they are in a unique position to influence technological changes that improve both worker health & safety and the environmental consequences of production. They are often silent partners with community groups in the latter's negotiation of "good neighbor agreements" with local industry (Lewis 1993)[16]. Labor contributes technical knowledge about plant technology to the local environmental groups who in turn press industry for improvement. Unfortunately, mechanisms improving access to information concerning toxic substances, known generally as right-to-know laws and policies, have not provided either labor or the community access to information about alternative technologies of production (Orum, undated). Thus, both informational avenues concerning technology options and the means to act (the right to act) upon that knowledge is required in order to empower workers and communities to press for technological change.

---

15 EPA continues to pursue cooperative approaches. EPA's National Performance Track was launched in June 2000. It consolidates and builds on several previous "reinvention" initiatives. It promotes "beyond compliance" by rewarding firms, depending on their (voluntary) placement in a tiered approach to enforcement. Firms are placed in one of three levels: status quo, achievement track, and stewardship track (Speir 2001). The higher the classification, the greater the firms are relieved of intense regulatory scrutiny.

16 See also Lewis and Henkels (2000).

In Europe, two comprehensive studies of "voluntary agreements" conclude that where there are no regulatory requirements to "back up" cooperative, negotiated agreements, little real progress at improving environmental and energy efficiency performance has been achieved (COWI 1997; EEA 1997). The exception is noted to be the "Dutch Covenant," which is much more than a voluntary agreement between industry and government. It is, in its best form, an enforceable contractual promise by the firm, with participation by environmentalists, and milestones and oversight with legal power to back up the agreement. But these approaches can be seen in line with more sophisticated ecological modernization ideas, where intelligent combinations of consensual negotiations and direct government regulation are made.

6. Conditions for Adequate Innovation

It is clear that firms need to adopt or develop technologies (and work practices) different from those currently being used in order to significantly improve their environmental performance. Depending on the particular environmental challenge, the needed technological change could be off-the-shelf available technology, technology available in a different industry, technology that needs to undergo minor development and adaptation, or major new approaches. These different types of changes are known as diffusion, technology transfer, incremental innovation, and either radical or disruptive innovation respectively[17]. In any case, there are three elements that are necessary and sufficient for technological change to occur: *willingness* to change, the *opportunity/motivation* to change, and the *capacity* to change (Ashford 1993, 2000).

Current ecological modernization approaches focus mostly on enhancing the capacity to change though cooperative efforts that also influence willingness and opportunity. When implementing ecological modernization for capacity-building, the question of "capacity building for what?" must be addressed. Neglecting this question has been a major omission in several branches of ecological moderniza-

---

17 In this paper, radical innovation is a major change in technology along the lines that technology has been changing historically, for example a much more efficient air pollution scrubber. As noted previously, in the context of product markets, Christensen (1997) calls this type of innovation "sustaining" and documents that it is usually pioneered by incumbent firms. Major innovation that represents an entirely new approach – characteristic of a 'technology regime shift' to use Kemp's terminology – even if it synthesizes previously invented artefacts, is termed "disrupting" and it almost always is developed by firms not in the prior markets or business. The replacement of Monsanto's PCBs in transformers by Dow-Silicone's dielectric fluid is a stark example. The new transformer fluid was based on an entirely different molecular model and pioneered by a firm not formerly in the dielectric fluid business. Unfortunately, the terminology used in the literature is not uniform; by the term 'radical' both Freeman (1992) and Kemp (1997) mean 'disrupting' as defined by Christensen (1997) who reserves 'radical' for major innovation *within* a technology regime.

tion – especially in failing to take into account the degree of innovation and the distinction between radical and disrupting innovation (Christensen 1997).

Willingness, opportunity/motivation, and capacity affect each other, of course, but each is determined by more fundamental factors. Therefore, policy approaches need to be chosen and designed for their ability to change these more fundamental factors. Willingness is determined by both (1) the firm's attitudes towards changes in production technology and products in general and by (2) its knowledge about what changes are possible. Improving the latter involves aspects of technical capacity building, while changing the former may be more idiosyncratic to a particular manager or alternatively a function of organizational structures and reward systems[18]. The syndrome "not in my term of office" describes the lack of enthusiasm of a particular manager to make changes whose benefit may accrue long after (s)he has retired or moved on, and which may require expenditures in the short or near term.

Opportunity/motivation involves both supply-side and demand-side factors. On the supply side, technological gaps can exist between (1) the technology used in a particular firm and the already-available technology that could be adopted or adapted (known as diffusion or incremental innovation, respectively), and (2) the technology used in a particular firm and technology that could be developed (i.e., major or radical/disruptive innovation). On the demand side, four factors could push firms towards technological change – whether diffusion, incremental innovation, or major innovation – (1) regulatory requirements, (2) possible cost savings or additions to profits, (3) community or public demand for a less polluting and safer industry or products, and (4) worker demands and pressures arising from industrial relations concerns.

Technical capacity or capability can be enhanced by both (1) increases in knowledge or information about cleaner and inherently safer opportunities, partly through formal Technology Options Analyses[19], and partly through serendipitous or intentional transfer of knowledge from suppliers, customers, trade associations, unions, workers, and other firms, as well as reading about environmental and safety issues – all leading to increased technological diffusion, and (2) improving the skill base of the firm through educating and training its operators, workers, and

---

[18] In an excellent discussion of capacity building, Weidner (2002) explores the conditions and requirements for changing the attitudes and practices of incumbent polluting (problem) firms through learning, interactions with cooperative networks, etc. to undertake changes that vary from incremental to radical innovation. In the context of sustaining innovations, and for encouraging the diffusion or technology transfer to developing countries, his insights are invaluable. What his discussion of capacity building does not capture is the regime-shifting, disrupting changes that may be required for sustainable development. But also see note 10 as well as Jänicke and Jacob (2002).

[19] Technology Options Analysis, as distinct from Technology Assessment, identifies what technologies *could* be adopted, or developed, to address a particular health, safety, or environmental problem (Ashford 1993, 2000). In a similar vein, also see O'Brien (2000) for a discussion of the need for "alternatives assessment" in responding to environmental challenges.

managers, on both a formal and informal basis – leading to technological innovation.

Capacity to change may also be influenced by the inherent innovativeness (or lack thereof) of the firm as determined by the maturity and technological rigidity of particular product or production lines (Ashford et al. 1985; Utterback 1987). The heavy, basic industries, which are also sometimes the most polluting and unsafe industries, change with great difficulty, especially when it comes to core processes. It deserves emphasizing that it is not only technologies that are rigid and resistant to change. Personal and organizational inflexibility is also important (Coriat 1995).

Finally, it should be realized that those policies that work to maximize win-win outcomes using (1) diffusion of presently available technology, might be different than those needed to stimulate (2) incremental innovation or those necessary for (3) radical innovation or (4) disrupting innovation. Policies of the first and usually second type strive for static efficiency; leveraging the firm's self-interest through consciousness-raising, continuous learning, and other techniques of ecological modernization may be helpful here. Other policies aiming at creating new dynamic efficiencies require much more than incremental learning and technological change (Ashford 2000).

7. Resolving the Apparent Policy Dilemmas

Recalling that a sustainable future requires technological, managerial, and social/cultural changes, it is likely that an evolutionary pathway is insufficient for achieving factor ten or greater improvements in eco- and energy-efficiency (McDonough and Braungart 1998), and reductions in the production and use of, and exposure to, toxic substances (Ashford 2000). Such improvements require more significant and revolutionary changes (Andersen and Massa 2000; Reijnders 1998). The capacity to change can be the limiting factor – this is often a crucial missing factor in optimistic scenarios.

Significant industrial transformations occur less often from dominant technology firms, or in the case of unsustainable practices, problem firms' capacity-enhancing strategies[20], than from new firms that displace existing products, processes and technologies. This can be seen in examples of significant technological innovations over the last fifty years including transistors, computers, and PCB replacements (Ashford 1994, 2000; Ashford and Heaton 1983; Strasser 1997).

Especially in industries which are "flexible" and always changing their products, we may be justifiably enthusiastic about existing firms' ability to move towards sustainable production. In this case, closer relations with customers and NGOs may be particularly helpful. But where the product line is "rigid" or mature

[20] Such as continuous learning, using life cycle analysis, change and niche management, and environmental management systems.

– as was the case of PCBs, and is the case with several other unsustainable technologies – change is not easy, and Schumpetarian revolutionary "waves of creative destruction" replace the product via new entrants to the market.

Christensen (1997) discusses the relatively rare successful management of disruptive product innovation by the dominant technology firms. In these disruptive product innovations:

- managers align the disruptive innovation with the "right" customers
- the development of those disrupting technologies are placed in an organizational context that is small enough to get excited about small opportunities and small wins, e.g., through "spin-offs" or "spin-outs"
- managers plan to fail early, inexpensively, and perhaps often, in the search for the market for a disruptive technology
- managers find new markets that value the (new) attributes of the disrupting technologies

Since, this is rarely done in the commercial context of *product* competition, it is unlikely to occur for many sustainability goals without either strong social demand or as a result of regulation[21]. This reinforces the view that disrupting innovations are necessary and the policy instruments chosen to promote sustainability need to reflect these expectations.

Rigid industries whose processes have remained stagnant also face considerable difficulties in becoming significantly more sustainable. Shifts from products to "product services" rely on changes in the use, location, and ownership of products in which mature product manufacturers may participate, but this requires significant changes involving both managerial and social (customer) innovations. Changes in socio-technological "systems", such as transportation or agriculture are even more difficult (Vellinga and Herb 1999). This suggests that the creative use of law is a more promising strategic instrument for achieving sustainable industrial environmental transformation, than the reliance of the more neo-liberal forms of ecological modernization on firms' economic self-interest.

This is not to say that technical assistance by government; enhanced analytic and technical capabilities on the part of firms; cooperative efforts and improved communication with suppliers, customers, workers, other industries, and environmental/consumer/community groups are not valuable adjuncts in the transformation process. And that is of course the value that ecological modernization scholars have brought into the discussion on major transformations in product, processes and socio-technical systems. But in most cases these means and strategies are unlikely to be sufficient by themselves for significant transformations, and they will not work without clear mandated targets to enhance environmental, safety, and health performance of the private sector. Nor will streamlining regulatory processes by itself be sufficient for the transformations that are needed.

---

[21] For a more optimistic view that large firms in established product markets can sufficiently transform, see Hart and Milstein (1999).

Government has a role to play in providing the opportunity for technological transformation/sustainable development through the setting of clear standards and policy goals, while allowing flexible means for industry to achieve those goals. Care must be taken to avoid dominant technological regimes from capturing or unduly influencing government regulation or negotiation processes. New entrants and new technologies must be given a chance to evolve to address environmental problems. Direct support of research and development, tax incentives for investment in sustainable technologies, and other technical assistance initiatives that fall under the rubric of "industrial policy" are other areas where government can make a difference (Nelson and Rosenberg 1993). Ideally, an "industrial policy for the environment" would include provisions relating to not only production and the environment, but also consumption, employment, and trade. Regulatory and other policy design and implementation are largely in the hands of government. The government can not simply serve as a referee or arbiter of competing interests because neither future generations nor future technologies are adequately represented by the existing stakeholders.

8. Final Commentary

Two different approaches are vying for the preferred pathway to address environmental problems. Ecological modernization approaches ask the question, How can we best encourage the creative forces of different sectors of society to make the necessary changes through cooperative involvement of stakeholders, continuous learning, innovative governance, regulatory streamlining, etc.? A technology-focused regulatory approach asks, How do we identify and exploit the opportunities for changing the basic technologies of production, agriculture, and transportation that cause damage to environment and health? In the latter approach, a policy choice has to be made for each environmental problem of (1) whether we want to effectuate a transformation of the existing polluting or problem industrial sectors or (2) whether we want to stimulate more radical and disrupting innovation that might result in technology displacement. Considerations of risks, costs, equity, and timing are relevant to all these questions.

Historically, the US EPA and most economists, scientists, and risk analysts have explored avenues of implementing the first approach. On the other hand, activists and others interested in significant industrial transformations have focused on the second approach and argued for application of political will and creative energy in changing the ways that industrial systems are constructed. The first effort promotes rationalism within a more or less static world; the second promotes dynamic transformation of the industrial state as an art form.

In a January 1994 report, the US EPA reveals a clear evolution of thinking, from a preoccupation with risk, to a concern for fundamental technological change. That report's introduction states:

Technology innovation is indispensable to achieving our national and international environmental goals. Available technologies are inadequate to solve many present and emerging environmental problems or, in some cases, too costly to bear widespread adoption. Innovative technologies offer the promise that the demand for continuing economic growth can be reconciled with the imperative of strong environmental protection. In launching this Technology Innovation Strategy, the Environmental Protection Agency aims to inaugurate an era of unprecedented technological ingenuity in the service of environmental protection and public health...This strategy signals EPA's commitment to making needed changes and reinventing the way it does its business so that the United States will have the best technological solutions needed to protect the environment. (EPA 1994)

Unfortunately, this article of faith has not been followed up with action, and neither the US nor Europe has come to grips with just how much major technological innovation should be encouraged, especially if it means the displacement of dominant technologies and even firms. If factor ten (or greater) is what is desired in pollution or material/energy use reduction, limiting policy initiatives to those involving cooperation with existing firms could limit success – especially if the targets, as well as the means and schedule for reaching the targets, are negotiated between government and those firms.

Finally, it must be realized that the choice of approaches are context-specific. It matters in a particular national environment whether there are (1) strong regulatory traditions and institutions, weak traditions and/or institutions, or (complete) absence of regulatory structure and culture; (2) strong trusteeship vs. arbitration traditions on the part of government; and (3) whether government is independent of capture or undue political influence by incumbent regulated firms. Current ecological modernization approaches might be best applied in regimes where diffusion, rather than innovation is likely to occur, as for example in the context of some developing countries[22]. But these same approaches could limit needed advances in industrialized countries. Involving a broader group of stakeholders and encouraging minor structural changes may not suffice; more radical and far-reaching institutional changes are needed, within the framework of "command-and-control" environmental governance.

[22] This contradicts to some extent the findings in most branches of the ecological modernization literature (cf. several contributions in Mol and Sonnenfeld 2000b).

Ecological Modernisation and the Creation of Lead Markets[1]

Martin Jänicke and Klaus Jacob

1. Introduction

The paper discusses the potential role of lead markets in the global process of ecological modernisation, here conceived as innovation and diffusion of environmental friendly technologies, including the innovation and diffusion of supporting national policies. This includes the question whether and how national pioneer roles in environmental policy can be played in times of economic globalisation.

Global Environmental Change in the direction of sustainability strongly depends on international markets for environmental friendly technologies. Such markets need national "lead markets" as a starting point. A lead market is the country that introduces an innovation that subsequently is adopted worldwide (Beise 2001). Lead markets are empirically characterised by high per-capita income, demanding and innovative buyers, high quality standards, political pressure for change and flexible, innovation-friendly framework conditions for producers and users. Unlike lead markets for normal technical innovations, *environment-friendly* technologies are specific insofar as they are problem-oriented and depend strongly on political influences. The problem dimension constitutes a potential global demand in terms of global environmental needs. It is mainly the role of pioneer countries to stimulate both, environmental innovation and their global diffusion, often in co-operation with international institutions and organisations. The interplay of innovation and diffusion of technology and policy takes place in different forms.

It is the high income countries which are able to afford the necessary investments in R&D for the development of new technologies. Many of them have also the demand conditions that enable environmental lead markets. These markets have to deal with the teething troubles of innovations, and they have to provide the pay back of R&D investments. They demonstrate the feasibility of technologies on a large scale application. Lead markets are not only related to potential first mover advantages, they also can attract foreign investors for environmental friendly technologies.

The paper is explorative in nature, as there is a lack of research on the mechanism and conditions for the successful making of global markets for environ-

1 The publication is partly based on a research project funded by the German Ministry of Education and Research, grant number 07RIW1A. However, the responsibility for the content of this publication rest with the authors only.

mental innovations. It should be understood as an overview and systematisation of aspects of global ecological modernisation with special regard to lead markets for environmental innovations.

2. Ecological Modernisation

By "ecological modernisation" we understand the innovation and diffusion of marketable environmentally friendlier applied technologies, including the innovation and diffusion of supporting policies. The concept describes the wide spectrum of possible environmental improvements that can be achieved through innovations beyond the purely end-of-pipe approaches. "Ecological modernisation" is not only the headline of the environmental policy of the present red-green government of Germany. It is a concept that at present rapidly diffuses, especially in the field of social sciences. We use it in its narrower technical-economic sense. Other authors - such as Hajer (1995) or Mol (2001) - tend to use a broad definition which includes institutional, structural and cultural changes of all kind. The main reason for our choice is that policies based on technologies and innovations not only represent a large potential of environmental improvements within the market system but are also easier to introduce and implement than those policies requiring intervention in the established production, consumption, transport, or lifestyle structures. We need a special term to denote these types of marketable solutions. If the concept of ecological modernisation is restricted in this sense it may also be easier to take the limits of the "technological" strategy into account, which cannot be ignored.

Ecological modernization starts beyond end-of-the pipe approaches or clean-up technology and way beyond merely reparative measures (see table 1). The scope of environmental friendly technologies varies from incremental improvements to radical innovation, where innovation means the initial market introduction of a new technology. The latter may improve some or all of the phases of a product's life cycle and thereby maybe labelled as "clean" or "cleaner" technology. Incremental improvement affects different dimensions such as efficient use of resources, efficient use of energy, efficient use of land, transport intensity, or risk intensity regarding plants, substances, products on waste intensity, i.e. waste materials and harmful emissions (Jänicke 1985).

Modernisation in its economic core is a process of continuous improvement of procedures and products. It is a compulsory necessity in capitalistic industrial societies driven by the forces of competition which generates innovative or efficient technologies. Typically, technological progress is a market based process. It is however possible, and for ecological modernisation a prerequisite, to influence the *direction* of modernisation. The spread of environmental concern can be an important motor for economic modernization (Brickwedde 1997). Recently, also public agencies adopted this view, as for example the Swedish Ministry of the environment (1996) argues that "environmental policy is contributing towards the modernisation Swedish enterprise" or the European Commission: "...high environmental standards are an engine for innovation and business opportunities" (European Commission 2001). This argument is extended to competitiveness: "How an industry responds to environmental problems may, in fact, be a leading indicator of its overall competitiveness ... Successful environmentalists, regulatory agencies, and companies will ... build on the underlying economic logic that links the environment, resources productivity, innovation, and competitiveness" (Porter and van der Linde 1995; cf. Wallace 1995). This argument, however, has also been disputed and certainly needs more empirical evidence and more differentiation.

Table 1. Model and Examples of Environmental Policy Approaches

Curative approaches			Preventive approaches	
Repair: Reduction / compensation of damage		*End-of-pipe treatment*: Clean-up technology	*Ecological modernization*: Clean(er) technology / Eco-efficiency	*Structural change*: Decrease of "dirty" industries / activities
Examples	Payments for noise damage	Passive noise protection	Less noisy motors	Alternative traffic modes, less traffic
	Ex-post measures against forest damage	Desulphurisation of coal power stations	More efficient power production and consumption; CHP; cleaner primary energy	Less power-intensive modes of production and consumption
	Measures against damage caused by industrial waste	Waste incineration	Recycling	Reduction of waste-intensive sectors

Jänicke 1985

Unlike the interpretation of Ashford (this volume) - ecological modernisation conceived as the transformation of the 'problem industries' into 'green' industries - our view on ecological modernisation, focuses on innovative technologies. The technologies which are put on the markets to substitute the environmental harmful technologies are not necessarily produced by the same firm or the same industry. For instance firms producing technologies for renewable energies constituting a new - booming - industry. Our view on ecological modernisation is concerned with technologies which are marketable. For certain environmental problems there is, indeed, the necessity of a structural change, e.g. the phasing out of nuclear energy or lignite coal, which cannot be effected via market mechanism. But the difficulty of this political task is so different that we should use a different term: The dichotomy between ecological modernisation on the one hand and structural change of the phasing-out type on the other hand may be helpful here. Ecological modernisation – using the logic of modernisation and markets - is the easier policy. It is at best a continuous process leading to win-win situations. However a decrease of an industry in its core technologies creating losers and e.g. regional employment problems requires huge political endeavour and is therefore possible only exceptionally.

As we will show later ecological modernisation, too, needs government policy overcoming market failures. And the policy may include the "big stick" as a final resort against laggards in the diffusion process. But this is, at least from the point of view of policy sciences, a quite different task compared to structural solutions.

3. The Political Dimension of Environmental Innovations

What are the driving forces of this process and how can they be reinforced? As a starting point for the analysis, the approaches of innovation economics for the explanation of innovations may be utilised. However, in addition, the special characteristics of environmental innovations must be considered. Traditional R&D policies do focus on the provision of infrastructure needed for the generation, transfer, and application of knowledge by the state and on the amount of subventions for R&D activities. Financial aids and research institutions are considered as the adjusting screws to explain success or failure of National Innovation Systems (OECD 1999).

For environmental innovations additional aspects have to be taken into account, too. It is not only the supply of technologies which is supported by public R&D policy, but also environmental regulations, frequently having a strong influence on the demand side that is of special importance. Environmental technologies which become obligatory as e.g. BAT standards (see Hitchens et al., this volume) do have a well protected market. But there are many other means to support and increase the demand for environmental innovations beyond command and control measures such as e.g. tax exemptions or reductions e.g. for unleaded gasoline, labelling schemes e.g. Blauer Engel, public procurement, or EMAS.

Innovations both in environmental technology and in environmental policy can nowadays count on a broad spectrum of transfer mechanisms beyond the market which - from the OECD, by way of the World Bank, right through to Greenpeace - help their diffusion on the world market. Pioneering measures taken by states and the international orientation along "best practice" lines serve to further reinforce these mechanisms.

To conclude, political strategies, aiming at a creation of markets for environmental innovations, can be build on three different approaches: 1) The improvement of the infrastructure for the supply with environmental innovations; 2) the safeguarding of demand by means of environmental policy and 3) the utilisation of transfer mechanism to speed up the diffusion of policy innovations into other countries.

There are, however, less favourable conditions for environmental innovations. First, there is a short-term and most often even static perspective on the technological possibilities of enterprises both by regulators as well as managers. Technologies, products and preferences are taken as granted and possible changes are perceived as provoking additional costs. Thereby, environmental policy is oriented on the state of the art rather than on the potentials of technologies. Second, there are many persistent reasons for an end-of-pipe orientation of environmental technologies. These technologies are more easy to control, they usually do not require a change in central processes, there are standardised solutions at hand. Furthermore, if substantial investments in EOP technologies have been done, sunk costs have to be depreciated before considering more innovative technologies. Third, due to the externalities of innovations and especially environmental innovations, there is an undersupply of R&D activities. For innovation in general, there is an incentive for free riding and to obtain second mover advantages. Klaus Rennings (2000) has pointed out, that there is a second externality of environmental innovation: There is an incentive for free riding on the environmental benefits of environmental innovations because these benefits are a public good. Environmental policy therefore has been promoting the diffusion of existing technologies rather than the stimulation of innovations (s.a. Hübner and Nill 2001).

Considering these peculiarities of environmental innovations it is a task of considerable difficulty for politics to implement a policy which is likely to foster environmental innovations. Whereas with "normal" innovations state and politics form only one factor among many influencing the framing conditions of the potential innovator, environmental innovations benefit from socio-political actors including NGOs. It is important to note, that environmental innovators often orientate their decisions on the early phases of public problem definitions and the early phases of policy formulation rather than wait until a suitable policy has been passed and enacted (Jacob and Jänicke 1998). Innovations cannot be explained by a single governmental instrument, but many other factors have been taken into account such is the policy style, the actors configuration and the instrumentation (Jänicke et al. 2000).

A political strategy should strengthen the ecological motivation of potential innovators, improve their situation regarding the available information, and above all cut their investment risk by providing calculable perspectives. A strategy of ecological modernization will begin with clear target data but with "soft" instruments and regard regulations and official directives as the very last resort (Wallace 1995; Jacob and Jänicke 1998). The guiding axiom is: The more credibly the government threatens specifications and sanctions right from the outset, the more effective the "softer" instruments will work. This rather management-oriented approach is likely to be effective particularly if *targeted* environmental innovations are at stake, for which potential innovators and target groups can be addressed directly. For a broad stimulation of unspecific environmental innovations it is necessary to address a wider spectrum of potential innovators less specifically and directly. For the latter type of innovation oriented environmental policy more traditional means of regulation and stimulation seem more appropriate.

Recently, with a broadly effective set of instruments applied as part of innovation-oriented environmental policy, it has been above all environmental levies and energy taxes that have gained in significance. Of course state provision of the necessary infrastructure for research, development, and knowledge transfer - as innovation research has always stressed - is also critically important. State-run "green" R&D programs play an important and more specific role in innovation-oriented pioneer countries (e.g. the Netherlands, Denmark, and Sweden).

Another important aspect is cooperative environmental planning as defined by "Agenda 21". This encompasses elements of classical regulation and control and of public management systems. The use of strategic targets in environmental plans and strategies reduces the insecurities involved in suitable innovation processes and offers innovators more reliably calculable investment conditions. If, for example, a hazardous substance has to be withdrawn from the market before a specified deadline, the potential supplier of a substitute substance has greater certainty with respect to the profitability of his research and investment planning. Moreover, sustained environmental planning can create motives for innovation and marketable solutions insofar as it is linked to a broad target-oriented debate on specific problem situations. Strategic environmental planning is usually associated with the formation of networks, among other things favouring the exchange of information so important for innovations.

For a comprehensive explanation of environmental innovations, it is not sufficient to look for the political management of single innovations only. The overall capacity of nation states, or even regions for innovations as well as market demands have to be reflected, too. This capacity has been conceptualised as "national innovation systems" where innovative firms are part of a network which encompasses actors from other firms, research institutes, universities, etc. (e.g. Freeman 1987, Lundvall 1992, OECD 1999). The concept of a national innovation system is, however, not a consistent theory, but it tries to combine a wide range of influencing factors which possibly explain national and regional differences in innovation activities.

So far, we have dealt with policy factors supporting innovation. But ecological modernisation is both innovation and - hopefully - rapid and complete diffusion of available and marketable solutions. Therefore, we discuss in the following the conditions for their diffusion.

4. Globalisation and National Environmental Policy Capacity

The greening of international markets strongly depends on national pioneers in environmental policy (table 2 and 3). But is pioneer behaviour of nations possible in the context of globalisation? Before we turn to the global diffusion of environmental innovations we should clear this important point.

Table 2. The Pioneer Countries in Environmental Policy. Policy Innovation or Early Adoption 1970 - 2000[2]

Country	1970-1985	1985-2000
Sweden (11):	7	4
USA (10):	8	2
Japan (9):	8	1
Denmark (9):	5	4
Finland (8):	4	4
France (7):	5	2
Germany (7):	5	2
The Netherlands (7):	3	4
UK (6):	4	2
Canada (6):	2	4
Total	51	29

Busch and Jörgens (FFU) 2001

Table 3. General Characteristics of the Present Pioneer Countries in Environmental Policy (PCEP)

Definition: PCEPs are innovators or early adopters of new environmental policy measures that diffuse into other countries (thereby contributing to the development of global environmental policy).	
Characteristic (Indicator/Measurement):	**General Hypothesis:**
Environmental policy innovations (Policy monitoring, FFU data)	Pioneering environmental policy is possible
Strict environmental regulation (e.g. Environmental Regulatory Regime)	Strict environmental policy is possible (s.a. Porter)
Innovation or early adoption of environmental technologies (Monitoring of environmental technology diffusion)	PCEPs having the capabilities for technology based environmental strategies and are by this candidates for becoming lead markets
High economic income (GNP/cap.)	High income means both, high (perceived) pressure **and** high capacity for environmental policy
High competitiveness (e.g. Competitiveness Report)	Environmental issue is important for the competition on innovation
Open economy[3] (export/import ratio of GNP)	Economic globalisation is no impediment for active environmental policy
Strong role of government[4]	No general "withering away" of governments in times of globalisation

According to a broad review of literature conducted by Bernauer (2000), there are three distinct understandings of globalisation. From a constructivist/sociological perspective globalisation encompasses political, economic, cultural phenomena of diminishing importance of national borders (e.g. Giddens 1990). Political scien-

2 Introduction of 20 new environmental policy institutions, laws or instruments: innovation plus first 3 adoptions. Preliminary data.

3 Mainly true for the present PCEPs (forerunners within the EU, and Canada).

4 See footnote 3.

tists which also stress the diminishing importance of boarders between nation states and using the term of denationalisation (e.g. Zürn 1998). A neo-marxist perspective stresses the globalising of capital, the dominance of transnational firms, or the unleashed world market (e.g. Altvater and Mahnkopf 1996). Finally, from the point of view of economists, globalisation refers to the extension of international markets both in terms of trade and investment activities. This is accompanied by an easier mobility of factors movements, especially the international movement of capital. For this paper we are interested in the possibilities and obstacles for the establishment of international markets for environmental innovations. Since *environmental* innovation/diffusion as a rule need to be stimulated by policies it is crucial to understand not only the economic, but also the political dimensions of globalisation and its implications for policy making as well.

There is an ongoing debate about the implications of globalisation for national policy making. Regarding social policy, economic policy, but also environmental policy it has been argued that a free movement of production factors limits the possibilities for national regulations which lead to rising costs for firms and to a competitive disadvantage. Globalisation therefore leads to a *"race to the bottom"* or to de-regulation to attract foreign investments. This phenomena of deregulation became known as the Delaware effect of globalisation (Vogel 1995). It was in Delaware where competition on deregulation of corporate chartering began. In the US charters are granted by individual states, but all states are required to recognize each other's charter. In the course of this competition, a race to the bottom was won by Delaware by lowering the level of protection for employees, shareholders, and customers.

According to Vogel (1995, 1997, 2001) economic integration and strict regulation is not as antagonistic as it can be expected. High standards in important markets may force foreign producers to adapt to these standards by which foreign governments react by raising their own standards. Furthermore, due to scale effects in production but also to obtain the image of an innovative firm, it may be sensible for firms to adapt to the higher standards for other markets as well on a voluntary basis. A prominent example of this *race to the top* are environmental standards set by California which lead to a world wide adaptation by car manufactures which became known as the California effect.

It is a question open to empirical investigation if this example of a successful convergence of environmental standards on a high level of protection may be generalised. It has been argued that this mechanism may apply to product regulation only (Vogel 1997; Scharpf 1999). The distinction between products and processes is not selective since all process technologies are products as well (e.g. wind mills). Empirical evidence is given for a spread of industrial pollution standards to developing countries (Hettige et al. 1996).

Regarding the expected decline in competitiveness by environmental policy the race-to-the-bottom hypothesis suffers from several highly questionable assumptions: It assumes that environmental regulations impose costs for producers that affect location, regardless of differences in labour productivity. It also assumes that governments react exclusively to the preferences of the international capital, ignoring the preferences of voters or interest groups (Drezner 2001). Last but not least, the race-to-the-bottom hypothesis not only overestimates the importance of environmental costs and the differences in regulatory costs but also the general role of prices, thereby ignoring the role of innovation in the global competition. The rising importance of the environmental issue in the competition on innovations may be the most interesting counter argument.

5. The Porter Hypotheses on Environmental Regulation and Competitiveness

The Porter hypothesis argues that a strict environmental policy can improve competitiveness of firms and sectors (Porter 1990; Porter and van der Linde 1995; s.a. Ashford 1979) may be split into two distinct parts (s.a. Taistra 2001): First, a competitive advantage might be achieved in case of a strict environmental policy which, at a later stage, diffuses internationally. If there has been a development of technologies in response to strict environmental standards, industries -not necessarily the polluting industry itself- might be able to export their technologies. Their competitive advantage may be based on learning effects or patent protection of their innovation.

Second, strict environmental policy might lead to innovation in the polluting industry itself which is able to compensate or even overcompensate for the costs of adaptation. This part of the Porter hypotheses has been labelled the "free-lunch" or even "paid lunch" hypotheses.

This second case refers to inefficient patterns of production. The existence of considerable inefficiencies is not expected by conventional economic theory. Possible explanations for the broad empirical evidence supporting this part of the hypotheses, might be seen in the fact that both regulators and enterprises most often have a static view when evaluating the expected costs of environmental regulation. Strategies for environmental protection are usually are developed on the basis of given technologies, products and preferences. Policies are most often formulated in a short term perspective only. All this leads to a policy which is based on the state of the art, instead of being oriented on the technical potentials.

According to Porter, environmental policy should choose instruments stimulating innovations which are able to take advantage of the potentials of technologies rather than stimulating the diffusion of existing technologies. Furthermore, national environmental standards should be a slight precursor for other countries. However, a wide gap between the different national standards should be avoided in order to beware of idiosyncratic solutions.

The "Porter hypothesis" has been supported by policy science research on environmental pioneer countries (Wallace 1995; Jänicke and Weidner 1997; Anderson and Liefferink 1997). There have been always national pioneer countries in environmental policy. In the context of globalisation these countries have gained additional importance - just in opposite to the "race to the bottom" hypotheses. They are - possibly more than international institutions - the paramount protagonists of the development of international environmental policy. While environmental policy is mainly based on technologies, they are at the same time supporters of a global ecological modernisation. It is mainly a few highly developed OECD pioneering national states which pushes technology based measures for environmental protection. For these countries the competition on quality which is based on innovation - rather than competition on costs - seems to be the primary push.

Governments do not have an exit option but must react to functional imperatives of their countries. In the global competition the nation state is coming under pressure in areas such as employment, financial policy, social security, infrastructure, R&D policy, and last but not least environmental policy. Here the *Cameron hypothesis* in political science may be remembered stating that open OECD economies tend to a higher share of public expenditure (Cameron 1978). The underlying causalities for this phenomena may be disputed. We expect, however, that open economies need more government activities, both to enable international competition e. g. by providing the infrastructure or an effective innovation system and to counteract its problems e. g. by compensating its losers. This pressure for action operates contrary to a diminishing importance of national boarders. For policy researchers it is no surprise that well developed OECD countries which are highly integrated into the world market are also more active in environmental policy (see Bernauer 2000). The nation state also remains the most competent and best organised actor in the global arena. According to these authors countries with an open trade regime do have more stringent environmental regulation.

While there has been a transfer of sovereignty to international institutions, nation states gained partially additional opportunities by concerting globally their actions (e.g. nature conservation, Basel convention, Rio process, but also the consolidation of national budgets). Therefore, the decline of national sovereignty should not be confused with a decline of capacity to solve national problems.

There is an ongoing debate in economic research about the question of whether a pioneering role in environmental policy influences the competitiveness of firms, sectors or nations. A number of empirical studies on technologies, sectors, firms and countries have been published which in general support the expectations of the Porter hypotheses (for an overview: Taistra 2001; furthermore Jaffe et al. 1995; Hübner and Nill 2001; Sturm et al. 2000; Esty and Porter 2000).

All of these investigations are not able to model the *causal* relationship between economic and environmental performance. At least, the correlations giving further evidence for the thesis that an ambitious environmental policy doesn't harm competitiveness. Furthermore, they giving evidence once again, that a well developed economy is a prerequisite for the development of a successful environmental policy. It is the highly developed countries which are characterised both by high environmental pressure (both objective and subjective, induced by high education and income) and high capacity (encompassing the institutional basis, administrative competence, economic/fiscal resources, knowledge, and the strength of NGOs) to react on it.

A key mechanism for an integrated approach which utilises the economic forces of globalisation, might be the establishment of *lead markets* for environmental innovations.

6. Diffusion of Policy Innovations and the Globalisation of Environmental Policy

As mentioned above, the international diffusion of clean(er) technologies strongly depends on the diffusion of their supporting policies. Therefore, the role of environmental policy diffusion is relevant in our context. Recent comparative research on the spread of environmental policy among countries reveals an astonishing international convergence in the development of national policy patterns (Kern 2000; Jörgens 1996; Kern et al. 1999). It is possible, by way of policy monitoring, to treat innovations in environmental policy as indicators and evaluate these accordingly (from the establishment of an environment ministry right through to the introduction of a CO_2 tax). It is also possible in the same way to assess the significance of pioneer countries and the role of certain strategic countries without which rapid diffusion would not succeed. This procedure also allows us to deduce, from the diffusion rate, the level of difficulty involved in solving a problem. Monitoring individual policy measures in this way (as policy output) is of course not a proper policy outcome evaluation; but the method of empirically describing national and global policy developments with the aid of policy indicators can still be considered a step forward in environmental policy research.

The result shows, for example, that the globalisation of environmental policy, insofar as this is reflected at state level, can indeed be described using the analytic concept of innovation diffusion: Standard solutions in certain pioneer countries are diffused worldwide, thus causing a substantial measure of convergence in policy formulation at national state level - irrespective of extremely different capacities for action. Unlike in the 1970s, when for example the USA or Japan had a major innovative function in global environmental policy, nowadays innovations in environmental policy emerge strikingly often in small EU countries tightly integrated in the global market (Jänicke 1998).

The - reformed - institutional fabric of the EU seems comparatively favourable both for innovations and for their diffusion (Héritier et al. 1994). The EU must firstly, at least in principle, accept a "high level of protection" in member states; it must secondly seek to harmonize innovations in environmental policy implemented at national state level. Pioneer countries, for their part, often have an interest in anchoring their policy innovations within the EU framework in order to thus minimize their subsequent need to adapt to European policy. It is also often a matter of "Europeanizing" certain national pioneer measures favouring the particular country's domestic industry. Policy diffusion within the EU, however, takes place not only by way of EU harmonization but also from country to country. In the latter case the policy innovation in question will often need first to be introduced by one of the more influential EU countries before it achieves the necessary widespread impact. For example, the CO_2/energy tax was already introduced in the Netherlands and the Scandinavian countries in the early 1990s - but it seems that the decisive push towards European diffusion has been the adoption of a green tax by the red-green coalition government in Germany in 1998. The CO_2 tax is an example of "horizontal" diffusion. It has yet to be established as a European measure.

The diffusion of innovations in environmental policy thus takes place both directly from one country to another, i.e. by way of imitative policy learning or "lesson drawing" (Rose 1993) and by way of international institutions (e.g. OECD, UNEP, World Bank), organizations (e.g. Greenpeace), or expert-networks (e.g. the International Network of Green Planners). It is striking how rapidly many innovations in environmental policy are diffused. Environment ministries have, in a period of just under 30 years, clearly asserted their position in the industrialized countries. Environmental plans, as defined under "Agenda 21", just ten years after the Rio Conference (1992), are going to be more or less in place worldwide - though in extremely disparate quality. However, in other cases (e.g. soil protection legislation) the diffusion rate is clearly curbed by the difficulty of solving the problems involved.

It can be expected that a high *capacity for environmental policy* is needed both for policy innovation and the adoption of innovations. The OECD defines it broadly as "a society's ability to identify and solve environmental problems" (OECD 1994). While the term capacity and capacity building was used previously by numerous institutions such as UNEP, FAO, World Bank, OECD, and others in connection with less developed countries only, it has been fruitfully extended to industrialised countries as well (Jänicke and Weidner 1997; Weidner and Jänicke 2002). It refers to the structural preconditions for successful environmental policy and encompasses the collective actors (esp. environmental institutions and organisations). The structural preconditions include (a) the institutional set-up (e. g. open and effectively integrated political institutions, administrative competence), (b) the system of creation, transfer and application of knowledge and (c) the economic-technical basis.

7. The Interplay between the Diffusion of Environmental Policy Innovations and Environmental Technology

There is a highly symbiotic fabric of interwoven interests between innovators in technology and policy makers. Suppliers of environmental technology seek the support of politicians and politicians are always looking out for technological options, because precisely these are much easier to implement than any sort of structural intervention.

However, the interplay between environmental policy and environmental technology in the case of innovation diffusion is characterized by a wide variety of possible constellations. Theoretically it is possible to distinguish between the following diffusion scenarios (figure 1), depending on the factors leading to the political and technological innovations:

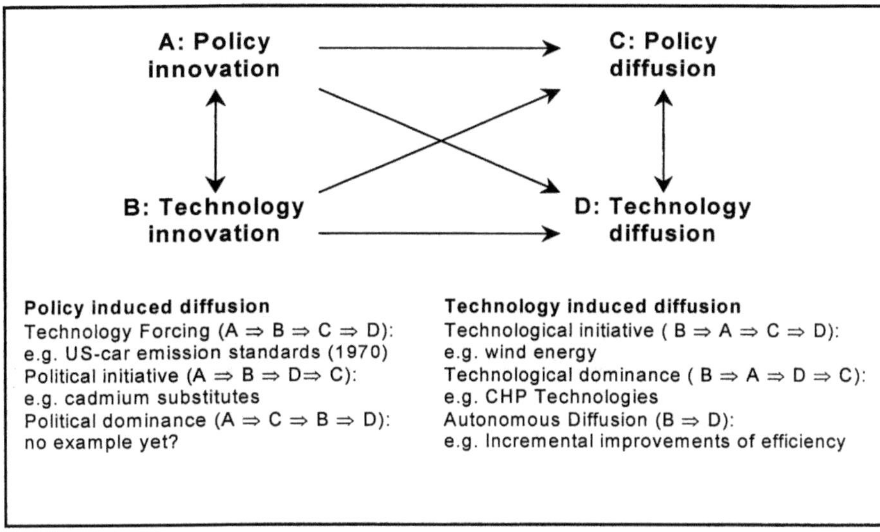

Fig. 1. Diffusion Patterns of Environmental Innovation

Technology forcing (A⇒B⇒C⇒D): A national environmental policy innovation in one country forces a technological innovation which diffuses if also the policy innovation is diffused (e.g.: catalytic converter technology in cars).

Technological initiative (B⇒A⇒C⇒D): A new but already existing environmental technology induces a political innovation whose diffusion in turn encourages the diffusion of the technology (e.g.: wind mills).

Political initiative (A⇒B⇒D⇒C): A national environmental policy leads to technological innovations whose diffusion in turn encourages diffusion of the policy innovation (e.g.: cadmium substitute[5]).

Technological dominance (B⇒A⇒D⇒C): An innovation in environmental technology is successfully diffused and as a result receives political support both nationally and internationally (e.g.: combined heat and power in industry[6]).

Political dominance (A⇒C⇒B⇒D): The innovation in environmental policy is successfully diffused before a corresponding technology is available (this scenario is, symptomatically, very rare in ecological modernization).

Autonomous technological development (B⇒D): An innovation in environmental technology is successfully diffused without political influence; this case, beyond incrementally increasing energy efficiency in companies, seems to be rather rare.

---

5 The use of cadmium was regulated in Sweden in the early 1980s with their standards for substitutes being adopted by European industry. Not until the early 1990s, however, were these standards made binding by the European Commission (Bätcher et al. 1992).

6 Combined heat and power in industry spread largely autonomously, even though regulatory measures were intended to encourage its use in public power stations.

Technological innovations do provide additional options for policy makers. For other cases, policy factors have been the major driving forces in the stimulation of environment-friendly technical innovations. The case of technology forcing has, however, been exceptional for environmental innovation (cf. Conrad 1998; Jacob 1999). So far, environmental policy has its merits in the promotion of the diffusion of technologies. It can be observed, however, that policies promoting the diffusion do support incremental innovations.

There is considerable plausibility for the assumption that autonomous emergence and diffusion of innovations in environmental technology is the exception rather than the rule and that such developments usually remain limited to incremental increases in efficiency in companies. The reverse border-line case is innovation in environmental policy where policy clearly exceeds the given technological possibilities.

The limits of ecological modernization (in the "technocratic" sense) are thus defined by the limits of technology. However, these limits are dynamic. They can be extended by research (and by backing for research). For example, research into the development of procedures for reducing CO_2 emissions, if successful, could substantially widen our room for manoeuvre in climate politics - even if only in the sense of end-of-pipe measures. The rapid diffusion of suitable policy innovations will then be as similarly predictable as the difficulty and slowness of a structural climate policy which de facto places restrictions on established energy markets (coal, oil).

The variants of this interplay between policy and technology in any case are a central theme in research on the diffusion of environmental innovations, especially when it comes to selectively optimising such innovations.

To summarise our main assumptions:

- Ecological modernisation can be conceived and has its strength as a market compatible strategy of technical environmental innovations and their policy based diffusion. It is the nation state, which is playing a crucial role in this context.
- The necessary pioneer role is a possible option for highly developed countries, many of them being especially open economies - there is no race to the bottom in times of globalisation.
- There is no general contradiction between competitiveness and demanding environmental policy, on the contrary, highly developed countries tend to integrate the environmental issue into the competition on quality.
- Global diffusion of best practice in environmental policy takes place and is a major driving force for the diffusion of marketable, technical solutions for environmental problems, that typically exist on a global scale.

If we are right, the creation of lead markets for environmental technologies would be a feasible global environmental strategy.

8. Lead Markets for Environmental Technologies

Lead markets are the geographical starting point of global diffusion processes. We understand lead-markets for environmental technologies as regional or national markets, which were stimulated by higher preferences for environmental goods in a given country, specific supporting measures, or policy interventions, which are able to influence the markets in other regions effectively, trigger reactions of adjustment and finally lead to an international diffusion of the new technologies. By this, we take again into account, that *environmental* innovations have to be largely ascribed to governmental (or NGO) activities.

For a targeted ecological modernization of international markets the potential of nation states for a framing of national markets might gain considerable importance. The history of environmental protection is rich in examples for lead-markets: it encompasses the legally enforced introduction of catalytic converters for automobiles in the USA, desulphurisation technologies in Japan, the Danish support for wind energy or the CFC free refrigerator in Germany. Another impressive example is the global diffusion of chlorine-free paper, from the political activities by Greenpeace and the EPA in the USA, by way of the introduction of chlorine-free paper whitener in Scandinavian countries and various Greenpeace campaigns in Germany and Austria, right through to effective political market intervention in south-east Asian countries like Thailand (Mol and Sonnenfeld 2000).

Lead markets for environmental technologies

An environmental lead market is the core of the world market for a product or process where:

- national policy or non-governmental influences successfully have created an incentive structure for users to adopt an innovation relating to a (manifest or latent) global environmental problem and
- the global dimension of the problem creates a *potential* demand also in other geographic markets.
- As a rule environmental lead markets are created by national policy innovations (e.g. standards) which potentially diffuse into other countries. There is a close interrelationship between policy innovation/diffusion and technical innovation/diffusion.
- The diffusion of environmental policy innovations is supported both by horizontal imitation ("benchmarking", "lesson-drawing") and by international organisations.

By setting up increasingly demanding environmental standards, pioneer countries in environmental policy may send out a twofold signal beyond the boundaries of their national market:

1. A national market for environmentally-friendly technology acting as a basis for subsequent expansion to bigger markets. The pioneer country demonstrates the feasibility of its standards, regulations. Subsequently the innovative regulation is adopted by other countries. For example, the German tax preference for fuel-saving cars (1997) has supported suppliers in that country (Volkswagen, Mercedes). The diffusion of this instrument, e.g. throughout the EU, can bring appropriate market expansion. Frequently, the international diffusion is supported by the national producers, if they were able to adopt successfully to the new standards (examples in Jacob 1999). A diffusion of regulations will be more likely if a country has attained the image of being a pioneer. It is only a few countries nowadays, mostly member states of the EU, which serve as the benchmark for the development of environmental policy.
2. The pioneer market with its demanding environmental regulations can, however, also send out signals to the supply side outside the domestic market. For example, California, with its stricter emission rules compared with the rest of the USA, was able to exert a general influence on the car industry word-wide (Vogel 1995). Similarly, Denmark, in 1994, with its targeted promotion of energy-efficient refrigerators, was able to prompt European suppliers to offer such devices there. In cases like these, competitive companies can advertise their ability to supply such demanding market areas as a sign of their technological competence. It can be cost efficient to orient the production on the highest standards, if there are scale effects.

An ongoing research project carried out for the German Ministry for Research and Education BMBF on "ecological lead markets" (conducted by the DIW, FFU, IÖW, and ZEW) aims at identifying both framework conditions and policy measures for the establishment of lead markets on a more systematic empirical basis. Lead markets generally are "geographic markets which have the characteristic that product or process innovations, which are designed to fit local demand preferences and local conditions, can subsequently be introduced successfully in other geographic markets as well and commercialized world-wide without many modifications. In the model of international diffusion of innovations a lead market is the core of the world market where the local users are early adopters of an innovation on an international scale" (Beise 1999). The U.S. as lead market for the internet, Japan as lead market for fax, or Finland as lead market for mobile phones are well-known examples. Empirically lead markets are characterized as follows:

- General characteristics of lead markets (see also Meyer-Krahmer 1997)
- High per-capita income, low price elasticity
- Demanding, innovative buyers, high quality standards
- Problems, pressure for change and innovation
- Flexible regulation, innovation-friendly framework conditions for producers and users
- Product standards are acknowledged also in other countries

Lead markets for environmental technologies, however, are characterised by additional factors. They typically are not only stimulated by higher environmental preferences of consumers in that country, but also by special promotion measures, or by political market intervention.

They provide marketable solutions for global environmental needs, offering at least improvements for environmental problems which are mostly encountered worldwide or at least in a great many countries. Thus technological solutions to environmental problems enjoy, right from the outset and by their very nature, potentially larger markets. Lead markets affect competition in other market regions, trigger appropriate responses and adaptations, and thereby lead to the international diffusion of the new technology. The creation of lead markets for an environmental technology takes place in two stages, the first being the most important:

1. Struggling for success on the national market: This includes the establishment of a national market (not only a niche market), successful incremental improvements of the product and its production. Government instruments may be standards, subsidies, charges, labels, public procurement, network management, or EMAS (demand of firms).
2. Government support for technology transfer by activities within international organizations (e. g. diffusion of the supporting policy pattern), bilateral actions with strategic countries (e. g. the environmental co-operation between Germany and China), special international conferences, use of the international media, cooperation with international NGOs. More important may be - on the demand situation - the diffusion motor of benchmarking and search for best practices which in many countries is an institutionalised mechanism, today. In addition, the cooperation with multinational companies may be a relevant transfer mechanism.

Possible functions of environmental lead markets

Global Functions:

- Problem solving function regarding global environmental needs
- Return function for R & D and learning costs (possible in high-income countries)
- Technological demonstration function (benchmarking)
- Political demonstration function (lesson-drawing)

National Functions:

- Competitive function, potential first-mover-advantages
- Potential attractiveness for foreign direct investments
- Increased market value of environmental **and** technological reputation
- Political legitimation function (for environmental policy, national policy actors as global players)

If successfully established, such markets may fulfil a range of functions: From a *global* perspective they provide marketable solutions for typical environmental problems. Lead Markets in high-income countries are able to raise the necessary funds for refinancing the costs for development and "learning". This is true for environmental innovations in particular since there is a need to survive the teething troubles of new technologies. They are demonstrating both the technical and the political feasibility and thereby giving a stimulus for other countries and enterprises to adopt to their pioneering standards. From a national perspective ambitious standards or support mechanisms might safeguard the first mover advantage for the own industries. Furthermore, ambitious policy measures can attract foreign investors which are interested in the development and marketing of environmental innovations. (It is not by chance that there have been recently some prominent investments for the production of solar cells or for fuel cells in Germany.) Finally, a demanding policy which holds economic advantages additionally legitimates the national policy makers, sometimes providing them also with attractive roles in the global arena.

9. The Limits to Ecological Modernization

We use this "technocratic" concept of ecological modernization, in its narrower sense, to describe the spectrum of technical, system-compliant solutions for environmental problems. Ecological modernisation in this sense, however, comes up against its limits where potentially marketable technological standard solutions are not available. The so far unsolved environmental "persistent problems" - urban sprawl, loss of bio-diversity, soil erosion, groundwater pollution, final storage for nuclear waste, or the deterioration of global climate - all, so far, show up these limits. The modernization approach is also no viable option where the risk is acute and immediate defensive action is needed.

If incremental increases in ecological efficiency are not a causal, sustained solution, the environmental relief might be compensated by subsequent growth processes. In this case, the effects of ecological modernisation are compensated by growth. A reduction in pollution tends to be followed by a resurgence. These facts were recognized as early as the late 1970s as the "dilemma of the N curve" (Jänicke 1979). This dilemma applies not only to clean-up environmental protection and end-of-pipe treatment but even to efficiency improvements. For example, Japanese industries, between 1973 and 1985, succeeded in saving energy and raw materials in a remarkable way but the high industrial growth in those days simply devaluated this effect (Jänicke et al. 1997). The overall growth rate must thus always be accompanied by equivalent progress in (compensatory) technology providing environmental relief. This "*hare and tortoise-dilemma*" of ecological modernization is even tightened if there are losers of modernisation: If it is not the polluting industry itself which finds new opportunities in environmental friendly products, the sector often seeks for new sales opportunities for the old product. For example, the successful campaigns of environmentalists against using chlorine in applications free to the environment leading to a considerable reduction in production and consumption have since been compensated by the expansion of chlorine uses in other areas (Jacob 2001).

What is needed in the long term therefore is, firstly, a transition from incremental to radical innovations in which ecologically problematic procedures and products are substituted by unproblematic ones (Kemp 1997). An example is the transition from efficiency improvement in coal-fired power plants to variants of solar energy. In between lie the border-line cases, a variety of incremental improvements which together represent a radically new quality (e.g. the zero-energy house).

What is also needed are structural solutions, i.e. solutions of a non-technical nature, changes in the structure of demand and of industry, and, based on these, an ecological industrial policy. Finally for the areas that are difficult to control, namely life-style, the level of personal mobility, and residential and housing structures, etc. have to be tackled by other means than technical approaches. Unlike the economic-technical variant of ecological modernization there are no marketable technical solutions to problems of that type.

The much higher degree of political difficulty for an even cautious ecological industrial policy aiming at a restructuring away from the environmentally intensive "chimney industries" is indicated by the fact that there are hardly any examples for such a far reaching policy. Examples so far, namely the running down of coal mining in the Netherlands or of crude steel works in Luxembourg, were hardly suitable for or capable of diffusion and are unlikely to find imitators. Often, environmental objectives haven't been the driving force in these cases, although there has been a considerable environmental relief (Binder et al. 2001).

Finally, an innovation oriented environmental policy is limited to those sectors where the target group has sufficient capacities to fulfil the expectations of environmental policy. A strategy based on innovation is more likely to be successful if the target group is small but encompasses potent actors.

Innovation, Time and Sustainability

Georg Erdmann

1. Introduction

From business economics it is known that the market success of innovative activities depends, among others things, on the timing of appropriate initiatives. Evolutionary economics uses the 'windows of opportunity'-concept for analyzing the varying effect of innovations in/over time. The application of this idea depends on whether and how windows of opportunity can be identified. This paper addresses this question. It discusses some of the working theses of a joint research project "Innovation, Time, Sustainability – Timing Strategies to disseminate Ecological Innovations", which is sponsored by the German Federal Ministry of Education and Research and began in 2001. This project takes an evolutionary definition of the term "sustainability", according to which it has to be specified in terms of temporally varying goals responding to problems changing in time. While the joint project further distinguishes inhomogeneities of the techno-economic, the socio-ecological and the political time that determine the success of innovations, the present paper takes only the techno-economic time into consideration.

2. Innovations and Environment

When discussing innovations for solving particular problems, technical as well as non-technical approaches should be included. With respect to ecological problems some institutional innovations have been applied, for example the organization of a market for sulfur dioxide emission certificates in the US. Other social and behavioral innovations are proposed – for example a different organization of cities to reduce the distances between living, shopping and working facilities or to ease the use of public urban transports instead of private vehicles.

So far most ecological challenges have been addressed by technical innovations. For example, particle filters, fluid bed combustion technologies and catalysts had been introduced to solve the local air pollution problems. According to some views this technology focus will not be able to solve present ecology related problems, such as the global warming issue. Instead institutional and social/behavioral innovations may be required (sufficiency instead of efficiency). However I am not quite sure whether this position underestimates the future potential of CO_2-sequestration, technologies improving the efficient use of energy or for using renewable energies. However, according to the problem solving cycle model above it is likely that through these solutions new endogenous problems to our societies will be caused. A recent example is the highly efficient combined cycle gas power plant; it is regarded as a CO_2-friendly technology but has become less attractive today due to the recent increase in natural gas prices.[1] The price increase is seen as a warning not to rely unique on natural gas as the future fossil energy source.

This example explains how technical innovations contribute to the solution of environmental problems but can be at the origin of new ecological, economical or societal problems. Technology assessment, technology forecasting and other instruments may be applied to identify such problems in advance, but according to the nature of innovations – novelties – the ability to anticipate their problematic consequences and impacts is rather limited. From an evolutionary point of view trial and error concepts are unavoidable in the sustainability process.

Following this argument, skepticism vis-a-vis technological solutions to energy problems has some good arguments on its side. But on the other hand the implementation success of institutional and behavioral or social innovations has, until now, not achieved important contributions to the problem solving process, at least in free societies where the power of a central agent on individual behavior is limited. For example it is unlikely to significantly modify the energy consumption by changing the consumer attitudes towards low energy content products, environmentally friendly mobility and a general denial of particular energy services. It is also unlikely that in free societies governments, or other national or international institutions, will achieve an institutional setting in favor of energy savings and subsistence. One should therefore not overestimate the problem solving capacity of non-technical approaches.

In the political debate in Germany it seems that this pro-technology position is being more and more accepted, even by adherents of a strict ecological position. It was probably triggered by the participation of the ecologist party BÜNDNIS 90/DIE GRÜNEN in the German government in 1998. The consensus about the role and importance of technological innovations is also growing in other countries. Recent examples are the Green Paper of the European Commission "Towards a European strategy for the security of energy supply" (EC 2000) and the Energy Plan of the new Bush administration in the USA (NEPDG 2001). At the moment there might be a *window of opportunity* to achieve an international agreement that the greenhouse gas problem as well as other energy related issues should be addressed by strengthening technology oriented efforts.

Therefore – and because a deeper analysis of non-technical innovations as an approach towards sustainability issues is certainly more complex – the remaining part of this paper focuses on technical innovations.

---

[1] See the problems associated with the VASA-Lubmin power plant project in Germany.

3. Innovations and Sustainability

The next point to address is the relationship between innovations and sustainability. The term "sustainability" is quite en vogue but a precise definition of its meaning is still missing. Even worse: By including ecological, economic or social targets practically any claim of interest groups can be based on the sustainability postulate (which may be the reason for the popularity of this label). But as the original definition of sustainability in the Brundland report "meeting the needs of present generations without compromising the needs of future generations" (UNEP 1987) suggests, a time dimension (beyond the time required for solving present problems) is an essential aspect of the claim for sustainable development. Therefore present ecological, economic and social problems should not be addressed by looking exclusively to the contemporary situation but by including the future implications of present actions. However, the future implications can never be completely anticipated, implying some degree of uncertainty about whether the needs of future generations are compromised or not. It happens quite often that solutions to present problems are at the roots of new problems.

A typical example is the emergence of the coal industry in the 19th century as a reaction to the declining availability of wood – unfortunately with the consequence of substantial local emissions leading to thousands of deaths particularly in industrial areas and big cities. The damages caused by the burning of coal became expressed by the term *smog* that came up after a particular catastrophic climatic situation in London 1952 and stands for describing the combined atmospheric effects of *smoke* and *fog*. But the development of the coal industry is still today a response to the limited availability of firewood. It is a major fuel source for contemporary electricity generation and is an important cause for the reforestation in many areas in the neighborhood of industrial regions (which can be assessed by comparing old photographs with the present situation). While the associated local pollution problems had been solved in the meantime, the global greenhouse gas concentrations in the atmosphere are the next problem to be solved. Many other examples of this kind can be found.

The risk that by solving given problems new and other problems emerge leads me to approach the sustainability concept by the figure of a *problem solving cycle*. Problems may arise both externally, that is without a human cause or interaction, and internally, that is as an – usually – unintended and unforeseeable consequence of human activities meant to solve another problem. In this figure, the society's problem solving process consists of three related elements:

- the problem perception
- the intellectual generation of solutions[2]
- the implementation or diffusion of actions that are assumed to solve the problem

As a conclusion, "sustainability" should be understood as a continuous process which requires a particular "balance" between the emergence of new basic problems on the one hand and the capabilities of solving these problems on the other hand. If, for a given type of problem (such as the greenhouse gas problem), both the problem itself and the solution can be expressed in quantitative variables (such as tons of greenhouse gas (GHG) emissions per unit of time and tons of GHG emission reductions, respectively) this "balance" is satisfied if the integral over time of both variables remains *always* below a critical threshold value. As a consequence of this interpretation, the postulate to achieve a state of sustainability is quite misleading. Sustainable development therefore is rather a permanent challenge. I should admit that the problem solving cycle in the above figure is a rather simplified representation of a more sophisticated reality where a multitude of dynamic problem solving cycles with multiple interactions exist and govern the innovation process.

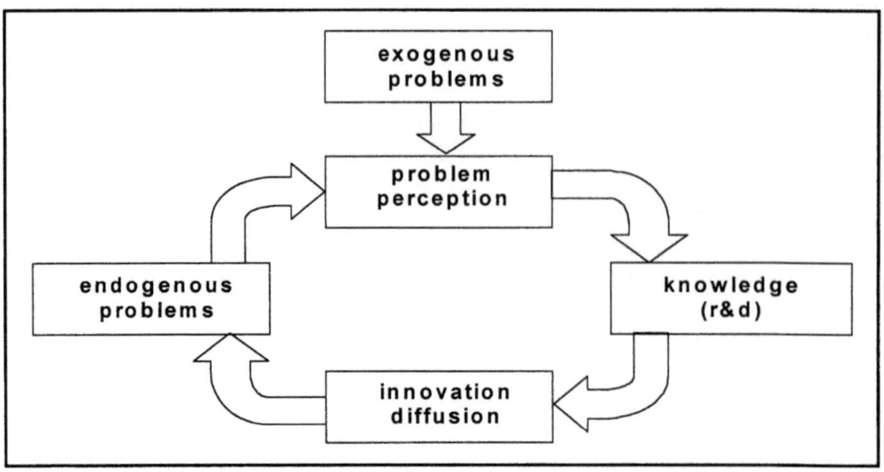

Fig. 1. Sustainability as a Problem Solving Cycle

One simple type of interaction between different problem solving cycles is the spillover of a perceived solution of one particular problem to other problems – sometimes by (re-) defining the other problems appropriately. Another interaction is the perception and relevance of a perceived problem that may grow with the ex-

2 In a recent discussion with the author, Rene Kemp claims that the perception of problems often follows the intellectual recognition of potential solutions and not the other way round.

istence of possible solutions and interest groups behind.[3] Examples for other and more complex interactions between separate problem solving cycles exist.

4. Innovation and Time: Windows of Opportunity

The third aspect of our model concerns the relationship between innovations and time. This relationship refers to the idea of modern evolutionary economics (in the tradition of Arthur 1988, David 1987, Dosi 1982, Nelson and Winter 1982 and others) according to which the success of innovative activities is time dependent. There exist windows in time ("window of opportunity"; David 1987; Erdmann 1993; Kemp 2001) in which an innovative activity is rather successful in terms of market penetration while outside it is not. The reasons for this inhomogeneous time is well understood by innovation theorists. The successful introduction of new technologies is reinforced by stabilised mechanisms, such as economies of scale, learning effects, indivisibilities, spill over effects within networks, and specific regulatory frameworks (lock-in effect; Arthur 1989), thus impeding alternative technologies to become successful (lock-out effect, Reichel 1998). However constellations might arise later in which the lock-in and lock-out can be reversed. These are phases in which other technological solutions have an opportunity to break the lock-in. Thus the creation and implementation of innovations is to be understood as an unsteady and path-dependent process, whereby phases of major innovations and breakthroughs alternate with phases of incremental improvements in given technological paradigms.

Innovations with positive ecological effects are thought to be especially handicapped during times of change, when the current technological paradigm can successfully be challenged by new basic technological approaches. The economic foundation for this is based upon the negative externalities of environmental damages, due to which the social benefit of environmental innovations cannot easily be transformed into benefits for the pioneering entrepreneur. But during times of open windows of opportunity even small political inducements might be sufficient to irreversibly change the direction of technological progress in an environmentally favorable way by achieving a lock-in of appropriate technical trajectories.

Accordingly timing is an important factor for both the private sector and the political regulator aiming at innovations to become solutions to sustainability problems. According to our initial examination two strategies for an appropriate innovation policy can be expressed that should be applied in accordance with the given situation. Inside windows of opportunities politics should apply temporal meas-

[3] A quite recent example is the greenhouse gas problem that had been pushed by the proponents of nuclear power in the eighties. These experts assumed to have a strong argument in favor of the nuclear technology against the raising social opposition after the *Three Mile Island* accident in 1979 and the *Tschernobyl* catastrophe in 1986 (see Häfele 1990). Later the GHG argument was adopted by the proponents of renewable energy technologies.

ures helping technical solutions to overcome the lock-out. Outside windows of opportunity the regulator could find it almost impossible to establish a new technical trajectory. The self-reinforcing nature of the dominant innovative paradigms and other lock-out-aspects of innovative options would exceed the capacity of the regulator to intervene. In such situations the regulator should stimulate a variety of alternative technological options that may later become possible technical trajectories once a window of opportunity is open: Because the *ex-ante* assessment of a particular technological innovation and its problem solving capacity (without creating too many other problems) will always be subject to possible errors, a reliable evolutionary approach is creating a variety of options. This strategy will be particularly appropriate if the regulator has limited *ex-ante* knowledge whether a proposed technical innovation will be able to contribute to the solution of a sustainability problem or not and what will be the nature and scope of sideline effects.

Thus, for applying the proposed time-dependent strategies the knowledge about the future potentials and capabilities of innovations is not absolutely necessary. Even if the anticipation of the consequences and impacts of innovations is not totally impossible – particularly in the case of environmental innovations the problem solving capacity is often quite obvious – (basic) innovations are novelties by nature and thus subject to surprise and non-predictability. But crucial for the implementation of time-dependent political strategies is the ability to anticipate the approaching of windows of opportunities or the time interval when a chance for breaking the lock-out for promising innovations exists.[4]

5. Examples

To find out how it might be possible for a political regulator to identify situations of approaching windows of opportunity I propose an example namely the technical transition in the chlorine production between 1975 and 1990. Chlorine (Cl_2) is one of the basic products in the chemical industry and is produced from sodium chloride (NaCl) and water (H_2O) according to the following chemical reaction

$$\mathbf{2\,NaCl + 2\,H_2O \rightarrow \quad Cl_2 + 2\,NaOH + H_2}$$

with the by-products sodium hydroxide (NaOH) and hydrogen (H_2). For exploiting this reaction in chemical industry the process engineers have to solve the task of avoiding contact and reaction between the by-products and the produced chlorine. During the period under concern three processes had been used:

The *amalgam technology* is the oldest chlorine production process; it is comparatively energy intensive and causes toxic mercury emissions because this substance is used on the cathode side to separate the products. These emissions are

---

4 It also requires that the innovation policy can be applied in the described time-flexible manner, which requires an appropriate policy setting which again may be implemented more or less easily depending on some sort of policy windows (Kingdon 1995). This question will not be discussed in this paper.

reduced, but not eliminated, by end-of-pipe-technologies. The advantage are products with high purity, that can be processed with minor additional treatment.

The second process, *diaphragm method*, is about twenty percent less energy intensive than the amalgam technology, but more costly. The higher investment costs result from the necessary purification of the chlorine and the sodium hydroxide. The merging of the products is avoided by a partition, the so called *diaphragm*, that is covered with a layer of asbestos material which is selected for its porosity and chemical stability. Therefore this technology causes cancerogeneous asbestos emissions.

The third method uses a semi-permeable *membrane technology* for separating the products at the cathode and the anode side. There are strong requirements for the material to be used for the membrane. Thus only in the mid 1970s a reliable technical solution was found. The products are of high purity, and the investment costs are about 30 percent lower, because no additional equipment is required for protecting the environment from dangerous emissions. Finally the energy requirements are again lower than with the diaphragm method. The only disadvantage of this process is a higher purity requirement for the sodium chloride, as otherwise the membrane gets destroyed.

At the beginning of the 1970s the dominant process in the chemical industry was the amalgam technology. But in 1989 the European industry mostly still used the amalgam technology, while in Japan the market share of the cheaper and environmentally advantageous membrane technology exceeded already 70 percent.

A closer look into the development of the market share shows, that, starting around 1975, the European regulator forced the chemical industry to invest in end-of-pipe-technologies in order to reduce the mercury emissions. In fact the mercury emissions had been reduced sharply and in a short period of time, but these investments into an old technology hindered the European industry to an early shift to the new membrane technology, which became available around 1980. Even in the 1990s the shift to the membrane technology didn't start because of overcapacities in the industry. In contrast to the situation in Europe the Japanese government put a ban on the amalgam technology that became effective from 1985 on. Thereby the Japanese regulator accepted the use of the amalgam technology in its current form until the new membrane technology was available on the market.

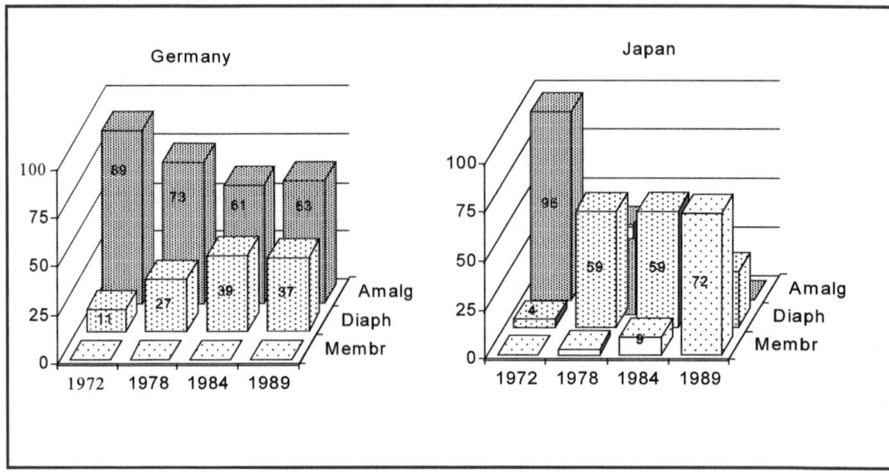

Fig. 2. Market Share Evolution of Chlorine Production Technologies

This example shows that while the Japanese government applied an appropriate time strategy for pushing a rather advanced and environmentally favourable technology into the chemical industry, the European regulator failed. The result have been higher production costs and higher emissions during the last 20 years. The Japanese regulator was able to anticipate the coming market maturity of the new membrane technology, and the same should have been possible also for the European regulator. But for reasons that I will not analyse here the European regulator had not been able to apply the appropriate instruments at the right moment to favour the market diffusion of the membrane technology. Instead, the mercury emission reduction policy triggered a lock-in of an inferior technology in the chemical industry for another investment cycle.[5]

This example is not unique. A closer look at the technological competition and the role of the governments herein shows a large number of similar cases where regulators missed to exploit windows of opportunity for the market introduction of new technologies and thus caused sub-optimal results. For example the power industry in Germany had to retrofit their coal fired plants with SO_2- and NO_x-filters; these power plants now contribute to the over-capacity in Europe and contribute to the lock-out of advanced power generation technologies. Many governments in Europe favor, without significant success, the market introduction of natural gas vehicles by fiscal incentives – just after the accomplishment of a costly gasoline filling station retrofitting program. Fuel cells are assumed to be a favorable heating technology which will be ready for the market within the next five years. But

[5] In the meantime the membrane technology succeeds also on the European market where a ban of the amalgam technology is scheduled for 2010. It should be noted that once the last amalgam process had been dismantled such a ban is superfluous, because there will be no voluntary return to this more expensive and environmentally problematic technology.

if the German government puts into force the planned energy use regulation before that date, it will induce retrofitting or marginal improvements of the existing technology rather introduction of the new technology. Thereby it will risk to create a lock-out of this technology. Examples of dysfunctional timing decisions outside the energy industry are the passenger vehicle recycling (see Aggeri 1999; with the opening of the iron curtain a lot of used passenger cars are exported to East-European countries instead of being recycled in domestic factories), the liberalization of the German railway system (the result is a company without a visible chance to become competitive under present conditions), the reforms of the retirement funds (as this happened with much delay and in rather small steps the future challenges due to the aging society will probably become serious) and some aspects of the economic policy in East-Germany (for example with respect to the wage increase).

This is only a short selection of examples but it demonstrates that in many cases the regulators are not even aware of the chances that an appropriate timing of environmental and other political instruments offers: Sustainability targets might be achieved at lower social costs and with a larger degree of political acceptance.

6. Modelling and Forecasting Windows of Opportunities

It is not only the regulator that fails, also the mainstream economic theory does. Until now, it has little to say about how to develop timing strategies for the use of environmental and other instruments. Particularly environmental economics is a static theory focusing on given technologies, preferences and the resulting marginal external cost curves from which the static welfare equilibrium is derived. That everything may change and that these changes might be triggered in favor of solving sustainability problems is yet the view of a minority of (evolutionary) economists.[6]

The theoretical analysis of timing issues in economic policy must try to develop a model that incorporates indicators identifying windows of opportunity and indication of their arrival. Even if the research is in its infancy, the examples presented above give some hints about how to proceed. As usual the first step is to carefully specify the relevant market. This is particularly important in markets governed by networks and spillover effects. The next step consists in quantifying the size of the *sunk costs* that would occur if the given technological trajectory was given up. As these costs are not constant but vary over time it is crucial to capture their full dynamics. They decline when the technology becomes old and the appliances amortized, but increase with capacity extensions and retrofit investments. The quantifi-

---

6 Business science has yet a better understanding of dynamic processes in economics and quite a number of models how to describe it. An economic research program of how to design and time strategies in favor of solving sustainability issues would benefit from these approaches.

cation of the sunk costs is a standard task in business economics. The new aspect here is the aggregation over all relevant markets linked by network effects and spillovers. A particular treatment is necessary if indivisibilities play a role in the technological competition.

In a similar way the phases of the product cycles can be assessed and quantified according to standard economic indicators such as the elapsed time since the first market introduction, the market growth along the logistic curve or the degree of market saturation.

A more difficult part of the analysis is the identification and assessment of new technology paradigms. Usually there exist thresholds for the minimum market share of new technologies that, once reached, will imply the self-supporting diffusion of the technology (for energy technologies see for example Marchetti and Nakicenovic 1979). However, this threshold is a function of several variables and thus difficult to identify. In addition the threshold may depend in a complex way on political market interventions.

But even if these threshold values are identified the question remains how to assess the market readiness of new technological paradigms (examples from energy technologies are fuel cells and photovoltaics). The problem is that the properties and costs of infant technologies can never be fully anticipated. The predictability remains limited even if scenario approaches, Delphi methods, and other instruments for technology forecasting are used.

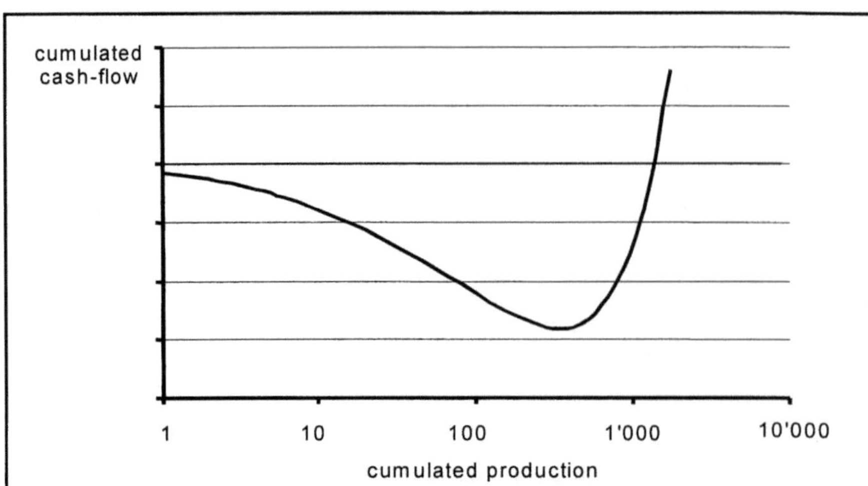

Fig. 3. Learning Curve Effects on Cumulated Cash-flows of an Innovation

A most promising approach is the use of learning curves as a formal representation of relative costs as a function of the cumulated production – a research field where much progress has been achieved during the last few years. At the beginning of the market introduction of new products or processes their costs are usually much higher than the market price of competitive products – even if the development costs are not included.[7] With learning by doing the costs decline and will reach the price level of the competitive products or technologies – at least for successful innovations. To make things simple one may assume that the market does not accept prices that exceed those of competing products, thus the innovative pioneers have to cover the difference. The high costs in the first phase of the market introduction lead the cumulated cash flow to grow into negative values. Once the cumulative production has been growing to a level that the costs drop below the price level, the pioneering entrepreneurs begin to create positive cash-flows which are used to pay back for the cumulative losses. If the cumulative production is further expanded the breakeven point will be reached and from this moment on the innovation becomes profitable to the entrepreneurs.

The figure above shows the expected cash flows as a function of the cumulated production. Its minimum and the breakeven production give two indicators concerning the market entry of new technologies. The smaller the cash flow minimum (in absolute terms) and the breakeven production, the more likely the successful introduction of the new technology. Again one has to set this figure into a dynamic context and assume that both indicators vary – through successful development efforts of the new technology or improvements of the old technology. From a declining cash-flow-minimum (in absolute terms) one concludes that a window of opportunity of the new technology is approaching.

Of course the learning curve effects and the two indicators derived cannot be known with certainty in advance, thus the approach to identify windows of opportunity must include methods of economic risk analysis – in particular the real options theory.

The application of learning curve models can further be improved if multi-use options and market differentiation effects are included; these allow innovative entrepreneurs to increase the cash-flows in the early stages of the market introduction. An actual example is the possible use of fuel cells as a vehicle drive train. It may still take some time until this becomes ready for mass market applications, but if fuel cells are used as auxiliary power units for electrical devices in conventionally powered vehicles, the industry may generate some premature positive cash flow.

Our examples show the general approach that may be used towards identifying the windows of opportunity for new technologies, products and business concepts. By applying the methodology to a larger number of cases from different periods and industry sectors the relevance of the mentioned aspects can be assessed. In some cases additional aspects should be taken into consideration, for example the nature and form of the competition in the relevant industry (the argument was first introduced by Josef Schumpeter 1912), openness of the relevant market for newcomers (as pointed out by W. Baumol 1982), the expected market growth, or exogenous developments such as changing world energy prices, exchange rates, in-

7 The case of innovations without competitive products or processes is not regarded as this will not be rather relevant for environmental innovations.

terest rates, the role of the government as first buyer, political regulations and incentives and so on.

7. Conclusions

The ideas about identifying windows of opportunities developed in this chapter represent the working hypotheses that are used as a starting point for our research project and they are preliminary as the elaboration of a workable model is still under way. But I hope I could make obvious that such identifications should be possible, not only in principle, as the project proceeds. Accordingly our project should also arrive at proposals on how to exploit windows of opportunity in favor of sustainability goals by applying time-sensitive political instruments such as temporal prescriptions and temporary support of new technologies. It will also include proposals how to proceed if the government has limited knowledge about innovations and their problem solving capacity.

The project will not only focus on the techno-economic time axis with its irreversibility and time-varying effects discussed in this paper. It includes likewise the social environment (with its time-varying ups and downs of public awareness of ecological problems). And it discusses aspects of "political time" according to which there are time frames in which policy is able to act, while some time earlier or later it is not.[8]

We assume that the calibration of such a model can be accomplished by analyzing case studies. The first step is the assessment of a broad number of cases from different technologies, business fields and industry sectors that are used to become more sensible for structural elements to be incorporated into the model. Once our approach seems workable we will evaluate our framework by a smaller number of in-depth studies. In these case studies[9] we will discuss, together with the protagonists in the field, how the concept of time-varying innovation policy can practically be applied to sustainability problems and how concepts and strategies laid down in our model would be feasibly. Thereby we have, among others, to face the complexity of innovations and the limited ability to anticipate their economic and ecological performances. Therefore governments might often be unable to intervene in an effective way. We have developed some ideas on how to deal with this issue, and these will be discussed in the in-depth case studies. Finally we will address the question on how the implementation of time-varying political instruments and intervention is legitimised regarding the postulate that governments

---

8 A quite trivial example explaining the inhomogeneous political time axis is the legislative period. Depending on the concrete settings in the constitution, the success of major political initiatives is improved if they are scheduled in the first half of a legislative period.

9 The selection of these case studies is inspired by the suggestion that there is, or predicted to be, soon to occur a deep-rooted change of technologies and/or institutional arrangements.

should not deeply be involved into the selection of technologies, but rather focus on defining the general rules of technological competition.

8. Acknowledgements

This paper was written within the project "Time dependent Strategies in order to disseminate Ecological Innovations" sponsored by the German BMBF. I have to thank Christian Sartorius, Stefan Zundel, Jan Nill and René Kemp for their helpful comments and suggestions.

Integrated Long-Term Strategies to Induce Regime Shifts towards Sustainability: The Approach of Strategic Niche Management

Remco Hoogma, Matthias Weber and Boelie Elzen

1. Introduction

For a number of years now we have known that there is something wrong with the ways we live and produce. Industrial societies are accumulating a heavy burden on future generations by depleting scarce natural resources, polluting the environment, and by reinforcing major economic and social imbalances. The Schumpeterian logic of *creative destruction* that drives modern capitalism gets a very particular and almost cynical connotation in this context.

It seems that in many debates an inherent conflict is supposed between the economic interest orientation of modern capitalism and the ecological, social and economic requirements of future generations. The notion of sustainable development was coined as a conceptual response, but it has proven to be a banner for many causes. In its simplest form, sustainable development is defined as *development that meets the needs of the present generation without compromising the ability of future generations to meet their own needs* (WCSD 1987). This definition assumes that it is possible, indeed necessary, to make trade-offs between economic growth, social balance and sustainability of the environment. It conveys the hope that the development of new modes of economic growth can dramatically reduce pollution and resource use. Furthermore, it presupposes that it is possible to deal with issues of equity and democracy on a world scale. Finally sustainable development requires that society, business and governments operate on a different time scale than they do now. Long-term aims must not be sacrificed for short-term gains. It seems evident that achieving the objectives of sustainable development requires a radical departure from the principles of operation of our economies, i.e. an economic *regime shift* is needed.

While on the one hand some place science and technology at the roots of current sustainability problems, on the other hand S&T promises to offer many opportunities to smoothen, if not solve, the conflict between economic, ecological and social interests. We thus argue that science and technology can indeed be beneficial for solving problems of sustainability, but that much depends on the way technology and innovation are actually shaped, how they are organisationally and institutionally embedded.

Technological change has also played an important role in policies directed at improving the environment. Technological innovation can change the energy and material basis of economic and societal processes and result in drastic improvements of resource productivity (Weaver et al. 1999) At the same time, it is clear that most technological change consists of incremental improvements, often not going beyond the control of particular pollutants. Ecological restructuring of production and consumption patterns will not so much require a substitution of old technologies by better ones, but radical shifts in technological regimes (this notion will be introduced in section 2), including a change in consumption patterns, regulations and artefacts. In other words, a technological and an economic regime shift need to go hand in hand.

This leads us to the question how to make progress with the design of policy strategies for moving towards sustainability. We argue that technological innovation is one of the key levers for policy to induce a regime shift. Until now, existing policies have relied on setting economic incentives and regulations, and these approaches allowed to tackle a number of environmental problems in the past. However, they are not particularly suited to induce technological regime shifts. Typically, they have resulted in so-called end-of-pipe technologies but not in the emergence of radically new types of production and consumption systems (Kemp 1997).

We will introduce the concept of Strategic Niche Management (SNM) as one promising approach to address this problem. SNM offers a framework for designing integrated long-term technology-based policy strategies to induce regime shifts to sustainability. Our particular attention goes to the question who could be or should be the manager(s) of a strategic niche management process. There are many situations where market actors will be in the better position to develop a sustainability-oriented niche, but other situations in which a very long-term perspective is needed to perceive interests in a regime shift may require a strong public role in leading the transition process.

The paper is structured as follows. In section 2, we discuss the notions of technological regimes and regime shifts and their dependence of the emergence of technological niches. We present a conceptual framework that addresses innovation, sustainability as well as policy making. Section 3 deals with the evidence of regime shifts that have occurred in the past. It provides the foundation for section 4 which addresses the question how the conditions for regime shifts can be influenced by policy – here in particular by policy strategies to induce sustainable innovation. We discuss the science & technology push approach, the setting of boundary conditions, and the demand-driven approach. Strategic niche management will be proposed in section 5 as a modulation policy which combines and optimizes the workings of the other three approaches. This section describes the approach and presents SNM as a modern tool of governance. Then in section 6, the role of niche managers is discussed. This will relate to the multilevel model of niche development from section 2: different types of activities to be performed by various actors employing a range of instruments will be discussed. The paper ends with conclusions with respect to both national and European policy and conceptual advancement of innovation policy.

2. Regime Shifts and the Role of Niches

The idea of a core technological framework guiding innovation activities has gained wide recognition in modern innovation theory, in economics as well as in sociology of technology. Concepts like technological trajectories and paradigms, technological guideposts, innovation avenues, techno-structures, technological momentum, etc. may all differ in important details, but they share the conviction that technological change does not follow a random walk[1]. The range of likely future development paths is narrowed down by what has been conceived before. Similarly technological, organisational and structural interdependencies and their role in stabilising and reinforcing development paths of systems, are highlighted from the perspective of socio-technical systems research[2].

In this contribution we use the notion of technological regime, defined as: the whole complex of scientific knowledge, engineering practices, production process technologies, product characteristics, skills and procedures, established user needs, regulatory requirements, institutions and infrastructures[3]. A technological regime incorporates a cognitive and normative framework and a set of (functional) relationships between technology components and actors throughout the production chain, which forms the basis for individual and collective action. A technological regime is the context for technological and economic practices within a production chain, which pre-structures the problem-solving activities that engineers are likely to undertake and the strategic choices of companies.

Technological regimes are a broader, socially embedded version of technological paradigms. The focused nature of socio-technical change largely results from the *embeddedness* of existing technologies in production practices and routines, in consumption patterns, in organisational structures and cultural values, as well as in mental frameworks, beliefs and practices of engineers, managers and scientists. This embeddedness creates economic, technological, cognitive and social barriers for new technologies.

This notion of technological regime helps to explain why most change is of the non-radical type, aimed at *regime optimisation* rather than *regime transformation.* It helps to understand why so many new technologies remain on the shelf, especially systemic technologies with long development times that require changes or

---

1 Technological trajectories and paradigms: see Dosi (1982), for technological guideposts: see Sahal (1985), for technological momentum: see Hughes (1983).

2 The role of such interdependencies is elaborated by looking at several in-depth case studies in for example Hughes (1983), Summerton (1994) and Weber (1999).

3 Rip and Kemp (1998) have accentuated the structured nature of a technological regime by defining a technological regime as the *coherent* complex of scientific knowledges, engineering practices, production process technologies, product characteristics, skills and procedures, and institutions and infrastructures that are labelled in terms of a certain technology (for example, a computer), mode of work organization (for example, the Fordist system of mass production), or key input (like steel or hydrocarbons). Since the accommodation between the elements in the complex is never perfect, it is perhaps better to talk about a *semi*-coherent complex.

at least major adjustments in the context of the technology in question, e.g. in terms of regulations, consumer preferences, infrastructures, etc.

In addition to constraining forces in the application context, radically new technologies also require transformations inside the organisation in which they are produced. Firms vested in the old technologies will be more inclined to reformulate their existing products than to do something radically new that may involve great risk to the firm. As noted by Rosenberg and Fransman, firms have a restricted technological horizon and a bounded vision, which serve to focus their exploratory activities upon problems posed by the existing product (Rosenberg 1976; Fransman 1990). This means that there are path-dependencies that act to contain radically new technologies.

These arguments show that both supply-side and demand-side changes are needed to successfully introduce radically new technologies and thus induce technological regime shifts. These changes consist of new ideas, production and user practices, the development of complementary assets and institutional change at the level of organisations and markets. These findings are confirmed by historical studies of technological transitions (see section 3).

2.1 Dynamics of Regime Shifts

In this paper we are interested in processes of technological regime shifts which raises the issue of the dynamics of past regime shifts or 'technological transitions'. Kemp has identified the following elements as key aspects of such transitions (Kemp et al. 1994):

1. Long periods of time. It often takes 50 years for a new technology system or regime to replace an old one.
2. Deep interrelations between technological progress and the social and managerial environment in which they are put to use. Radically new technologies give rise to specific managerial problems and new user-supplier-relationships; they require and lead to changes in the social fabric and often meet resistance from vested interests.
3. New technologies tend to involve 'systems' of related techniques; the economics of the processes thus depend on the costs of particular inputs and availability of complementary technologies. Technical change in such related areas may be of central importance to the viability of the new regime.
4. Perceptions and expectations of a new technology are important, including engineering ideas, management beliefs and expectations about the market potential, and, on the user side, perceptions of the technology. These beliefs and views of the new technology are in constant flux, and the progression of these ideas may either be a barrier or a catalyst to the development of a particular technology.
5. Specialised applications in the early phase of technology development when there is usually little or no economic advantage of a radically new technology help making innovations more robust. Moreover, the existing technologies tend

to improve during the development phase, rendering open market competition even more difficult.

Technological regime shifts thus entail a number of structural changes at different levels. This implies the co-evolution of several factors: technological options, user preferences and institutional changes need to be jointly created and shaped. Users, for example, do not have fixed demands that are fulfilled with a new technological option, but learn about and discover new application possibilities. Several case studies have shown how user demands are developed, articulated and expressed in the process itself, in interaction with the technological options available[4].

This process also works *vice versa*. Producers learn new ways to view their own technology. In the absence of an articulated demand, which is typical of regime shifting technologies, producers underestimate the market potential of a radically new technology. The interactions between users and producers of new technologies are just one example of a range of complex mechanisms and feedback loops within the systems underlying radical innovations with a regime-shifting potential.

If a regime-shift is such an encompassing and complex process with long lead-times, how then can we understand its emergence, not to speak of influencing it politically? Historical evidence shows that the start can be very modest. *Regime-shifts often start at the periphery of existing dominant technological regimes in small, isolated application domains.* Schot, in a six-volume history of Dutch technology in the nineteenth century consisting of 24 case studies of various innovations covering many sectors in the Dutch economy, concluded that technologies often first appear in niches[5]. In the historical literature many other examples can be found. The steam engine was developed by Newcomen to pump up water from mines. Clocks were first used in monasteries where life was arranged according to a strict timetable. The wheel was first used for ritual and ceremonial purposes. The railways introduced the telegraph, and the rapid press was first employed in the production of newspapers[6].

In these first niches commercial viability might well be absent. The first applications of electricity at world fairs, theatres and public events had symbolic value; they brought excitement. The first applications of aeroplanes and cars in races were never commercial successes; indeed, the motivation to engage in such activities was not primarily economic in nature.

Expectations play a crucial role in early phases of technical change (van Lente 1993). Technologies in the making have yet to prove themselves (in terms of technical, social as well as commercial viability). Parties that apply a new technology, therefore, often construct and communicate positive expectations in order to make actors (including themselves) *believe* that it will yield returns in the future.

---

4 See, for example, the work by von Hippel (1976), von Hippel (1988), Rosenberg (1982), Teubal (1987), Lundvall (1988), Habermeier (1990) and Slaughter (1993).

5 See Schot (1998a). An adapted version of this chapter has been published in English, see Schot (1998b).

6 See for more detailed analyses of these examples Mumford (1963) and Bassala (1988).

After successful application in one domain, the technology finds through a series of new applications that become attainable because of its progressive improvement in terms of performance characteristics and economics. This is the process of *niche branching*, which includes the emergence of new application domains and the creation of a bandwagon effect (that is, a wider diffusion) through replication of the niche elsewhere[7]. During these developments the technology is improved and further technical choices are made partly in response to new selection criteria. New reference points are created for evaluating the traditional alternatives. The technology links up with others, perhaps creating hybrid forms.

The changes brought forth by this process of niche branching culminate in the emergence of a new technological regime by coupling to developments on other levels. The model we use, based on work by Rip and Kemp (1998) and Schot (1998b), is a *multi-level* model that stresses developments at various levels which then could *couple* (i.e. create positive and negative feedback) at specific times[8]. This coupling is, in turn, a major source of bandwagon effects. Although this is an unplanned and contingent process, actors can enhance the *probabilities* for coupling to occur, and this is precisely one of the underlying aims of strategic niche management. Our hypothesis is that a regime shift requires three types of coupled developments (Hoogma et al. 2002):

- Niche developments of novelties followed by increasing returns of adoption;
- Erosion of opportunities to make progress within the dominant regime; and
- Emergence of new external opportunities and constraints which challenge the problem solving capability of the existing regime. Such external development can be events (such as a war or a scientific breakthrough that allows for new technical developments) or broad trends such as urbanisation.

These developments play out at various levels (see figure 1). Niche development is based on innovation processes at the local level where experimentation and piloting play a critical role. Of course, successful innovation implies that the innovation will become more broadly known and adopted, hence more global. This means that technological niches emerge at a supra-local level, composed usually of a number of piloting experiments which give rise to a first round of learning processes between individual sites.

---

[7] Rip (1995) has summarised the process of niche branching as follows: "*Technological change is not a continuous process along dimensions of increasing functionality. It is more like a patchwork quilt, or if one prefers a different metaphor, the way yeast cells grow. Developments branch off in different directions, cross-connections and interactions occur, and niches, that is limited and relatively easy/or advantageous domains of application and further development, strongly determine what steps can be taken productively. The eventual shape of a technology, its usage and the way it is embedded in society can be very different after 5, 10 or more years than it looked at the beginning*".

[8] See also Schot et al. (1998).

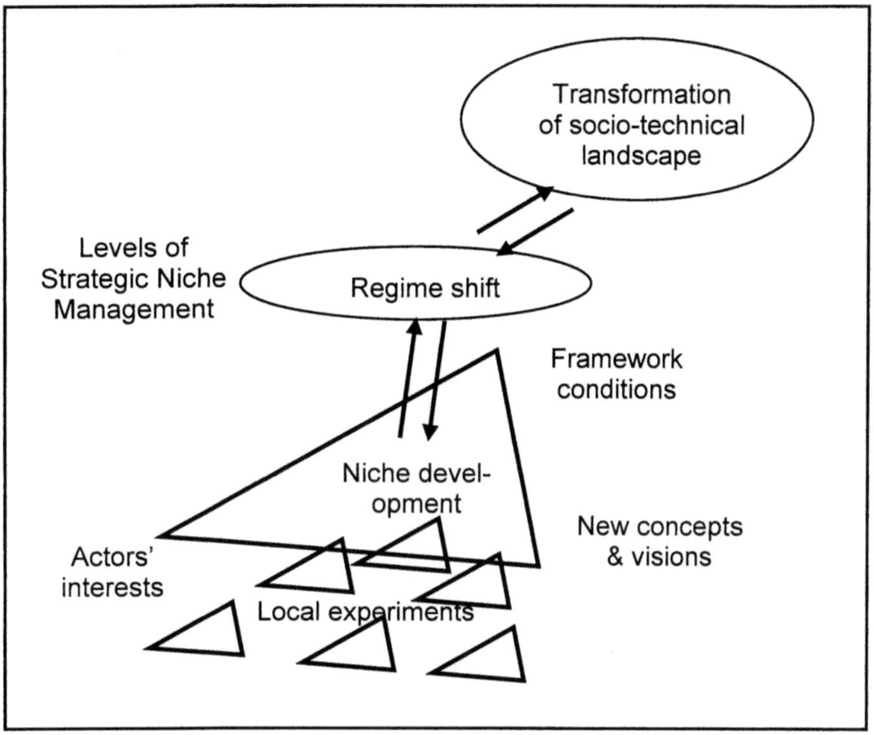

Fig. 1. Levels of Strategic Niche Management

The level of a regime is then an even more global level of shared understanding and rules that orient the actors' behaviour at the niche level. At some point actors might start to rethink the viability of technological options in the existing regime. A classic example from the history of technology is the development of aerodynamic theory in the 1920s, which suggested that the traditional regime of piston engine and propeller combination in aeroplanes would never be able to compete with turbojets. Consequently, the design constituency lost faith in incremental improvements to the existing regime. It is important to note that this process did not occur overnight, and was heavily influenced by World War II (Constant 1980).

The fourth level, referred to as a sociotechnical landscape (Rip and Kemp 1998), is also global but its local effects can differ widely. Others may call it the wider context or the boundary conditions of the socio-technical system under study. The socio-technical landscape can be defined in two ways, first as a set of connected technological and societal (hence socio-technical) trends, deep structures and major events that influence the development of technologies embedded in regimes as well as new promising alternatives. These factors resemble a landscape that accommodates some developments more easily than others. Second, the landscape is not directly influenced by the success of local innovation processes and only to a limited extent by individual regime shifts.

These factors influence processes of regime shifts, but regime-shifts in specific sectors will not affect the landscape itself dramatically. Of course, if a number of regime-shifts occur, the landscape will be changed. For example, the emergence of electricity led to changes in factory regimes, transportation regimes, and household regimes and thus to a new kind of electrified society and economy. This is a special case of a pervasive technology[9]. Such a landscape change may also currently be ongoing in connection with information and communication technologies.

Niches are a necessary component of a regime shift by creating a pathway to a new regime. Of course not all niches will be instrumental in this respect. The niche technologies must have ample room for improvement that allows for cost efficiencies and for branching out. They also must have a synergetic relationship with other developments in technology and markets in order to find new users and capture new domains of applications. This is a second necessary condition. A third condition is that the gap between existing domains of application and new ones should not be too big. And a fourth condition is that the rate of progress of the emerging technology system offering particular services should be larger than that of existing technologies with which it must compete. We thus have four conditions for a regime shift to occur.

2.2 Successful Niche Development

What makes niche development processes successful? We would like to introduce a first set of approximate measures for evaluating the success of early niche development, namely the *quality of learning*. Learning refers to a range of processes through which actors articulate relevant technology, market and other properties and thus contribute to the establishment of a niche. This is a learning process because outcomes are not known beforehand, but actors have to work hard to define outcomes. Learning concerns a number of aspects:

- Technical development and infrastructure: this includes learning about design specifications, required complementary technology and infrastructure;
- Development of user context: this includes learning about user characteristics, their requirements and the meanings they attach to a new technology and barriers for use they encounter;
- Societal and environmental impact: this entails learning about safety, energy and environmental aspects of a new technology;

[9] For an elaboration of pervasive technologies see Freeman. and Perez (1988). They distinguish four types of innovations: incremental innovations, radical innovations, changes of technology system (this is what we call a regime shift) and changes in techno-economic paradigm (following from pervasive technologies).

- Industrial development: this involves learning about production and maintenance networks needed to widen diffusion; and
- Government policy and regulatory framework: learning needs to take place about institutional structures and legislation, the government's role in the introduction process, and possible incentives to be provided by governments to stimulate adoption.

Learning can be limited to single-loop learning, meaning that various actors in the niche learn about how to improve the design, which features of the design are acceptable for users, and about ways of creating a set of policy incentives which accommodate adoption. However, for niche development to result in a regime-shift, another kind of learning process is needed – double-loop learning[10]. In such learning processes, conceptions about technology, user demands, and regulations are not tested, but questioned and explored. Room emerges for co-evolutionary dynamics, that is, mutual articulation and interaction of technological choices, demand and possible regulatory options. Co-evolutionary learning will also allow for what Brian Wynne has called collective value learning, that is clarifying and relating of various values of producers (designers), users and third parties involved such as governments (Wynne 1995). Successful niche development consists of single-loop learning on a whole array of aspects, along with the occurrence of double-loop learning.

The emergence of a new technological regime implies a change in the selection environment for innovations. This change is prepared through processes of niche development, usually in a sense that allows mutual adjustment process between niche and regime level. This process is called *institutional embedding*. Next to the quality of learning, the quality of a niche development process with respect to a compatible regime shift can thus also be expressed by the extent to which the new niche technology becomes embedded[11].

Three crucial aspects of institutional embedding can be identified. First, institutional embedding implies the creation of complementary technologies and infrastructures, a factor that is also highly important for increasing return to adoption dynamics in later diffusion phases. Second, embedding produces widely shared, credible (i.e. supported by facts and demonstration successes) and specific expectations about what might be feasible under the new regime. Third, embedding enlists a broad array of actors aligned in support of the new regime. This network includes producers, users and third parties, especially government agencies.

Numerous studies of product innovation and failure have shown that the involvement of users is important for successful market introduction, and that a lack of user involvement is a major cause of failure[12]. Especially for innovations that serve broader societal goals, like more sustainable transportation, third party in-

---

[10] The distinction between single-loop learning and double-loop learning was introduced in Sabatier (1987).

[11] Cf. societal embedding of a new technology, which implies that it becomes integrated into the structure and culture of society, according to Deuten et al. (1997).

[12] For an overview see Leonard (1995), especially chapter 7.

volvement representing interests in a sustainable development (i.e. often environmental organisations) is crucial. The alignment of the actors supporting a new regime refers to a situation in which these actors have developed a stable set of relationships and can easily mobilise additional resources in their own organisations because the network is seen as an important, credible and strategic operation. In such situations, as Rip has suggested, often so called macro-actors are available who have a separate responsibility to realise and maintain a high degree of alignment (Rip 1995). Thus, successful niche development requires the development of complementary technologies, more robust expectations and a broad, highly aligned network.

The third criterion of success, next to learning and institutional embedding, concerns the *direction* of regime shifts and niche development processes. So far, the issue of regime shifts and niches has been discussed independently of the direction of change. These processes do not necessarily lead to more sustainable outcomes, unless additional measures are taken to define the corridors of development in which experimentation and niche development take place. These corridors can be defined at several levels. Firstly, there are fundamental and cross-cutting principles that should characterise any research and development effort that is geared towards sustainability, such as the precautionary principle or the participation of a wide range of stakeholders in the definition of research fields for public funding. Secondly, regulations and incentives can help orient R&D efforts and experiments towards the goals of sustainability. And finally, there is a need for forward-looking information provision (e.g. using foresight or delphi techniques) to improve our knowledge and understanding of emerging issues and technologies. Strategic Niche Management, which will be introduced in more detail in a later section, aims to incorporate these different elements in order to ensure that niches and regimes are established that help improve the sustainability of the systems in question.

3. Evidence of Regime Shifts

With the three criteria of success for niche development processes – learning, institutional embedding and a sustainable direction of change – at hand, we will analyse some likely candidates of past niche creations and regime shifts. This is important because for our perspective to serve as a foundation of policy strategies, we first need to show that indeed regime shifts have occurred in the past and that they were in fact built on niche development processes. As far as relevant, we will also point to the criteria of a sustainable direction of change. In this contribution, we can present only selected evidence from energy supply and some other fields.

3.1 Regime Shifts in Energy Supply[13]

Energy supply has undergone earlier regime shifts in the past, for example with the introduction of the steam engine and electricity supply. Both even transformed the sociotechnical landscape due to their pervasive impacts on the organisation of industrial production processes and living patterns.

Currently we are also going through a phase that may well prelude a regime shift. The liberalisation and (at least in some countries) privatisation of energy supply has already led to transformations that match several of the elements of the definition of regime shifts. It led to the establishment of new engineering practices, production processes, skills, procedures, institutions and infrastructures. At the structural level, new organisational patterns emerged that are associated with corresponding changes in the technical systems. For example, traditional electricity supply companies lost control over their networks and moved into other, neighbouring businesses. And the era of large-scale centralised power stations is being replaced by one with more decentralised systems, at least in some countries.

The process of liberalisation is not driven by technological development, but is motivated politically in a top-down manner. At its origin stood the recognition that several of the monopolistic features of energy supply could no longer be justified. This process raises also many important questions with respect to technological development. The – mainly economic – regime shift favours certain new technologies while others appear less promising under the new conditions. So, the "dash for gas" and the success of gas-based combined cycle plants cannot be explained on the basis of technical arguments, but was very much stimulated by the new framework conditions. In any case, liberalisation affects the direction of technological change in the field of energy supply and has in fact led to a co-evolutionary process of change in economic structures, organisational structures and technological structures.

The technological regimes that emerged in response to liberalisation differ significantly across various countries. Several models have been tested, with the UK pioneering a very radical transformation of its supply system. Other countries, such as Germany or the Netherlands, have also opened up and liberalised their electricity supply markets, whereas France is very reluctant to adopt a far-reaching liberalisation approach. The access to the French power market continues to be a controversial issue, and EdF continues to be under government control.

In terms of the technological dimension of the new regime, the British liberalisation and privatisation led primarily to a fast uptake of combined cycle gas turbine technology. The Dutch reforms at the end of the 1980s (which led only to a partial opening up of markets) clearly favoured decentralised combined heat and power systems. As regards Germany, it is still too early to assess the impacts of full liberalisation that took place in 1998.

The regime shifts in the UK, Germany and the Netherlands also differed significantly in terms of the respective contributions to a more sustainable supply of energy. Environmental issues were clearly more prominent in the Netherlands and

13 Based on Weber (1999) and Weber (2000).

Germany than in the UK, where the economic dimension was of primary importance. This is not to say that the liberalisation and privatisation in the UK did not have a positive impact on the environment , but it was not a major issue. If more attention had been paid to the environmental consequences of liberalisation, it is very likely that more sustainable results would have been achieved. The Dutch strategy of forcing combined heat and power (CHP) was certainly very successful when looking at the diffusion curves of CHP, and due to the high energy efficiency of CHP also in terms of CO_2 emissions and resource consumption. In Germany, a less technological inroad was chosen; in essence it was left to industry how the environmental objectives were met, but framework conditions and technology policy were geared towards improving the environmental characteristics of energy supply while leaving the final choices to business. With liberalisation in 1998, however, the balance seems to have shifted towards the economic dimension of the regime shift.

This regime shift in energy supply was mostly policy-driven, as a top-down measure to induce change. However, in all countries it could be observed that technological niche development processes in industry followed this drive. The Dutch government made targeted attempts to establish processes that come close to the core of SNM. Experimentation was supported in many ways, framework conditions were adjusted and new carrier organisations of key actors established that played a major role to organise and enable the processes of learning and societal embedding.

In the UK, the definition of – mainly economic – framework conditions determined the niche development processes. However, as little attention was paid to specific technological alternatives under the new regime, these niche developments remained comparatively marginal. Such learning processes took only place under very conducive circumstances, e.g. when a major potential user of CHP systems joined forces with a system developer. The German system with its highly decentralised structure results in a high degree of variation of framework conditions; these differ between regions as well as between municipalities, thus offering a wide range of slightly different spaces for learning. There have been failures as well as successes, and the lessons learnt were taken up by the engineering and industrial associations until the design of technologies such as CHP or similarly windpower was technically standardised.

Whether the niches developed more or less successfully was then a matter of rather small differences in preceding energy supply histories, in targeted support measures and in framework conditions. The Dutch complemented their policy to open up electricity supply markets with an active technology policy in support of CHP as well as by other constituency-building measures. In fact, one could argue that their approach of Strategic Niche Management *avant le mot* was very successful in establishing a new, decentralised and highly efficient electricity supply regime.

3.2 Other Examples of Regime Shifts

Historians of technology have presented many cases of technological regime shifts, but most often not in these terms. Schot has extensively studied the history of artificial light and has looked at it through the lens of regime shifts (Schot 1998). A process of niche branching led to the subsequent penetration of gas lamps, incandescent lamps and paraffin lamps which prevailed over candles and oil lamps in ever more applications, first of all in textile factories in England at the beginning of the 19th century. It spread from there via street lighting, the illumination of factories and domestic lighting to become the dominant form of lighting until electricity made its appearance; gas was then pushed to new niches for cooking and heating.

Historical studies warn us that we should beware of the simplistic image that a regime shift occurs by a new regime replacing the old. In practice, the existing regime usually transforms gradually by incorporating elements from niches. In the popular story, for instance, steam ships 'of course' replaced sailing ships because they were faster and more reliable. Geels has shown, however, that in reality the operators of sailing ships opposed steam ships, partially because they had to pay for coal while the wind was for free and partially because the coal storage went at the expense of cargo space. What subsequently happened is that rather small engines and limited coal storages were installed on some ships as *auxiliary power sources* to be used in the case of lack of wind[14]. Next, in a gradual process engines and coal storages got bigger and the balance between the use of steam and windpower shifted (Geels 2002). Thus, the sailing ship regime did not change by a 'head-on' attack from an emerging steam-ship regime but it was gradually transformed from the inside after adopting specific elements of the steamship niche. Hence the notion of a technological *transition.*

4. Policy Strategies to Induce Sustainable Innovation and Regime Shifts

The preceding section has empirically underpinned the theoretical argument that regime shifts are nothing unusual and that they are induced and given direction by niche development processes. For a policy perspective aiming at a more sustainable provision of services like energy or transport this raises the question where and how policy could apply its levers to reinforce, guide or even lead such processes.

When looking at the fundamental policy options available, we can recur to a differentation into three basic strategies:

---

[14] This process is called 'hybridisation'.

- A science & technology push approach, supported by development of visions and *leitbilder* (e.g. the 'global village' which boosts developments in the area of telecommunications and transport). This provides, in terms of evolutionary theories, a broad set of 'variations' (Schot 1992).
- The definition of (regulatory) framework conditions. This has been a very important approach for dealing with environmental issues over the last two decades.
- Demand-oriented approaches that target the users' interests to exert an indirect effect on the innovative efforts made, e.g. by means of taxation, labelling, etc.

The latter two strategies have in common that they try to bend the development process by judiciously applying economic and/or social incentives and disincentives, so as to make some possible paths more, and others less interesting and feasible. In terms of evolutionary theories, these strategies work on the 'selection environment'. Policy makers have tried to do so, for example, though the use of gasoline taxes.

These three approaches have been applied in different situations over the last two or three decades, with varying degrees of success with respect to inducing a more sustainable development. In fact, many political initiatives have been pretty successful in terms of improving the environmental (and in some cases also the economic) performance of technologies. In most cases, however, the developments have remained within the frame of the dominant technological regimes. It should be noted here that these generalisations are based on the empirical evidence from the fields we are most familiar with, i.e. transport and energy supply.

This tendency to remain within the dominant regime is not surprising, first because the dominant regime often had a significant improvement potential and secondly because many of the elements of a regime cannot be influenced by social and economic (dis)incentives alone. Moreover, regime shifts simply take a long time. They are the outcome of a myriad of decisions over a long period in a changing landscape. They are not a linear process but involve processes of co-evolution that give rise to new "configurations that work" (Rip and Kemp 1998), combining old and new elements in novel ways. Technological regime shifts are associated with structural change at different levels – of companies, production chains, users and government policies – and are connected with new ideas, beliefs and even new norms and values.

This is why integrated approaches are needed to stimulate technological regime shifts. However, a full orchestration of a wide range of different measures would require a very knowledgeable and wise planner to coordinate and predict the joint effects of all these forces. Historically there has been a lot of planning in transport as well as in energy. Most infrastructures were built based on planning decisions. For example, planning has an important role to play in making the current transport regime more sustainable, by reducing the need for transport, providing for transfer spaces and special infrastructures for cycling and collective transport means to substitute for individual modes of travel. But there are clearly limitations to a planning approach. When there is high uncertainty about the end states and best means to meet these, as in the case of sustainable travel, you can not use a planning and implementation approach. Consequently, policy strategies are not about detailed planning any more, at least since the end of the planning euphoria in the early 1970s.

Increasingly, a different interpretation of integrated approaches is adopted. It builds on a broad systemic model of the innovation process and takes social, technological, economic and political aspects into account, with the underlying dynamics being captured by means of concepts derived from complexity research[15]. Such a perspective points to a set of relevant determinants of innovation that in principle all represent potential inroads for policy intervention, both at the structural/macro level and at the behavioural/micro level. These include scientific and technological factors, determinants on the demand and user side of innovation, various types of framework conditions, institutional and organisational settings, and the visions and *leitbilder* associated with regimes and technologies. Obviously, in a dynamic perspective, these potential inroads for policy cannot be looked at in isolation. Reinforcing mechanisms between them, processes of self-organisation, co-evolution and learning need to be taken into account.

An 'integrated strategy' aiming at inducing regime shifts, should pragmatically combine elements from the different approaches and react to new developments occuring. The objective is to *"float with the co-evolution processes and modulate them"*[16]. This means the aim is to exercise some influence, or leverage, to *modulate* ongoing dynamics and open up corridors of future development. This approach is more directly oriented at learning and adaptation, and the creation of visions and plans to guide private and public decision-making. Here policy makers engage in a kind of process management, exercising some leverage to socially beneficial developments and putting constraints on less desirable developments in order to bend these into more advantageous directions. Modulation policies are thus in many respects more modest than the other approaches, but on the other hand more demanding because they require taking a wider range of aspects and factors into account.

The approach of Strategic Niche Management which we introduce in the next section is an elaboration of such a modulation policy.

5. SNM as a Modulation Policy for Inducing Regime Shifts

Central to the Strategic Niche Management (SNM) policy perspective is the view that technology policy must contribute to the creation and development of niches for promising new technologies through experimentation. A niche can be defined as a discrete application domain where actors are prepared to work with specific functionalities, accept teething problems like higher initial costs, and are willing to

---

15 See for instance the complex innovation systems as described in Weber (2003).

16 See Kemp et al. (2001). The approaches are not mutually exclusive. Planning and policies that change the frame conditions will be *part* of the third approach, which is more inclusive. The distinction is not so much based on the instrument choice but on the management or governance philosophy.

invest in improvements of new technology and the development of new markets. If successful, a new technology might move to follow-up niches resulting in a process of niche branching. Subsequently, it may start to eat into markets covered by the existing regime, either because the regime starts to adopt specific elements from the niches or because the new niche technology competes head-on with the existing regime technology.

SNM is an example of an "evolutionary" policy, aimed at deliberately shaping paths, creating circles of virtuous feedback through carefully targeted policy interventions, rather than at correcting perceived market failures. It thus helps to overcome the weakness of current environmental policies that have been found to have a marginal influence on innovation[17]. Of course, it is not a panacea and does not guarantee success, but this holds true for all instruments.

The proposition that it is possible and productive to engage in SNM rests on two fundamental assumptions. The first is that the introduction of new technologies is a social process that is neither an unavoidable deterministic result of an internal scientific and technological logic, nor a simple resultant of the operation of the market mechanisms. The notion of co-evolution or co-production captures this assumption. The second assumption is that it makes sense to experiment with this co-evolutionary nature of technology. Such experiments can be envisaged as (part of) a niche in which technologies are specified and consumers defined and concretised. Experiments make it possible to establish an open-ended search and learning process, and also prepare societal embedding and adoption of new technology.

There is no lack of experiments, top-down ones organised in demonstration projects, or bottom-up ones developed in small market niches. In the fields of energy and transport, in biological farming and in the multimedia area (users experimenting with new forms of information and communication technology), there are many actors exploring new options through experiments. Although they are abundantly present, most of them do not get beyond the experimental stage, in spite of very promising characteristics. SNM goes beyond individual projects and we thus propose a new kind of technology policy for sustainable development based on experimentation[18].

We prefer to use the notion of experiment rather than more common notions such as demonstration or pilot projects to stress the idea that *learning* is central in SNM. This learning goes beyond technical learning; it involves learning about user needs, societal benefits and negative effects, and regulation. And not just learning to specify existing user needs, technological options and regulatory requirements (i.e., forecasting), but also learning to question the existing preferences and find ways of building new ones. SNM is not just about testing user acceptance but tries to find ways for tinkering with user needs, i.e., double-loop learning (cf.

---

17 A discussion of the pros and cons of different environmental policy instruments, especially the choice between the use of economic incentive and standards, is offered in Kemp (1997, 2000).

18 SNM is part of a broader set of possible policies; this whole set is called constructive technology assessment. A first casebook is Rip et al. (1995). See also Schot and Rip (1997).

section 2). In the transport case, this implies experimenting with new technologies connected to new mobility forms and regulatory incentives. Visions or *Leitbilder* play an important role here as "glue" and common orientation for the different actors involved. This process goes beyond user-producer relationships because it aims to involve also those in charge of defining framework conditions. This aspect is crucial to learn about the requirements and the conditions for a wider uptake of the technology in the making.

SNM thus makes flexible use of different fundamental policy inroads: experimentation with new science and technology developments, learning in close interaction between producers and users to take the demand-side fully into account, and adjusting framework conditions to accomodate the developments in the experiments and niches. SNM provides an integrated approach that is based on the modulation perspective.

As argued in section 2, niches are a necessary component of a regime shift because they help to create a pathway for a transition. Four conditions were identified for a regime shift to occur. Although it is difficult to tell beforehand whether these conditions pertain, they can help to identify technology systems that are eligible for support through niche management. Of course, the choices may be wrong in the sense that no new path gets created and the project fails to bear fruit. The attractiveness of SNM is that one finds this out in a bottom-up, non-distorting manner by carefully choosing a domain of application for which the technology is already attractive. The costs of discomfort are thus minimised (or carried by a local actor with a special interest) while useful lessons may still be learned. Here SNM as a probe-and-learn strategy differs from strategic planning or control policies based on the achievement of set goals in the sense that it is more reflexive and open-ended. It is aimed at the exploration and creation of new paths by building on developments at the local and supra-local level. Thus, initially, SNM is *not about pushing desired winners but about identifying, testing and training possible winners*. After a period of testing and learning, pushing may follow which involves an element of control, of limiting side effects. SNM thus combines elements of push and control.

Concerning the type of technologies to explore, we have to acknowledge there is no way to decide up-front which ones will be the winners. The best we can do is identify a range of technologies, each of which has a promise in the sense that it might solve a specific problem in the current regime but each of which has problems and faces barriers as well. Also, it is quite unlikely that a single alternative will be able to solve all the problems of the current regime. Furthermore, future developments will, probably in an unforeseen way, combine bits and pieces of different alternatives that are currently developed in separate niches. This implies that we should not just look at one promising solution but that we need to look at a variety of new alternatives and, through SNM, train various 'athletes' for the contest to follow.

Different actors have different preferences or different expectations of the potential of these. The promise of clean vehicles is a clear case in point. In such a case, it is not sensible to try and get consensus beforehand on what is the most promising option (electric, natural gas, improved petrol). Instead, we need to acknowledge that there is uncertainty that needs to be clarified on the basis of practical experience rather than on the basis of desk research or expert opinions. The result of this exercise is a definition of what we call the 'portfolio of promises'. By this we mean a spectrum of new (in this case, transport) technologies and concepts, each of which entails a certain promise to solve an important problem of the current traffic and transport regime, but each of which is associated with uncertainties and potential new problems as well. One of the objectives of SNM should be to ensure the portfolio is kept sufficiently large and varied.

It should be stressed that the alternatives in the portfolio of promises are chosen because each has a specific advantage over a current technology (e.g. very low emissions) rather than that they can be made economically competitive in the near term. It has to be acknowledged that the economic viability of new technologies is something that cannot be reliably assessed in such early stages. Economic viability often is the effect of sustained efforts of specific actors that believe in a new technology or concept, at times working against the odds, until they have managed to get all the elements in place to attain market success. To induce a regime transition, it is more important to get the right partners together that believe in new options than to do detailed marketing analyses in these early phases.

Through SNM, user experiences can inform private investment as well as government policies. By carefully choosing an appropriate domain, the costs may be kept low. New options are explored at the local level while at the same time a transition path may be created to a new and more sustainable system in a non-disruptive way. SNM helps actors to negotiate and explore various interpretations of the usefulness of specific technological options and the conditions of their application. The outcomes of experiments may be used to fine-tune government support policies and to change the framework conditions. Technology experiments should be supplemented by niche management policies aimed at stimulating the diffusion and further development of niche technologies.

As discussed in section 3, SNM is not a substitute for existing policy strategies for sustainability, but a useful addition. It renders other strategies more effective by improving the functioning of the variation-selection process. It gives an enlightened, conditional and socially embedded push to technology by increasing the variety of technology options upon which the selection process operates. This selection process remains dependent on policies that make sustainability benefits part of economic decision-making. Sustainability is a much weaker driver for change and path creation than economic benefit is. The two things have to be reconciled: there should be an economic benefit in activities that produce sustainability benefits. Subsidies and other types of positive rewards (such as prices) might help to achieve this; taxes, standards and other penalties are another route. Infrastructure provision is a third route. All routes have a role to play, depending on the circumstances.

Following the standard diffusion model of innovation we can make a distinction between a niche-phase, in which the emphasis is on learning, and a diffusion phase in which the emphasis is on market penetration of new technologies. This suggests that policy approaches should also be phased, starting with a period of strategic niche management and followed by a period using market-based instruments. In practice, however, there is considerable overlap between the phases. The picture becomes even more complicated when we acknowledge that niche developments concerning a specific technology or mobility concept are influenced by developments in the dominant regime and vice versa.

To account for this, it is necessary to make an inventory of barriers in the dominant regime that make it difficult for various alternatives to be able to compete eventually. These barriers need special policy attention and attempts should be made to lower them. For example, there are many examples where physical (infra-)structures or existing rules very much favour use of a private vehicle above the use of shared means of transportation. The general problem is that public policies in many cases do not adequately reflect the negative impacts of specific characteristics of the traffic and transport regime. Doing this in a more consistent way would not only affect the dominant regime but would also create more space for alternative technology niches. Well-chosen policy approaches may exploit the dynamics between niches and regimes to speed up the pace of innovations towards sustainability.

The question then is what policy options are most promising to stimulate such synergies under specific circumstances. This question is beyond the scope of this paper but it is clear that certain policy approaches can have noticeable effects on the development of niches as well as the dominant regime by exploiting their dynamic interaction. Clear examples of this in the area of traffic and transport are the federal energy policy act (EPAct) in the U.S. and the California ZEV (zero emission vehicle) mandate. Both can be seen as examples of 'technology inducement'[19]. In the late 1990s EPAct required American owners of large vehicle fleets to buy increasing numbers of alternative fuel vehicles, including methanol, ethanol, natural gas and electric vehicles. This created a large stimulus for the developers and suppliers of these technologies to enhance the performance of these fuels and technologies. This is even more the case for the ZEV mandate that has provided a worldwide stimulus for further development of EV-technology. As these alternatives gained international acclaim the American major automakers, which had strong reservations about the potential of these alternatives, increased their efforts to make conventional vehicles cleaner. They even went as far as offering to supply cleaner conventional vehicles than they were obliged to by federal law (the so-called 49-state programme) in exchange for not having to supply electric vehicles in specific states.

The point to stress here is that certain policies can have rather strong impacts on both the development of alternative niches as well as the dominant regime. This stresses the point that we need a *portfolio of policy approaches* to induce a regime transition, one of which is strategic niche management aimed at learning on how a variety of alternatives can be made to have a practical impact on the existing regime.

SNM clearly involves a dilemma. On the one hand, it is driven by the concern to introduce new technologies and practices to alleviate existing transport prob-

---

[19] Technology inducement is a form of policy (using mandates or incentives) that induces industry to develop technologies with specific desired characteristics. Cf. Elzen (1999).

lems. In this sense, SNM prescribes rough corridors of technical solutions to current problems, informed for example by forward-looking Delphi or Foresight exercises. Diffusion is not possible without making choices, fixing certain technological and demand (mobility) options and subsequently creating new path-dependencies. So creating fixes is part of SNM. On the other hand, learning is central in SNM, implying a change (unfixing) of present day mobility patterns and connected technologies and the articulation of new patterns and technologies. To allow for learning, flexibility must be preserved to limit the possibility of premature selection of inappropriate solutions. Thus, a tension between learning and institutional embedding is inevitable and perhaps desirable. It is precisely the learning processes encouraged in a SNM approach that lead to early identification of negative impacts and explorations of technological, regulatory and user needs, which allow one to work towards solutions beyond technical fixes offering a particular solution to problems of unsustainability.

The SNM approach puts learning processes at the forefront, with the result that it becomes difficult to be specific about outcomes beforehand. What kind of sustainable transport future will be pursued then through SNM? In a way applying SNM demands that we avoid trying to answer this question directly. We do not know in detail what a sustainable transport regime looks like, but we have well-founded insights into possible building blocks of such a regime, which include technologies but also new regulations, consumption patterns etc. To make a step towards sustainability these elements need to be investigated – not separately but in connection. The SNM approach leads to a detour, to a process in which a sustainable transport future is explored. Still, implementing the SNM approach does include picking a set of technologies for experiments, so an assessment *ex-ante* of the potential of new technologies is necessary. This assessment is not, however, focussed on gains in terms of resource productivity or social equality of individual technologies. Rather, SNM tries to answer the question: which technologies might open up a pathway towards a more sustainable technological regime including a new set of consumption patterns (user preferences), regulations and artefacts?

We can also put it another way. SNM concerns *changing change*: introduction processes might be designed differently. The long-term goal of SNM policies is to create new routines – "institutions" as neo-institutional economists as well as sociologists would call them – that would anticipate impacts, user requirements and related technical choices earlier and more frequently. At the same time, the aim is to set up introduction processes to stimulate learning and reflexivity, and thus, to create space for experimentation. In the long run the ability to deal with difficult and complex processes such as the introduction of more sustainable technologies and mobility concepts will become more widespread.

Having said this, we do not claim SNM guarantees sustainable development. Uncertainty is an intrinsic part of both technical change and sustainable development and cannot be lifted through moral enterprise for a certain course either.

6. The Role of Niche Managers: Theory and First Experiences

Above we positioned Strategic Niche Management as part of a third model of governance, which we called modulation policies, which try to utilise the "winds of change" and seek to exploit windows of opportunity. Such policies are especially suited when end goals are not clear (because manifold) nor is it clear how these goals can be reached. SNM helps to deal with uncertainty about the desirability and costs of new technologies and with opposition from vested interests that often stand in the way of doing something new. SNM may actually enrol companies vested in the status quo in the process of niche development. These companies should not be allowed to control the process, though, given their interest in the status quo. For radical change, outsiders and entrepreneurs are crucial.

SNM is not something completely new. It has been attempted by companies that introduced radical innovations such as cellular telephones, optical fibres, aspartame (Lynn et al. 1996). But although some attempts like the Swiss large-scale demonstration program for lightweight electric vehicles[20] could be labelled as de facto SNM policies, it is a new approach for policy making. In our view there is a need for policy makers to go beyond demonstration projects and to promote user experiments with new technologies.

Table 1. Actors' Reasons for Engaging in or Supporting Technology Experimentation

Type of actor	Reasons to engage in or support technology experimentation
Companies	Learn about the current state of a technology either for supply or use and inform company policies Be prepared for a shift in market conditions creating a demand for a new technology Influence public policy by offering a solution to an environmental, economic or other type of problem
Local authorities	Learn about a new technology and about sociotechnical arrangements that may solve a local problem (pollution, nuisance, employment, congestion)
State authorities	Have society learn about new technology options and facilitate transition processes Create business Inform public policies to achieve socially desirable outcomes
Consumers and citizen groups	Learn about their own consumption patterns and needs Demonstrate to others sustainable life styles Contribute to a reduction of environmental impacts
NGOs	Demonstrate feasibility of sustainable lifestyles in order to get support for other policies Experiments are vehicles for campaigns

20 Discussed in detail in Hoogma et al. (2002).

Different people and organisations may thus be interested in technology experiments and SNM for various reasons: to seize a business opportunity, to alleviate a local problem of unsustainability or simply to learn. Table 1 above gives an overview of different actors' motivations to engage in technology experimentation.
The table shows that technology experiments allow for mutual benefits that help various parties to find a common ground to be involved in experiments. On the other hand, it shows that SNM involves difficult tradeoffs. It involves making decisions on the choice of first application domain, choice of partnership, choice of protective measures taken to shield the new technology from the selection environment, and more. Suggestions for how to implement SNM are offered in Weber et al. (1999), Weber and Dorda (1999) and Hoogma et al. (2002) Table 2 with an overview of phases and key issues of Strategic Niche Management is adapted from the former publication.

Referring back to the multi-level model in section 2, however, we can see that Strategic Niche Management cannot be limited to merely conducting local experiments. It also needs to include activities at the niche level and the regime level. It is important that experiences from different experiments with the same technologies are brought together for comparison and triangulation, to find out under what circumstances a technology option functions best. This can be done at the level of RTD programmes, which are found at the level of niches. Moreover, experiences from different programmes need also be compared and juxtaposed, to find out the best technology for the right niche applications as well as the potential of synergies between developments in different niches and synergies between niche- and regime developments. This should be done at the overarching level of the regime. This means that we need to introduce 'portfolio management' as an essential element of SNM – management of both portfolios of experiments which compose a niche and portfolios of niches.

Table 2. Overview of Phases and Key Issues of Strategic Niche Management

Phases of SNM-Proc.	Key Issues	General advice
Identifying a new technol-ogy/ concept	(1) Incremental or radical departures from the cur-rent regime	Choose a technology or concept which is as close as pos-sible to the existing regime, but which allows to induce more radical changes later on.
	(2) Path-dependency	Seek to keep open a variety of technological options; there-fore phase your experiments and organise them in mod-ules in order to avoid their becoming too complex
Designing an experiment	(3)The structure of net-works	Keep the experiment sufficiently broad in terms of partners (users, suppliers, government, operators) and have com-mitted partners in the team.
	(4) Successful experi-ments and successful niche formation proc-esses	A successful experiment needs not be conducive to niche formation, and vice versa.
	(5) Protection measures	Explore which types of market pressures could be opera-tional in the experiment.
	(6) Involvement of users	Create opportunities for the active involvement of pioneer users in the early phase of an experiment, and of mass us-ers in its later phase.
Implementing an experiment	(7) Communication with the wider public	Create opportunities for discussing results of the experi-ment with groups which are not actively involved in the ex-periment but are affected by it.
	(8) Broad expectations of partners in an experiment	Monitor the tacit and vague expectations and visions of participants and articulate them specifically.
	(9) Learning about the facets of an experiment	Seek broad coverage of opportunities for learning about new implications of a technology.
	(10) Learning about un-derlying assumptions e.g. regarding mobility	Reflect upon existing mobility patterns and changes which the new technology may bring about in relation to the mo-bility objectives pursued.
Expanding an experiment to a niche	(11) Changes in the net-work structure	Be aware of changing requirements in terms of network structure in the course of the progress and scale-up of the experiment.
	(12) Complementary pol-icy measures	Consider which kinds of complementary policies could be conducive, needed or detrimental to the experiment.
	(13) Transfer of an ex-periment	Look for opportunities to replicate an experiment and try to keep the experiences stored in a network.
	(14) Changing require-ments during niche ex-pansion	The technology or concept needs to be customised when the pioneer market turns into a mass market.
Review of the protection of an experiment	(15) Structure and timing of specific protection measures	Seek to establish productive and smart ways of protecting an experiment.
	(16) Generalised protec-tion of a niche	Seek to establish productive and smart ways of protecting a niche as part of the prevailing transport framework.
	(17) Continuation or ter-mination of an experiment	When phasing out a niche development process, try to en-rol the established network into the development of other options for addressing similar problems.

These activities at different levels call for different actors to assume a responsibility in SNM, in other words: to step forward as niche managers.

- At the *regime level*, the *European Commission* and *national policy makers* are responsible to guard the *portfolio* of alternative technology options, to prepare a 'level playing field' and ultimately make choices which options deserve support for market introduction and diffusion. They should ensure that the overall portfolio of options experimented with is sufficiently broad and try to organise experiments addressing 'blank spots' in the knowledge about technology applications. They should also assess the potential of specific combinations of experimental results at the programme level. In short, they should guard the overall 'portfolio of promises'. They also need to develop visions of how different technologies may be combined to bring about synergies. The portfolio management and vision development can be delegated to intermediary organisations, as these may be more acceptable to, and therefore more successful in involving non-government actors (research institutions, industry, and users) in these activities. Instruments may include TA, foresight, Delphi, scenario workshops, backcasting and cluster policy for portfolio management, and subsidies, taxes and discretionary measures for (dis)incentives. Creating a level playing field may require the modification or abandonment of existing tax structures and discretionary measures.
- At the *niche level*, the objective is to manage a 'portfolio of projects' in a specific area (e.g. biofuels, electric propulsion). The aim is to bring together and widely spread knowledge of local experiences and lessons and establish the optimum conditions for application and diffusion of a technology. This is an often-underdeveloped aspect of technology programmes. There may be international fora for the exchange of experiences and information in specific niches (like the annual electric vehicle symposia for the EV domain) but there are virtually no attempts to systematise this information and experience and make it widely available[21]. This is a crucial task in niche management, though, on both levels. At the regime level, it would facilitate an assessment of 'niche progress' and evaluate this in relation to progress in other niches to identify potential synergies and potential impact on the regime. At the niche level, systematised information is crucial for people with plans for projects to be able to base their design on what has been learnt elsewhere. To create such overviews, we need co-operation between niche managers at two levels. At the national levels, program managers from executive agencies sponsoring concrete projects could produce and continuously update a national overview. This information should be brought together under the responsibility of international bodies, either linked to the EU or working independently. These bodies should keep in close touch with the 'portfolio managers'. Ideally, for a variety of niches a continuously updated *knowledge and experience centre* should be created on the Internet.

[21] An exception is the work under the auspices of the International Energy Agency (Implementing Agreements).

At the *project level* the first responsibility of local project managers is to run an experiment effectively and collect and evaluate the results in accordance with the objectives that reflect the motivation behind conducting the experiment. Extensive guidelines on how to do this have been elaborated in a recent EU funded project and can be found on the Internet[22]. Effective SNM, however, implies that the results should have a wider bearing, necessitating a broadening of the objectives of local experiments as well as making the results more widely available. In practice, local projects often require co-operation with national or international bodies for funding and these bodies could guard that this need is taken into account. Thus, project level SNM also requires a two-level management structure with a local manager taking responsibility for the actual conduct while a national (or EU-level) manager should ensure that the results add to the experiences from earlier projects elsewhere. After the completion of the project the latter should take responsibility that the results are made available at the knowledge centre for the niche in question.The SNM activities, the designated managers and the instruments applied are summarized in Table 3.[23]

Table 3. Strategic Niche Management Activities and Instruments at Different Levels Attributed to Managers

	Regime level	**Niche level**	**Project level**
SNM activities	Evaluation of progress on 'portfolio of promises', identification of potential synergies, portfolio management (identification of blank spots), create level playing field, vision development, economic/social (dis)incentives	Managing portfolio of projects, cross-project learning, institutional embedding, create and maintain 'niche knowledge centre'	Project management, network management, wide dissemination of results
Managers	EU, national governments, delegated intermediaries	Delegated intermediaries at national and international level such as programme agencies and interest groupings	Local actors (industry, users, government, interest groups) and national / international programme managers
Instruments	TA, foresight, Delphi, scenario workshops, backcasting, cluster policy, programme / project funding, subsidies, taxes, discretionary measures	Delphi, strategic workshops, backcasting, develop standards, benchmarking, identifying best practice, clearing house, project funding	Strategic workshops, project evaluations in 'niche knowledge centre' format

22 http://utopia.jrc.it/prog6.html.

23 This table is inspired by the report by Smits et al. (2001), which elaborates on the possible roles for one intermediary organisation, the Netherlands Organisation for Energy and the Environment, in transition management (cf. managing regime shift).

The discussion above and the table illustrate that SNM is not a matter of a central manager "pulling all the strings". It is typically a matter of co-operation between different types of actors at different levels with different interests and responsibilities and to try and create synergies in the mutual interest. Local actors, for instance, are primarily interested in specific solutions to local problems. At the national and EU level, however, people are interested in generalised solutions that can be transferred to a range of specific locations. Strategic Niche Management is very much an intereractive approach that tries to organise programs and projects such that both types of interests are taken seriously.

7. Conclusions

The SNM approach was initially developed in and for the mobility area, but we think it is useful for other technical systems too. Investigating this sets a challenging research agenda. For drawing our conclusions in this paper, we will again refer to traffic and transport, but we suggest that the conclusions have wider bearing.

To develop a sustainable transport system a regime transformation seems to be needed. A wide variety of alternatives have been experimented. They suggest that large improvements over the current situation are possible. It would not be difficult to design a sustainable mobility system behind a desk, even when only using demonstrated technologies. To realise this in practice, however, seems to confront us with unsurmountable barriers if only because different actors have completely opposing views on what can be realised in practice or what is desireable.

It is then impossible to move forward via some sort of centralised 'strong policy' approach that attempts to realise a blueprint of sustainable mobility. The opposition to such attempts would be too strong to be able to realise it in practice. A better way to move forward is to find out in practical situations what the potential of various alternatives is and to try and develop them further so that they can make practical contributions to sustainability.

The approach of strategic niche management tries to do this in a systematic fashion. It starts by acknowledging that nobody knows what the constituting elements of a future transportation system will be but that we can identify various alternatives that, at least on one or two dimensions, could theoretically solve problems of the existing regime. This range of alternatives is called the 'portfolio of promises'. SNM subsequently seeks to learn on the practical usefulness of each of these alternatives in a variety of practice experiments. The results of these experiments are fed back into the developmental processes in connection with these alternatives.

As a theoretical construct, there could be one enlightened central manager who organises experiments of all kinds of dimensions across the world. In the real world, programme and project managers at different levels and in different locations have their own priorities. A practical approach is then to create better mechanisms for co-ordination between experiments and make their findings widely available in a form that makes cross-project as well as cross-niche evaluations possible.

In the previous section, possible roles of niche managers at different levels were discussed. Government agencies at the national and EU level could, in close co-operation, take the responsibility to monitor progress at the 'portfolio of promises' level and ensure the portfolio is sufficiently wide (e.g. by initiating projects to fill holes). Evaluations at the portfolio level should be fed into the policy-making process so that other policy measures can be taken to create a 'level playing field' and, when appropriate, take policy measures to stimulate the market uptake of alternatives that are considered to have been sufficiently demonstrated. This point illustrates that SNM is not a stand-alone policy approach but should be tuned with other approaches.

At the niche level, an appropriate international body should work with national agencies to develop and implement programs in specific domains (e.g. electric vehicles), monitor progress and create and maintain a knowledge and experience centre accessible for anybody (e.g. as an Internet site). These knowledge centres (for a variety of niches) would provide the basic input for the portfolio assessments mentioned in the previous paragraph while they would also provide a great help to local people considering a local project since it could help them to take into account experiences obtained elsewhere.

At the local level, a local manager has to ensure proper conduct and evaluation of an experiment but care should also be taken that the local experience is relevant for and made available to others. The latter is probably best ensured by co-operation with a national programme manager who has better oversight at the niche level.

Thus, SNM is not a centralised policy approach but a strategy to facilitate co-operation between different levels to obtain synergies between different but partly overlapping interests.

Until now, SNM has been applied mainly to look at past experiences with experiments, demonstration and pilot projects. The effectiveness of policy approaches that take into account the principles of Strategic Niche Management have been convincingly shown at least for the area of mobility (Weber et al. 1999; Hoogma et al. 2002). Whether a systematic application of the approach would indeed help shift dominant regimes towards sustainability remains an open question, but the review of past regime shifts has shown how important successful experimentation with non-conventional alternatives is to enable them. Therefore, there are good reasons to assume that SNM as an instrument of RTD policy should be an important constituent in any policy portfolio to induce a regime shift towards sustainable mobility. The next step to assess the usefulness of SNM, and to elaborate and refine it would require setting up new experiments explicitly using the SNM framework. Initiatives of that kind have been proposed but have not been implemented yet.

As a final warning, we should not mis-interpret SNM as a simple "tool". It should be clear that SNM can never be implemented in an instrumental mode only. It is a perspective that facilitates specific kinds of communication and co-ordination processes, with specific contents. It helps to better align the technical and social dimensions of innovation processes, which is a crucial element that has been missing in many cases of promising, but finally failed experiments. SNM does not create a highway towards sustainability. What it does is that it provides a strategy to experiment with a range of alternatives increasing the chances that at least some of them will eventually get practical value. It thus broadens the range of possible pathways towards sustainability which thus eventually also increases the chances of arriving there.

Policies and Conditions for Environmental Innovation and Management in Industry

Nigel Roome

1. Introduction

This paper draws freely on ideas developed through the work of an expert group established by DG Research of the European Commission.[1] The group met in 2000 through to April 2001 with the task to examine research, technology development and innovation for a competitive and sustainable European production system. It sought to advise the EC on appropriate policies and actions for the period to 2020. It therefore specifically addressed future policies and conditions for environmental innovation and management in industry.

The aim, then, of this paper is to discuss policies and conditions for environmental innovation and management in industry. This is a critical issue. Industry harnesses technological innovations to provide artefacts or new materials in their products and services. Through this process industry generates wealth and contributes to the satisfaction of human needs but also adds significantly to environmental degradation. For example, to illustrate the scale of the direct environmental impacts of European industry. Industry contributed 26% of European N_2O releases and 23% of the green house gas, whilst manufacturing generated 26% of the waste produced by EU Member States (Environment in the European Union at the Turn of the Century: European Environmental Agency 1999: European Environment Agency; Copenhagen). While this gives a broad indication of the scale of industry's environmental impact it should be noted that these figures only represent the effects of industrial and manufacturing activity at production sites. The figures do not include the environmental impacts arising from the extraction of resources used by industry or the impacts and wastes associate with products in, or after, use.

European industry is then a significant actor in environmental degradation. Improving these environmental impacts, while remaining competitive, involves innovation. Only in this way can Europe move progressively toward more sustainable forms of development. The transformation required will involve industry to engage in change, with a mix of other societal actors, including the public sector and public policy makers. In particular, public policies are needed to support innovation and provide the conditions within which transformation, to a sustainable

---

[1] Although I will draw on the EG's work you should be aware that the comments in this paper are my own. I do not speak for the members of the expert group nor for that matter do I speak on behalf of the EC or DG12.

industrial economy, becomes possible. Not only do we need public policies that focus on industry by supporting and spurring innovation, these policies, and the structure developed for their implementation, can be viewed as a form of (social) innovation in their own right.

This paper explores critical issues about .what is implied by the need for innovation in public policy and industry. First, I want to devote part of the paper to addressing some of the key terms used in the title of the paper, in particular, the core notions of environmental innovation and environmental management. Frequently, these terms are used without recognising the wide variety of interpretations they can have. These introductory comments are then used to develop a conceptual model of the transition from innovation in environmental compliance driven industry to innovation in more sustainable forms of enterprise. This will be used as background to the work of the expert group. It leads into the second section of the paper where the main outcomes of the Expert Group's work are outlined. The final section sets out for discussion some key points that need to be addressed as we establish policies and conditions for environmental innovation and management by industry.

2. Definitions

Environmental innovation can take many forms. In an industrial context the notion of innovation is associated with purposeful, or designed change. These changes range in scope from modifications to production processes and technologies through new products/services and the technologies on which they are founded, to innovation in complex socio-technical systems. Examples of innovations in production processes include the move away from the use of CFCs in the manufacture of electronic components. Innovations leading to new products include the introduction of catalytic converters, or the advent of hybrid engines in automobiles. An example of innovation in complex socio-technical systems includes reconsideration of ways to meet societies' needs for mobility.

One of the key factors that underscores this hierarchy of innovations is the complexity of the issues and the number of actors who must engage in concerted change as the scope and boundary of the system addressed through innovation is expanded. For example, the introduction of CFC-free electronic components requires the identification of substitute cleaning agents and inevitably leads to changes in manufacturing processes. However, the consumer is left largely untouched by these innovations unless product functionality or price is affected. The introduction of catalytic converters, on the other hand, involves change to lead free fuels as well as the redesign of internal combustion engines, especially the development of hardened valves and valve seats, which no longer benefit from the protective effects of lead additives in fuel. This type of change means that automobile manufacturers and gasoline producers must collaborate, although they may have very different interests in existing and future combinations of technologies. However, this form of innovation has also obliged automobile users to adopt new habits and routines, buying lead free fuel and accepting the loss of power or higher fuel consumption that might follow from catalytic converters.

The purposeful redesign of mobility systems is even more complex. The overall architecture of the existing mobility system, with its vested interests and technological and social rigidities, has to be addressed. This involves many actors: producers and consumers; automobile and traffic engineers; town planners and many others.

Moving up the innovation hierarchy increases in the complexity of the issues, the number of actors involved in change, and the number of linked, multiple technological and social options, the innovations and new practices that need to be undertaken and the uncertainties that have to be considered.

Given the serious overuse of materials and resources arising as a result of our developed industrial life-styles and population pressures, the innovations demanded by any transition to sustainable development are likely to be at the more complex end of the hierarchy.

How does this relate to the issue of environmental innovation? This term has many meanings. A key issue is what makes an innovation environmental, rather than just an innovation? It is not simply that the drivers for change arises from the environment department of a company or an environmental ministry. I contend that a better perspective on what defines and innovation as environmental arises from the tautological, yet profound, idea that all innovations that involve resources, materials or social practices, which impact the quality or quantity of resource endowments or natural systems, are environmental. This perspective is profound because it means that virtually every economic actor - consumer, industry or service provider - is involved in 'environmental' innovation. Put another way, all industries that combine human ingenuity with materials and resources to produce products and services are engaged in the process of environmental innovation. Consequently there are no industries that can say they are not environmental industries.

Environmental innovations are also characterised by the demands that arise from the systemic nature of environmental impacts and the changes that result form those impacts. This implies that a necessary prerequisite for (environmental) innovation is the gathering of information about the systemic impacts of the innovation. However, gathering information tells us little about whether an innovation should be judged as good or bad. Indeed, the dualism between good and bad is not helpful because most innovations have distributed effects. That is innovations produce a range of effects, some regarded as good and others as bad. Moreover the distribution of these effects changes over time and space, from local to global. We know this from most studies aimed at determining the life-cycle impacts of new products.

Whether an innovation is, on balance, good or bad from an environmental point of view can be tested at two extremes. At one extreme is the question of whether, on balance, an innovation has more or less severe environmental impacts than the product or social activity it is designed to replace? Does product A have a better or worse environmental profile than product B? This is viewed as a weak test because it is only concerned to compare the environmental profile of a new product against an existing product. This weak test still involves a complex evaluation of the effects provoked by products in natural and resource systems. It also involves an assessment of whether the impacts, individually and in aggregate, are judged to be good or bad. There are many ways to form this assessment, from expert opinion to multi-criteria scoring systems or stakeholder consultation exercises.

A much more demanding test, however, is whether an innovation results in activities that can be conducted within the 'carrying capacity' of the environment - local, national, regional and global. This is more demanding because it requires an evaluation and assessment of the impact of the innovation in relation to the processes and sinks provided by environmental systems. This implies a sound working knowledge of the dynamics and fluxes of environmental systems together with an evaluation of how innovations affect the 'carrying-capacity' of those systems, at different spatial levels and over time.

Comparing these two tests helps to distinguish between environmental innovations, which on balance lead to less damage to environmental systems, from those that maintain or improve carrying-capacity. In my experience most innovations to date have been guided by the less demanding of these tests: do they cause less damage than the practices they replace? In this way innovations lead to environmental improvements or, what might more appropriately be seen as, reduced levels of environmental damage. I would contend that we have little way of knowing whether innovations of this kind are environmentally sustainable, without applying the carrying-capacity test.

Environmental management in industry can be characterised in a similar way. At it simplest environmental management involves the application of a (relatively) structured environmental information system to provide the basis for understanding and making decisions about the environmental consequences of industrial practices. At the next level this structured system may be used to make decisions about priorities for reducing environmental impacts. Often this involves attempts to integrate environmental assessment and choice with existing, conventional business processes – investment analyses, policies and mission statements and so on. We can, for example, distinguish between pollution control as an innovation where it is important to know the costs of control as well as the environmental returns from more advanced notion of pollution prevention. In pollution prevention integration has come to mean seeking out innovative designs that simultaneously reduce costs and/or improve productivity, and, lead to environmental improvements.

In its most extreme form environmental management can be directed toward environmental sustainability. True environmental sustainability implies that only choices and innovations that operate within the carrying-capacity of environmental processes and systems can be viewed as feasible (sustainable) options. Environmental management based on this concept of carrying-capacity is regarded in this paper as 'strong' environmental sustainability, whereas, environmental management based simply on improving environmental impacts is a 'weaker' notion. At best, this weaker notion only leads in the direction of environmental sustainability. It is acknowledged that sustainability also has an important social dimension but for the purposes of this paper the discussion will be restricted to the environmental aspects of sustainability.

The idea of an innovation hierarchy and the differences between environmental management and sustainability can be used to develop a simple conceptual model of the transition that takes place in industry as it moves from environmental compliance to sustainable forms of enterprise. This is shown as figure 1. The figure is based of the relationship between four variables. The complexity of innovation, the scope of change, the strength of environmental management, and the set of actors actively involved in design for innovation.

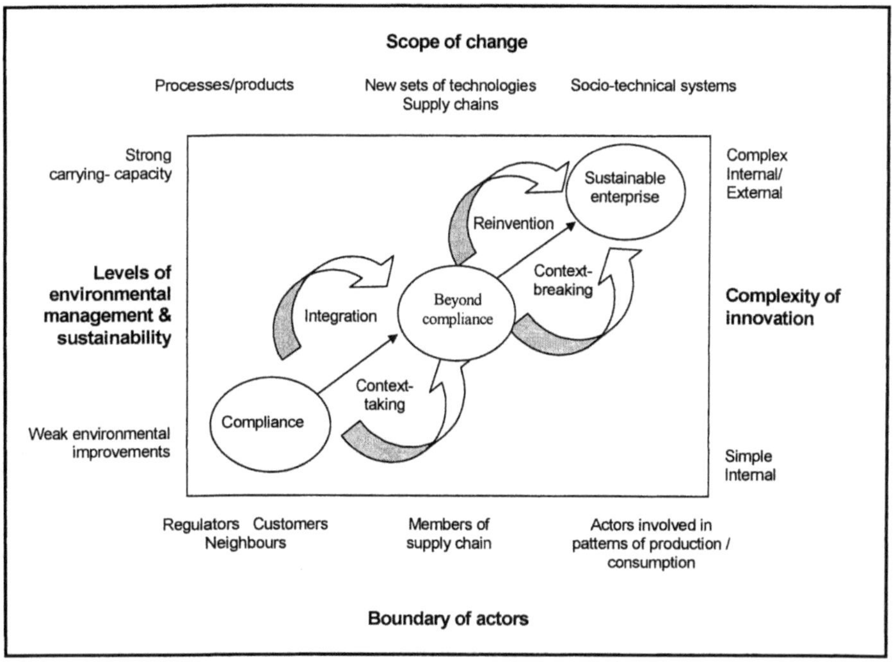

Fig. 1. Framework

The model suggests that in compliance-driven companies the scope of change is dominated by incremental improvements to processes and products, there are relatively weak environmental management systems in place and the main actors of concern to industry are regulators, customers and neighbours. The innovation process is mainly driven internally and is relatively simple. Sustainable enterprise, by contrast, has a scope of change that is based more on reformulating socio-technical systems. Here environmental management is very strong and highly connected with business processes, including the overall business strategy. The actors involved are those who shape and influence patterns of consumption as well as production in socio-technical systems. The innovation process is consequently relatively complex and involves linking actors internal and external to an industry in concerted change.

The model suggests that the transition from compliance to beyond compliance is centred on internal integration of environmental and business processes while accepting the context provided by markets. The bridges that contribute to the process of integration include notions such as quality, organisational learning, pollution prevention or full cost accounting. The transition from beyond-compliance to sustainability is centred on notions of reinventing the company, together with its relationships with others in the socio-technical systems in which its products/services are embedded. As a result innovative product offered by industry are a part of larger scale, context-breaking change. It is suggested that the bridge to this type of innovation is found in processes that involve industry in collaborative problem-finding and the identification and implementation of solutions involving technological and social change. This represents a form of organisational and social learning based on continuously reforming collaborative structures.

If the transition to environmentally sustainable industry is to be accomplished in the long term it will involve a shift from innovation built on internally integrated technological change to innovations based on integrated arenas for innovation. These involve highly collaborative, multi-actor processes, where industry is one of the key actors.

With this model in mind we can now turn to the main conclusions of the expert group.

3. Outcomes of the Work of the Expert Group on Policies and Actions for Sustainable and Competitive European Production Systems

Through its work the expert group developed a vision of a European system of industrial production which would guide more environmentally sustainable innovations for the prospective period to 2020. In this vision:

Human ingenuity (knowledge and technology), capital, resources and needs are harnessed and governed so people can live better lives while consuming less material resources and energy. This system is sustainable when production and consumption support the quality of individual and social life, in ways that are economically successful while respecting environmental limits within the changing context of local-global conditions.

The key condition to the realisation of this vision is a more integrated view of the arena of innovation, with a focus on sufficiency. The thinking that guided this orientation, fits with the notion that, innovations have to be designed by actors who know a specific production system. The report's central argument is that this orientation is the necessary pre-requisite for new policies, actions and practices customised to the needs of different systems.

A more integrated arena for innovation sees production and consumption as key parts of an overall socio-technical system. Socio-technical systems include technologies, products and materials 'in use' and human systems. Examples of socio-technical systems are information and communication, mobility, or household services, such as clean clothes or nutrition. Innovations arise in these arenas through purposeful processes of change that engage many actors (producers, consumers and others). An integrated arena requires other conditions. It brings together economic, environmental, social and scientific concerns. It requires solutions tailored to specific socio-technical systems and localities under the influence of national, European and global pressures. A particular issue that affects all such arenas is the way globalisation creates new axes for the governance of technological and social innovations. This means future innovation must be governed in ways that are responsive to global competitiveness and innovation, environmental concerns and social needs, while harmonising public policies and business strategies through collaboration and joint action. This is not easy. It runs counter to most existing policies and practices for RTD&I. It has implications for the orientation of environmental and industrial policies as well as industrial practice.

This outlook builds on a number of trends in manufacturing and adopts some known and simple principles. For example, production is becoming progressively more resource efficient, there is an increasing number of examples of closed-loop production on which to draw, and manufacturers are increasingly addressing consumers' needs for product performance. There is also a strong link between the development of knowledge and innovation, as the foundation for competitiveness, and, the achievement of environmental and social sustainability. This link is provided by ideas about learning organisations and multi-actor (social) learning platforms. These emphasise collaborative processes for developing visions, systems thinking, problem finding and problem solving, and resolving barriers to change and joint action.

The innovations arising out of this process are either context-taking or context-breaking and fit within one of two archetypes – efficiency and sufficiency. Efficiency is a linear concept seeking lower inputs for a given activity. Sufficiency is concerned with the search for, and implementation of, new ways to meet human needs. Sufficiency addresses the services required to meet needs and the performance of material products. Illustrations of sufficiency are selling flooring services not carpets, providing photocopied documents rather than photocopiers, or selling clean clothes rather than washing machines. In sufficiency solutions manufacturers retain ownership of the material in physical products and sell the service performance of those products to customers.

There are a limited but growing number of innovations that illustrate the sufficiency archetype. Implementing these innovations is beset with institutional, organisational and managerial obstacles. Some obstacles are generic others are specific to individual innovations and socio-technical systems. Development and implementation of sufficiency solutions requires these obstacles to be overcome as part of the innovation process.

A number of obstacles to sufficiency and, especially to the collaborative processes on which organisational and social learning for sufficiency is based, were identified. In a European context these include the many different management styles and cultures. Similarly collaborative processes are affected by different cultural and linguistic conditions. These differences should not be overcome by reducing the diversity of the European system, as diversity is a potential source of innovation. Rather, what is required is the move toward a more overarching level of management, which is self-reflective and knowledgeable about the contributions and values of different management styles and cultures.

There are structural problems due to lack of coherence and the perverse incentives in the overall mix of policies. For example taxes, subsidies, capital write-offs, and trade agreements are often contradictory and do not support economic and environmental efficiency let alone sufficiency.

At the institutional level there is weak participation by the private sector in public policy making and weak collaboration between private and non-governmental interests. This is associated with relatively low levels of co-operation and collaboration in research and innovation and an absence of initiatives that bring together potential partners across sectors and interests. In particular consumer groups and other actors outside of business and the public sectors have difficulties to engage effectively in learning and innovation programs. And when these do arise there are poor mechanisms to diffuse good practice in learning and knowledge development.

While at the organisational level there is risk aversion in committing resources and forging new organisational collaborations around innovations together with a low disposition of organisations to support life-long learning at all levels of enterprises.

At the level of individual managers there are shortages of capabilities in systemic thinking and systems integration. And, a scarcity of facilitation skills and the skills to support multi-actor, multi-disciplinary, multi-functional, multi-sectoral processes.

These deficiencies are underscored by the continuing confusion between environmental management and sustainable development.

Accelerating the shift toward conditions that foster sufficiency requires broader, more flexible, policy instruments than are provided by the present support for collaborative R&D projects. For example the group took the view that new conditions were needed. These included ideas such as: 100% funding of search exercises for key socio-technical systems to enable the generation of ideas about, and commitments to, sufficiency solutions through the moderation of material consumption and sustainable product service offerings. It is possible to envision funding support for the remodelling of R&D infrastructure and the innovation system so that it better reflects the new demands for knowledge, the new context for interaction with industry and the requirements for new skills and competencies.

An important new element of the innovation infrastructure would be the establishment of international competence networks as a basis for research and the dissemination of research results. These networks should be set up for a period of 5-10 years (maximum), equipped with a (relatively) stable budget and working to a remit that emphasises communications. These competence networks should form nodes in a broad Europe-wide communications and co-operation network.

In the same way RTD&I policy administration could be restructured better to reflect the participative processes that are being encouraged for RTD&I. What this means is that policy administration should be multi-disciplinary and participative, experimental and possibly open to continuous interaction with EU experts and supported RTD&I partnerships. This might involve experimentation with continuous process evaluation and mid-term corrections in projects.

In response to the deficiencies of the present system of RTD&I and the obstacles to sufficiency in Europe the expert group went on to advocate a 'design framework for innovation'. This would be based around six concurrent processes. These were:

1. Generating ideas for innovative approaches to sufficiency strategies for selected socio-technical systems.
2. Improving understanding of Socio-Technical Systems.
3. Resolving the barriers to change and establishing the feasibility of new solutions.
4. Supporting the development and adoption of enabling technologies.
5. Engaging a variety of relevant actors to participate in the process of organisational and social learning and change.
6. Demonstrating and disseminating these processes and their outcomes to others.

Each element is the overall process is discussed in more detail below.

Generating ideas for innovative approaches to sufficiency strategies for selected socio-technical systems would involve mechanisms such as 'foresight forums'. These would bring together societal groups to generate new ideas and learn about the expectations of any set of relevant actors for competitiveness within the framework of sustainability in relation to specific socio-technical systems.

Developing approaches that go beyond marginal improvements involves maximum encouragement for maverick, or wild card, approaches to RTD&I through a continuously open call within the theme Competitive and Sustainable Development. This to be matched by specific funds designated for innovations, which have merit but do not meet traditional criteria.

Improving understanding of socio-technical systems could be brought about through the development of participative forums that establish the key actors involved in specific socio-technical systems, and, identify and map the specific characteristics of those systems together with each actors' needs and interests. This would establish the basis for inputs and contributions by these actors to collaborative action.

Resolving the barriers to change and establishing the feasibility of new solutions means addressing knowledge transfer problems and organisational barriers for companies, which want to adopt competitive strategies for sufficiency. It also means devising appropriate cost-accounting and financial control mechanisms that reflect the true economics of material recovery and material assets held in product-related service performance systems. This would need to be supported by schemes that seek to develop competence in designing for service-performance rather than in producing products so workers can become service providers. The support of inter-firm co-operation using information technology & knowledge management and logistics, especially on reverse supply chains and take-back schemes. Establishing and resolving the barriers that arise from the demand for venture capital oriented toward competitive sustainable development projects. And, finally, identifying a policy mix that supports sufficiency in specific socio-technical systems.

Supporting the development and adoption of enabling technologies places an emphasis on basic science and research in technologies, which allow decentralisation of production systems. This highlights areas that include information and communication technologies, biotechnology and micro and nano-technologies. These represent important enabling technologies in the areas of dematerialization and resource productivity.

Engaging a variety of relevant actors to participate in the process of organisational and social learning and change would have to incorporate societal and environmental actors together with management of business in the programme committees of the Framework Programme. It would oblige a broadening of the knowledge base on sustainability innovation mechanisms in manufacturing practice, together with socio-economic research on sustainability management and innovation management within competitive frameworks. It might benefit from the introduction of a voucher system for societal groups, which would allow them – if collaborating – to give research grants.

These processes would need to be able to draw on improved professional support and function effectively. This could include the development of the contribution of socio-economic experts as a support input/vision to RTD&I on social needs. Training participants in effective multi-actor procedures and the facilitation of processes.

Demonstrating and disseminating these processes and their outcomes to others places a need on the assessment and development of a policy-mix that encompasses legislation and taxation allowing technical alternatives (through R&D for technology) to be examined in advance of drafting directives. It could harness R&D in the hard sciences in support of areas of public sector spending where the objective is the promotion of competitive and sustainable solutions.

It would benefit from action research that was able to demonstration of the principles that underpin future production systems demonstrate participation in action. Ideas that would enhance this approach include multi-actor Implementation Forums for RTD&I and sustainability combined with competitiveness at levels appropriate to specific socio-technical systems. Database and resource guide on good practices in SMEs involved in competitive approaches to sufficiency. This would be especially valuable for companies not in existing networks. Finally the group saw scope for socio-economic shadowing of the process of mainstream RTD&I research with monitoring in real time, with the express objective of presenting challenges, learning and disseminating rather than evaluation policy implementation.

This design framework should be guided by principles such as – lightness, flexibility, durability, adaptability, and closed material loops. The framework involves processes based on collaboration for mutual learning and action. Together these principles and processes provide the elements of a 'design guide for sufficiency'.

4. Conclusions

This final section draws out some key points that need to be addressed if we are to establish policies and conditions for innovation by industry that improves competitiveness within the framework of sustainability.

The most important points to emerge from the paper are that the conditions for environmentally sustainable innovations can not be defined in terms of a set of hard parameters. Rather conditions are seen more in terms of principles that circumscribe the innovation process and provide guidance on the boundaries to that process. These principles are seen as recursive. That is, the same principles apply to the interactions between all actors engaged in environmental innovation as a form of organisational and social learning, whatever their level in an organisation.

Three particular principles are identified to illustrate this point. These concern open-ended learning, multi-actor processes and flexibility.

In the case of open-ended learning the argument of the expert group is that innovation directed to problem-solving is not really appropriate for environmentally sustainable innovation given the present need to address fundamental flaws in the (un)sustainability of existing socio-technical systems. The current approach to incremental forms of environmental improvement is based on problem-solving that takes the existing system as a given. Technological bottle-necks are viewed as problems, which actors then set about resolving mainly through technological solutions.

In contrast what is proposed by the expert group is not problem solving but a more open, problem-finding approach. In this approach overall socio-technical systems are addressed. More environmentally sustainable socio-technical systems are envisioned through multi-actor processes. Agreed visions of future, more sustainable, socio-technical system are developed. Only then is it possible to establish the barriers and problems that need to be resolved in moving toward that vision, given present reality. This approach posses real issues for public policy because it demands faith in the success of an open-ended process that begins without concrete and measurable targets.

For example problem-solving may begin with a target such as the reduction in automobile carbon emissions through the introduction of an efficient catalytic converter. Success against this target can be assessed by various hard measures. Problem-finding requires support for a process, the outcomes of which are not clear at the beginning of the process. Indeed, the moment policy makers, or other actors, seek to define desired, hard outcomes, the more the process of problem-finding is circumscribed and tends to become less open to bringing about the breakthroughs that are needed.

In the case of problem-finding, targets and outcomes are 'soft'. They demand the adoption of processes (for example multi-actor search processes) where hard outcomes are defined through the process itself. In terms of public policy this may mean the continuous interaction between policy-makers and the other actors immediately involved in the innovation processes, which have been stimulated by public policy. Open-ended processes also imply the possibility of a series of iterative mid-course adjustments as the process unfolds and as the process shifts from envisioning the future to taking concrete steps to make that desired future a reality.

Secondly, if environmentally sustainable innovation is a multi-actor, multi-disciplinary, multi-sector, multi-functional process, then the principle of problem finding applies not only to the platforms used to identify the needs innovation must address, the same conditions apply to the administration of the policy process in support of those innovations. For example, if the aim of public policy is to foster multi-actor, multi-disciplinary collaborative innovation then the administrative mechanism used to review and assess proposals for public support must also have a multi-actor, multi-disciplinary character. This means that the administrative structure by which public funds are allocated should employ multi-disciplinary teams or individuals in proposal evaluation. Yet these teams are hard to manage and individuals with multi-disciplinary skills are hard to find. The development of multi-disciplinary skills, and skills in the facilitation of multi-disciplinary processes, runs counter to our existing system for education and research, which provides for the development of policy makers and industry practitioners engaged in innovation.

The third important consideration is that environmental sustainability is defined in relation to the specific demands of a socio-technical system and the local conditions under which those systems operate. At the same time these socio-technical systems innovations develop in the context of global and regional pressures. These two sources of influence – global/regional, on the one hand, and local demands, on the other can prove paradoxical. For example, innovation has to be sensitive to local environmental conditions and yet solutions are often influenced by, say national or regional tax structures and other elements of the policy-mix. More often than not, this means that the existing mix of public policies constrains the possibilities for innovation.

It is necessary, then, to move toward a framework for public policy that is more sensitive to local circumstances and the demands arising from local innovation rather than the demands for national or regional administrative efficiency and consistency. Yet this shift is hard to imagine. A clear example of this paradox is found in the debate in the WTO, between those who argue the right of nation states to establish environmental policies that are fitting for their local (national) conditions and the demands for global free trade, which is unencumbered by the variability of local restrictions. In the same way, tax regimes set at national or regional level are not normally open to modification in the light of local demands. Consequently, actors involved in socio-technical innovation often have their choices limited by the framework, or mix, of policies within which they operate.

The claim of this paper is not to that hard, instrumental top-down processes should be replaced by softer, more flexible bottom-up approaches based on problem-finding, conditioned by ideas of carrying-capacity and precaution. The real challenge is to develop a form of continuous iteration between the policy framework and the demands of local environmental and specific socio-technical systems. This demands more flexible and open bureaucracies and more open and flexible industrial commitment to learning and change with a mix of actors. Indeed I contend that these are hallmarks of the kinds of process of organisational and social learning that lead to the social and technological innovations that we need to secure sustainability.

The Need for Environmental Innovation Indicators and Data from a Policy Perspective

Yukiko Fukasaku

1. Introduction

Because of the market failure and the systemic difficulty involved in optimising environmental innovation, well-designed public policies are needed. These policies include both environmental policies and research and innovation policies. Environmental policies need to be designed to stimulate innovation. They need to be complemented by research and innovation policies so that the innovative responses are appropriate, adequate and timely.

Sound policy-making needs to be founded on clear understanding of what is going on in the real world. However, systematic information about what drives firms to innovate for the environment, or the knowledge requirements of firms to generate environmental innovation is not yet available. Two types of information are necessary for sound policy making to optimise environmental innovations. One is public and business expenditure in environmental R&D, and how and where the funds are spent. This is needed to assess whether we are investing enough to generate knowledge to understand and improve the environment and whether the funds are being used efficiently. Also, policy makers need to judge if public investments well complement private investments in research. The other is information about the determinants of environmental innovation in industrial firms, how firms assess the costs and benefits involved, how they acquire needed knowledge. Appropriate indicators and methodology need to be developed to collect these types of information to facilitate the design of environmental policies that stimulate innovation and innovation policies that can supply appropriate and adequate knowledge.

2. Harnessing Science and Technology

It is now widely recognised that technology and innovation play a key role in directing our development path toward sustainability. A recent OECD study on sustainable development concluded that harnessing science and technology was a key policy tool in moving toward sustainable development (OECD 2001). In the face of the urgency of many global environmental problems such as climate change, technology is even regarded as easy "fixes" that can bring about sustainability even in the absence of other policy measures. However, there is no easy "technological fix". Appropriate innovations need long gestation periods guided by public policies that define the demand for and supply of such innovations.

A growing number of studies point to the pivotal role that public policies play in enhancing environmental innovation. These policies range from environmental policy instruments, i.e., regulations, market based instruments, information measures, and voluntary or negotiated agreements to innovation policy tools, i.e., direct and indirect R&D subsidies, public/private partnerships, the use of national innovation system approach such as clusters and networking. Well-designed environmental policy instruments are needed to stimulate innovative efforts in the business sector to generate and take up cleaner options. Innovation policies are needed to complement environmental policies so that the innovative responses are appropriate, adequate and timely. Working out the right mix of policies to enhance environmental innovation is the challenge facing public policy makers[1].

Sound policy-making needs to be founded upon solid understanding of what is taking place in the real world and how it responds to public policy. Relevant information for effective public policy making to enhance environmental information is not easy to find. Moreover, there is no standardised methodology or indicators for compiling environmental R&D or innovation data that can be applied in any country. This results for one thing from the "diffuse" nature of environmental innovations and the scientific knowledge base that contribute to it. Environmental innovations draw upon a diverse knowledge base. For another, it is not always clear to what policy signals firms are most responsive. This could also differ according to the industrial sector, and the country. Business response to public policies can only be known through thorough firm/sector case studies or well-designed surveys. Either type of studies is still relatively few.

The paper first highlights some characteristics of environmental innovation that distinguish it from other types of innovation. It then discusses the issues policy makers face in making effective policies to enhance environmental innovation. The paper then turns to the discussion of the types of information needed to aid policy makers, namely, indicators and data on environmental R&D expenditures and the determinants, cost and benefits of environmental innovation in industrial firms.

3. Characteristics of Environmental Innovations and Obstacles to Stimulating It

Although environmental innovations share many of the characteristics of "innovation" in general, they do distinguish themselves in some aspects that make it more difficult to develop indicators of how the innovation process is taking place and to formulate policies to enhance it. These special characteristics are the market fail-

1 Need for a mix of policy instruments rather than the application of a single instrument is the conclusion of some major case studies on environmental innovation, for example in the German studies contained in Hemmelskamp et al. 2000.

ure and the systemic difficulties that environmental innovations are subject to and the "diffuseness" of the knowledge base relevant for environmental innovations.

First of all, innovation for environmental sustainability suffers from "double" market failure[2]. It is widely recognised that because of the spill-over effects of knowledge, private (business) investments in R&D remain sub-optimal. Also, because of the "public" nature of environmental qualities, private investments in contributing to improve the environment also remain sub-optimal. In other words, the market failure involved in environmental innovation is more serious than for other types of innovation; therefore, more intensive efforts at the public policy level need to be made to stimulate it.

Knowledge about environmental changes and their impacts as well as innovations that improve it arises from research and development in different scientific and engineering disciplines[3]. Knowledge advances in diverse fields of basic and applied sciences and engineering need to be combined to generate innovations that enhance environmental performance. Relevant knowledge is generated not only by innovating firms themselves, but may also be generated by upstream or downstream firms, universities or other public research institutions. Environmental innovations often require inter- or multi-disciplinary approach to research as well as inter-firm or inter-institutional co-operation in R&D. The research and innovation systems in many countries are still not well adapted to enhance inter-disciplinarity or inter-sectoral co-operation. This subjects environmental innovations to what may be termed as "systemic" difficulty.

The systemic difficulty implies that the knowledge base that potentially contributes to environmental innovation is diverse and diffuse. Any body of scientific or engineering knowledge and technology can be applied for environmental objectives. Also, a wide range of "organisational" innovations can enhance efficiency and hence improve environmental performance. The growing importance of cleaner processes and products as opposed to end-of-pipe technology is adding to the diversity of the knowledge base for environmental innovations. This also makes it more difficult to define the boundaries of environmental goods and services sector which now is defined to include cleaner processes and products (OECD/Eurostat 1999). Inclusion of cleaner technology implies that firms not necessarily producing environmental goods and services including those in emerging areas such as ICT, biotechnology and nano-technology, probably have a great

---

2 Some experts call this the double externality problem of environmental innovation. "...neither innovators nor those investing in environmental protection can automatically secure returns on their actions. There is a danger that the actual level of environmental innovation will lag behind that which is economically desirable..." (Lehr and Löbbe 2000).

3 For example, a recent article in Nature discusses the rise of bio-monitoring, the use of living organisms to scientifically assess the impact of environmental pollution and changes on living systems, which may be used as complements or substitutes to more conventional chemical monitoring (Whitfield 2001). It is clear that this branch of environmental R&D needs to combine biology, chemistry and ecology.

potential and do in fact contribute in an important way in supplying needed technology to user industries in enhancing environmental performance.

These characteristics of environmental innovation make it difficult to gather needed indicators and data for effective policy making. The boundaries of environmental innovation itself as well as the boundaries of the pool of knowledge that contribute to this type of innovation are difficult to define. However, because of the seriousness of the market failure and the systemic difficulties involved, there is a need for effective policies to enhance environmental R&D and innovation. There is a need to better understand how industrial firms generate environmental innovation and what their knowledge needs are. Only an adequate understanding of this could contribute to designing of policies (both environment and research/innovation) that can stimulate the demand for environmental innovation and assure the supply of useful knowledge for that purpose.

4. The Issues Policy Makers Need to Address in Formulating Effective Policies

Then what are the issues public policy makers need to address in formulating effective policies? In order to simplify discussion, it is assumed that environmental and innovation policies play different roles. Environmental policies largely define the *demand* for environmental innovations and determine the *direction* of technological and innovative change, whereas research and innovation policies define and manage the *supply* of knowledge for innovation, hence determine the *rate* of technological change and innovation. It is to be noted that this distinction is somewhat artificial, since to a certain extent each determinea both the rate and direction of innovative efforts. Also, policy coherence and integration between these policy domains is an issue in itself. However, the distinction has its own merits, in that in most governments, environmental policy making and research and innovation policy making remain distinct and separate policy areas.

4.1 How to Regulate to Stimulate Innovation – The Issues for Environmental Policy Making

Since environmental innovations are generated in industrial firms, policy makers need to know what drive business firms to innovate to enhance environmental performance. A closely related question is what are the barriers to environmental innovation encountered by firms and industries. Obviously, environmental regulations are the most direct drivers for firms to improve environmental performance through regulatory compliance. Then the policy issue is how to regulate so that firms are stimulated to search for the means of compliance through innovation.

4.1.1 Reglatory Design that Stimulates Innovation Flexibily, Cost-effectively and Continuously

In this context, studies undertaken so far have shown that regulations differ in their effects on innovation (OECD 1999; Kemp 2000; Hemmelskamp et al. 2000; Environmental Law Institute 1999). These studies find that regulations based on technology specifications tend to stifle innovation, although the diffusion of the specified technology option is stimulated. The instruments that are favoured from the point of view of stimulating innovation are performance standards and market-based instruments[4]. Also, regulatory stringency and enforcement strategies are considered important factors in stimulating innovation (Ashford 2000).

Whether or not policy instruments stimulate innovation is not the only relevant policy issue. Some more key policy considerations need to be addressed. These include whether the policy instruments stimulate innovation in a *flexible* manner (i.e., the choice of innovative response is left up to the polluter), whether they do this *cost effectively* and in the *long range*. The flexibility, cost effectiveness and continuity considerations imply that cleaner process and product innovations, where appropriate, are normally to be stimulated rather than end-of pipe solutions.

4.1.2 Where and How Large are the Win-win Opportunities?

These considerations also open up the debate that the well-known Porter hypothesis (Porter and van der Linde 1995) has sparked. This argues that regulatory compliance presents "win-win" opportunities for firms and stimulates environmental innovation that increases their competitiveness. Environmental innovations "offset" the cost of regulatory compliance through innovations that reduce cost to the firm by increasing resource efficiency. The study presents numerous case examples revealing such win-win situations. Because the cost of regulatory compliance is normally higher, the more stringent the regulation, Porter's study as well as others argue that the win-win pie is larger, and can stimulate more significant innovative response, the more stringent and focused the regulations (Ashford 2000).

The Porter hypothesis was in part supported by theories of competitiveness, but the evidence presented was for the most part anecdotal firm level evidence. It did not provide systematic, statistical evidence; consequently, it encountered criticisms both from environmental economists and other management researchers. These criticisms present statistical or anecdotal evidence that regulatory compliance incurs cost to the firms, and/or that these costs varied according to industries or plants (Palmer et al. 1995; Walley and Whitehead 1994; Environmental Law Institute 1999).

Therefore, the central policy issue of how to regulate so as to stimulate innovation requires an understanding of the interrelationship between changes in production costs, R&D inputs on one hand and process and product innovations on the other. This requires disaggregated data on costs and benefits of environmental in-

4 Product bans also stimulate innovation, but clearly the applicability of this instrument is limited.

novation, which in turn requires systematic work on indicators and data on drivers of environmental innovation, their costs and benefits. The debate surrounding the Porter hypothesis raises the issue of the general paucity of indicators on environmental innovation (Kemp and Arundel 1998). Also, the inter-industry or inter-plant differences in compliance costs imply that "win-win" opportunities are not distributed evenly. There is little doubt that significant "win-win" opportunities do exist, but how large they are in the aggregate or where they are found in the industrial sector are not clear (Norberg-Bohm 2000). These are policy issues that need to be addressed. Again, more systematic understanding of the cost and benefits of environmental innovation is needed.

4.1.3 Central Role of Incentives: Designing Effective, Market-based Instruments and Making Voluntary Agreements Work

The debate surrounding the Porter hypothesis highlights the importance of incentives in stimulating firms to innovate for the environment. The importance of incentives is the main *rationale* behind the increasing use of economic or market-based instruments (taxes, tradable permits, pollution charges, deposit-refund schemes), and the policy advice to enhance their use[5]. Empirical analysis of US situation in the past few decades demonstrates the strategic usefulness of properly designed and implemented regulation complemented by economic incentives (Strasser 1997).

However, the implementation of effective market based instruments has not been easy. The difficulty mainly stems from the frequent resistance to their adoption, especially energy or environmental tax, notably from the industrial firms. The main lesson to be drawn from past experience is that market-based instruments are more effective when applied in combination with other policy instruments, especially regulatory standards, in a policy mix, rather than independently. It may be noted that the optimum mix differs according to industry.

The industrial aversion to market-based instruments, seems to be inducing a proliferation of voluntary agreements in many industries and many countries[6]. Like market based instruments, voluntary agreements normally are applied in the context of existing or new policy mixes. Theoretically, they are flexible as they leave industry more freedom with regard to the method and moment of compliance. They have been criticised on the basis of the danger of free-riding and under-exploitation of opportunities on the part of the industry as well as the frequent

---

5 Some policy advice organisations recommend more extensive use of these instruments, such as the OECD (see OECD 2001).

6 There are numerous examples, such as the chemical industry's Responsible Care Program, voluntary agreements to reduce perfluocarbon compound emissions in the aluminium industry. These programmes involve several countries. An example of a major national programme is Japan's Keidanren Voluntary Action Plan to reduce CO2 emissions, which involves a major part of the industrial sectors through participation of industrial associations.

absence of technology forcing targets (Kemp 2000; OECD 1999). But recent experiences suggest that voluntary agreements do stimulate innovation[7].

4.1.4 Interplay of Drivers of Environmental Innovation, but Inter-industry and Cross Country Differences

The existence of win-win opportunities and the importance of incentives imply that for industrial firms, the regulatory driver is often translated into commercial driver to generate environmental innovations. Some recent environmental surveys show this correlation clearly (Malaman 1996; Green et al. 1994; Cleff and Rennings 1999). In addition, some of these surveys show that the social awareness factor is also an important driver, and interact with other drivers as well. Firms want to demonstrate social awareness by being innovative in environmental performance. This in turn improves the image and the performance of the firm in general. It may also be noted that these surveys demonstrate significant inter-country differences in the relative importance of the different drivers.

Numerous surveys of environmental innovation in business firms have been undertaken. These surveys have been conducted at the national level with diverse methodologies, and reveal the incentives that drive firms to innovate for the environment. Although the studies identify regulations, cost considerations and social awareness as drivers, there are clear differences between the relative importance of these drivers according to countries. For example, the UK study showed that anticipation of regulation, the fear of rival products, and the threat to market share were important drivers. However, a German survey revealed that "maintaining market share" or "expected future legislation" were drivers of relatively low importance (Green et al. 1994; Cleff and Rennings 1999). It is difficult to draw conclusions as to whether these results represent genuine cross national differences in firm behaviour, or the divergent results are caused at least in part by the difference in the survey method.

It is equally conceivable that there are significant inter-industry differences in the relative importance of drivers. For emission intensive ("dirty") mature sectors like steel, regulatory driver is likely to be the most important. But for further downstream industries whose products sell directly to consumers, the social awareness driver could well be more important. Environmental policy making need to take into account these important cross-country and inter-industry differences.

---

7 Such as Keidan Voluntary Action Plan to reduce CO2 emissions in Japan, and aluminium industry's efforts to reduce perfluocarbon compounds emissions in several countries. See OECD 2001.

4.1.5 Working out the Best Mix of Policy Instruments Adapted to Industry or Country Specificities

The most widely recognised conclusion about the design of environmental policies that stimulate innovation is that a mix of policy instruments need to be worked out within a regulatory framework adapted to the specificities of the national regulatory regime (Hemmelskamp et al. 2000; OECD 2000).

To work out this policy mix, there is a need to better understand firm behaviour and the ways they respond to signals that policies provide, since policy makers have to assess the relative merits of the different policy instruments from the viewpoint of stimulating innovation. The effective policy mixes differ according to the industrial sector, the country and over time. Policy makers need to be ready to tailor policies to diverse contexts. Hence, there is a need for better indicators and data on drivers and barriers to environmental innovation, and how these factors interact; also, the costs and benefits of environmental innovation for industrial firms. Also sensitivity of business firms to the various environmental policy instruments or the combinations of policy instruments need to be assessed and documented. These require more systematic surveys and information gathering of business attitudes and strategies towards environmental innovation. Standardised environmental accounting methods and standardised survey methods on environmental innovation would enable compilation of relevant information.

4.2 Issues for Research and Innovation Policy Making

If environmental policy making addresses the issue of *why* firms innovate for the environment, *how* firms do it is the central question for research and innovation policy. Once business firms decide to innovate for the environment, they would invariably turn to R&D to search for knowledge required for innovation. Some statistical studies demonstrate the correlation between environmental (compliance) expenditures and R&D expenditures or patenting (Lanjouw and Mody 1996; Jaffe and Palmer 1996). A recent survey by the World Business Council on Sustainable Development indicates that when firms decide to adopt "sustainable development" as part of their corporate strategies, links with R&D and innovation management becomes important. For these "sustainable" firms, improved technology and better engineering skills are the essential tools for supporting "sustainable development" strategy, and the considerations of sustainable development have helped them to launch new products and improve existing products and processes (Dearing 2000).

4.2.1 Making the Case for the Key Role of Research and Innovation Policy for Environmental Objectives

If research is crucial to environmental innovation, active research and innovation policies would facilitate business efforts in the search for needed knowledge. The seriousness of the market failure for environmental innovation discussed previously justifies public support of research. However, two counter arguments arise.

One is that if environmental policies are well designed they would be sufficient to induce appropriate research and innovation. This is probably not the case. For one thing, even if environmental policies send the right signals, it normally takes a long time for the appropriate innovative response to emerge. R&D efforts typically require long time horizons especially for radical innovations. Also, firms may favour less costly (in the short term) incremental innovation to more radical innovations which in the long range may be more cost effective. Network dependent technologies such as energy supply and transport are examples. In other cases, the assessment of the impact of an environmental issue may change constantly, resulting in considerable time lag for appropriate environmental policies to be implemented. Some environmental problems may become irreversibly aggravated by the time proper policies are in place. Climate change is the case in point. Reduction of CO2 came on the policy agenda more than ten years ago, but some of the key policies to address it, such as the Clean Development Mechanism, is yet to be fully designed and implemented. Finally, it is a widely accepted view in research and innovation policy, that demand side factors alone do not determine innovation. Supply side factors play a crucial role. For example, medication for infectious diseases must have been in great demand since the dawn of history, but effective drugs were only developed after advances in bio-medicine since the late nineteenth century. Policies to enhance research and innovation would be needed to address the time lag factor, and to facilitate the development and adoption of appropriate innovations.

Another argument against active public support of research and innovation that often arise in economic policy making, is that public support for R&D, especially public funding of technology programmes, tends to "pick winners". This is viewed as conducive to "locking in" technological development paths which may later be judged sub-optimal from the viewpoint of environmental sustainability or economic efficiency. Some power generation technologies and transport technology are cases in point. This view generally tolerates support to "basic" research at best, but opposes public support to the development of specific technologies.

4.2.2 Importance of Public Support to Broad-based Basic Research

While the market failure factor and the time lag factor involved in innovative response justify public support to environmental R&D, how best to do it is a difficult question that research and innovation policy makers face. The first question that needs to be addressed follows from the winner picking issue: Do we need more than support to "basic" research in order to enhance environmental innovation? If so, how should the focusing be done so as not to pick winners?

It was pointed out previously that the on-going general shift from end-of-pipe solutions to cleaner process and products approach broadens the range of innovation and technology that can be applied for environmental objectives. This first of all, indicates that environmental innovations would benefit from knowledge advances in many scientific and engineering areas, as well as social and behavioural science areas providing knowledge base for organisational and managerial innovations. In a survey of American firms in the environmental technology sector[8], a major R&D issue was the lack of long-term basic research. Two thirds of the companies indicated that at least 90% of their research has a short-term focus, because of economic pressure which make their research market-driven and oriented to developing specific products, with a correspondingly short time frame. Also, nearly half of the firms developed their technology from basic research not oriented specifically to solving an environmental problem (Environmental Law Institute 1997).

These results show the importance of the public funding of basic research to complement business environmental R&D and innovation. It also shows the importance of the serendipity factor in environmental innovation; hence, the basic research that needs to be publicly funded need also to be sufficiently broad-based. It is not easy to foresee which lines of research would lead to environmental innovations.

4.2.3 Inter-sectoral and Inter-country Diversity in R&D Requirements

The above survey also demonstrates differences in the R&D requirements between the different segments of the environmental technology sector. First of all, the share of revenue devoted to R&D differs according to the segment. Water and air technology firms spend an average of 2.5% and 3% respectively, instrument manufacturers 8%, and process and prevention technology firms invest 25%. Also, air and water companies finance 80% of R&D from its own capital and instrument companies 60%, but the share of government funding of R&D is larger for the process and pollution prevention segment.

These results show that the R&D requirements of firms differ considerably even within the environmental technology sector. The difference seems to depend on the maturity of the segment, with the relative importance and the public dependence for financing of R&D higher, the less mature the segment.

Also, another study shows that one sector, paper and pulp sector in US, Japan, Sweden and Germany take different approaches to R&D and innovation (Blazejczak and Edler 2000). The difference stem from the type of regulatory regime and the differences in approaches to innovation.

4.2.4 Ways to Focus Need to be Worked out According to the Sector and the Regulatory Regime?

These inter-country and inter-sectoral differences imply that it is extremely difficult to answer the question of how much R&D investment is adequate in either the public or the private sector as a whole. The answer probably differs according to

---

[8] This was an interview survey of 45 small to medium sized firms exclusively dedicated to developing environmental technologies distributed between four principal categories of air pollution control equipment, water pollution control equipment, monitoring instruments and process and pollution prevention technologies. (Environmental Law Institute 1997).

the sector and the country. The sectoral differences reflect the diversity in the technological base of industries, and the differences in the innovative paths that these technologies will take in the future, while country differences are likely to arise from the diversity in the environmental policy framework as well as national research and innovation systems.

4.2.5 Focusing without Picking Winners: Partnerships and Involving Multiple Stakeholders in Research and Innovation Decisions

It may be noted that recent changes in innovation policy are addressing the question of focusing public research investments without picking winners. This is seen in the shift away from large publicly supported technology programmes towards the use of networking approaches, especially public/private partnerships in funding and executing research. The important element is the involvement of both public and private actors in research and innovation decision making and taking part in funding of research. The partnership approach leverages private R&D funds, and pre-empts "free-riding" on public funds by making the private sector commit itself financially. The use of this approach has enabled reduced public R&D funding in some sectors such as energy (IEA 2000).

Public/private partnership approach is already used widely for the purpose of enhancing environmental research in many countries. These programmes involve a variety of public and private actors in collaborative research efforts. Partnerships approach can overcome institutional barriers to facilitate networking and address the systemic difficulties in R&D activities to enhance inter-disciplinary and inter-sectoral co-operation (Fukasaku 1998).

Another innovation policy that enhances networking and multi-stakeholder involvement in research and innovation decision making is the cluster approach. For example the recent Finnish Environmental Cluster Programme provides seed funding for research on new environmental technologies to be carried out by consortia of producers and suppliers, universities and institutes. Projects have been launched which aim at improving eco-efficiency through the application of life-cycle techniques in agriculture, forestry, basic metals and water management (Honkasalo 2000; OECD 1999). Collaborative projects enhance networking among researchers and users and facilitate innovation, without picking winners.

4.2.6 Moving towards Radical Innovations

Experts in research and innovation policy know that inducing radical innovations is much more difficult than incremental innovations. But in pursuing any direction of technological change, it is inevitable that at some stage, adoption of radical innovations becomes necessary. Environmental innovation is no exception. Environmental policies normally induce incremental innovations. Inducing radical innovations by environmental policies is not easy, since if it is to be done through regulations, they need to be extremely stringent. If it is done through market-based instruments, tax or charge levels need to be extremely high. It is unlikely that such stringent regulations or drastic market-based instruments can be negotiated as acceptable public policy.

In principle, stringent regulations and effective market instruments increase the size of the win-win pie. However, it is pointed out that it is not always easy for firms to exploit the win-win potential when it involves large investments. Public policy, not only in the form of environmental regulation, but also R&D support and other measures are necessary to focus firms on the win-win potential, especially in cases of possible shifts away from incremental process and product improvements and towards radical changes in processes and products (Norberg-Bohm 2000).

Then how can research and innovation policies facilitate the move towards radical innovations? Two innovation policy tools can be identified that can do this. In both the involvement of diverse stakeholders is a key. One is technology foresight. An increasing number of countries use technology foresight processes to set priorities in research. In the technology foresight exercises conducted during 1990s, identification of technologies that potentially contribute to environmental sustainability occupied a major place, and a broad range of future technologies have indeed been identified that in the long range are expected to contribute significantly to sustainability. The list includes applications of biotechnology and information and communication technologies, new materials and micro and nano-scale technologies, new energy technologies, innovative waste treatment and recycling technologies (Fukasaku 1999). Many of these may be categorised as radical innovations.

A recent trend in technology foresight exercises is the involvement of diverse stakeholders, including the research community, government, business and the civil society, in the process. The multi-stakeholder involvement clearly guards against setting priority on the basis of the interest of any one group or industry, and guarantees that where a choice is made, that choice is in the interest of diverse stakeholders in society. The involvement of the research community and business enables the matching of the supply of new knowledge developments and the market demand since long before the actual realisation of innovation. Technology foresight, as an innovation policy tool, can not only identify research areas and technologies that are likely to contribute in an important manner to environmental sustainability without "picking winners", but also enable matching the supply of technologies with business demand by actively stimulating networking and inter-sectoral collaboration.

Undesirable winner picking and lock-in effects may be avoided if radical innovations of entirely new technology systems are given a chance for experimentation. Such "systemic" innovations that lie beyond incremental innovations can transform large infrastructures that have been built up over the long term. Transportation and power generation infrastructures are the cases in point. It is clear that the existing infrastructures that accommodate increasing traffic and fossil fuel power generation with complex grid infrastructures are probably not sustainable in the long range. An innovation policy instrument that induces systemic innovations by allowing experimentation is strategic niche management.

SNM is a means of trying out new systemic technologies in a selected environment - niches - by real users. In niches the technology is temporarily protected from full selection pressures of the market and acts as a test bed and incubator for the new technology. This has been applied in introducing the use of light-weight vehicles in a Swiss town, and in developing organised car sharing, also in Switzerland (Kemp 2000).

5. The Type of Indicators and Data Needed

Then, what kind of information do policy makers need in order to address the issues and questions raised above? There is already a substantial amount of funds devoted to research and development for environmental objectives in both the public and the private sectors. Also, industrial firms have innovated to enhance environmental performance in various ways. The starting point for designing or reforming policies should be to find out what is going on, i.e., to gather systematic information and data on the inputs and outputs of environmental innovation.

5.1 Input Indicators and Data – Public and Private R&D Expenditures for Environmental Objectives

The most obvious input indicator of environmental innovation is R&D expenditures devoted to environmental objectives[9]. R&D is invested both in the public and private sectors. Public policy makers need to know both, so that policies can be designed to complement business efforts.

5.1.1 Public Environmental R&D

The availability of internationally comparable data on environmental R&D remains very limited. The only available indicator compiled by the OECD is government budget appropriations and outlays in R&D (GBAORD) for environmental objective (table 1). On average, OECD governments appropriate about 2 per cent of their R&D budgets to research for environmental objectives. This share rises to about 5 percent when environment-related research on other objectives is added, such as that on energy and agriculture. Also, since the early 1980s, growth in budget appropriations on environmental research has outpaced most other research areas so that its share of the total has increased by about half a percentage point.

However rapid the increase may be, it is dwarfed by the smallness of the budget appropriation itself on environmental objectives. An OECD study undertaken about ten years ago (OECD 1992) also pointed out that there was considerable "relabelling" of existing activities as environmental in response to the demands on

9 Another important input indicator is human resources, but this is not discussed in this paper since available data is non-existent.

public policy to respond to environmental issues since mid 1980s. The study also argues that publicly financed environmental R&D does not appear overall, to have increased in line with the increased recognition of environmental threats or even with the development of environmental policies and institutions. In spite of considerable general policy discussion in some countries about an appropriate level of funding, the small share attributed to environmental issues suggests that they have had little influence in wider discussions of S&T funding issues.

The scope of environmental research as defined in the Frascati Manual includes both the identification and treatment of pollution *and* prevention of pollution, but the latter may not entirely correspond to the concept of clean or cleaner technology. Also, data are not normally disaggregated into treatment and prevention. Reporting by objectives, therefore, masks trends at more disaggregated levels, the knowledge of which is important for policy makers.

Table 1. Government Budget Appropriations and Outlays for Environment R&D Levels in Millions of 1995 US$ PPPs and as % of total GBAORD

	Million 1995 US$		198	198	199	199	199	199	199	199	200	200
Australi	199	68	2.7	1.9	3.1	2.9	2.5	2.6	2.7	-	-	-
Austri	200	23	0.4	0.9	1.9	2.5	2.2	2.1	1.9	1.9	2.0	-
Belgiu	199	40	2.8	2.5	1.0	1.7	2.5	2.5	1.9	2.8	-	-
Canad	199	147	1.2	1.9	1.9	3.2	3.2	3.6	3.8	-	-	-
Denmar	200	16	1.8	1.5	3.8	4.4	4.4	2.9	3.7	3.4	2.8	2.0
Finlan	200	27	0.9	1.5	1.4	2.5	2.6	2.3	2.2	2.2	2.3	2.2
Franc	200	229	0.5	0.5	0.7	1.9	2.0	2.1	2.2	1.6	1.8	-
German	200	525	1.8	3.1	3.5	3.6	3.7	3.5	3.4	3.5	3.4	-
Greec	199	14	3.1	3.4	2.8	3.6	3.8	3.4	3.3	3.3	-	-
Icelan	200	3	-	0.1	-	3.4	4.0	4.6	2.9	3.6	3.7	3.9
Irelan	199	4	0.4	0.8	1.2	1.4	1.7	1.6	1.1	1.4	-	-
Ital	199	238	1.8	1.0	2.2	2.4	2.4	2.4	3.4	-	-	-
Japa	200	158	-	-	0.5	0.6	0.6	0.6	0.6	0.7	0.8	-
Kore	199	157	-	-	-	-	-	-	-	3.3	-	-
Mexic	199	19	-	-	1.4	0.6	0.7	0.8	1.0	1.0	-	-
Netherlan	199	118	-	3.2	3.4	3.9	3.7	3.8	4.0	4.0	-	-
New	199	3	-	-	2.6	3.3	-	0.8	-	-	-	-
Norwa	200	26	3.6	2.7	3.2	2.8	2.8	2.9	3.0	3.0	2.8	2.7
Portug	200	45	-	-	3.2	4.4	4.2	4.3	4.2	4.3	4.4	-
Slovak	200	3	-	-	-	2.0	2.4	2.0	1.6	1.4	1.3	1.3
Spai	199	102	0.7	0.4	4.3	2.6	2.7	2.2	2.6	2.6	-	-
Swede	200	21	1.8	1.5	3.2	2.3	2.3	-	0.8	1.6	1.4	-
Switzerla	199	2	2.7	-	1.8	-	0.9	-	0.2	-	-	-
United	199	201	1.2	1.2	1.4	2.3	2.2	2.3	2.5	2.4	-	-
United	200	472	0.8	0.5	0.6	0.8	0.7	0.8	0.8	0.7	0.7	0.6
European	199	1,81	1.4	1.7	2.2	2.9	3.0	2.9	3.0	-	-	
Total	199	2,76	1.1	1.1	1.3	1.7	1.7	1.7	1.7	-	-	

OECD S&T databases, July 2001

For the purpose of policy making other disaggregated data would be necessary. Since public sector research includes most basic research; also, most research done in the universities, indicators and data on how much is being spent on basic as opposed to applied research and how much spent on which areas of environmentally relevant research would be useful. Where the funds are used, universities or in public research institutions or contracts to business is another needed information. If there is significant amount of budget appropriated in environment-related research in other objective areas or generic technology areas, this needs to be more systematically assessed. Also, needed is information about where and by whom the funds are being spent.

5.1.2 Business Environmental R&D

The OECD has very fragmentary data on business expenditures for environmental R&D (table 2). This reveals that for some countries, business expenditures on environmental R&D are larger than government budget appropriations. This implies that business expenditures in the OECD area could well be much larger than public expenditures, since this data set does not include US, Japan and Germany which are the largest producers of environmental goods and services. The business sector in these countries is likely to be investing more than the government. In any case the unavailability of data for these and other larger OECD countries severely limits drawing conclusions about major trends in business environmental R&D.

Table 2. Public and Business Expenditures for Environmental R&D (in Millions 1995 US$ PPPs)

	Business expenditures for environmental R&D		Government Budget Appropriations and Outlays for environmental R&D	
Australia	1998	69.2	1998	67.5
Austria	1981	0.7	1981	2.5
Iceland	1999	0.3	1999	3.3
Ireland	1990	4.0	1990	1.7
Korea	1998	173.1	1999	157.3
Mexico	1995	17.2	1995	7.4
Netherlands	1991	63.8	1991	94.7
Norway	1999	9.0	1999	26.9
Slovak Republic	1999	1.4	1999	3.3
Spain	1999	108.8	1999	102.2
Sweden	1999	3.5	1999	23.9
Switzerland	1996	138.7	1996	12.4

OECD S&T databases, July 2001

For the needs of policy making, business sector R&D indicators and data need also to be disaggregated. Which sectors invest larger share of their total R&D for environmental objectives? How do firms use their R&D, on which environmental problems? How much is spent for research on prevention and how much on treatment? How much is spent on basic research? How much is being spent internally, how much R&D is contracted out, or spent on collaborative research? These are some of the questions that policy makers need to know from appropriate indicators and data on business environmental R&D.

5.2 Output Indicators – Indicators and Data on Environmental Innovation in Business

The second set of indicators and data needed for policy making is determinants of environmental innovation in industrial firms, and the types of innovation generated by firms. More work has been done in this area than for environmental R&D. The main problem regarding these output indicators and data is that methodology used varies considerably, and in many cases comparability of results across sectors and countries is very uncertain.

As discussed above, policy makers need to know the relative importance of the various drivers of environmental innovation in industrial firms, regulatory, commercial or social awareness. They need to know how firms react to various policy instruments. Policy instruments need to be assessed in terms of the extent of incentives for innovation they create for firms. This should include the assessment of industrial preference for voluntary agreements as opposed to regulatory or market-based instruments, even if voluntary agreements are not directly designed by the government.

These types of information can only be obtained by conducting innovation surveys. A number of environmental innovation surveys have been conducted in several countries. The problem is that the methodology used differs considerably as discussed above. Hence in many cases, it is difficult to determine if the differences in results are genuine or if they are caused by differences in methodology. Since policy makers do a lot of learning from experiences of other countries and sectors which they are not directly responsible, designing of standardised methodology is a genuine need for policy makers.

As discussed above, another dimension of environmental innovation in industrial firms that policy makers need to know are the costs and benefits involved in improving environmental performance. It is clear from the innovation surveys conducted so far that whatever the incentives for environmental innovation, firms are ultimately interested in reaping win-win opportunities. Information about costs and benefits of environmental innovation can be obtained if firms conduct environmental accounting. This is being encouraged by environmental policy makers, and in recent years many firms have adopted environmental accounting systems. Here again, standardisation of methodology is an issue that needs to be addressed. Also, it should be noted that an important part of the costs of environmental innovation is R&D investments. Patent data would constitute an important indicator of set of output of environmental innovation.

Innovation surveys also should elucidate the "knowledge value chain" for environmental innovation in various sectors. Who supplies the most relevant knowledge for environmental innovation, downstream or upstream firms, public research institutions? What kind of knowledge are the firms looking for? Addressing these questions are key for policy makers in designing appropriate environmental innovation systems.

6. Conclusion

There is clearly a need for better indicators and data on environmental innovation. At least two types of information is necessary, environmental R&D expenditures and determinants of environmental innovation. Existing indicators are far too insufficient. Indicators and survey methods need to be better defined and better standardised, so that the results can be compared across sectors and countries. There is a need for intensified and co-ordinated national and international efforts in building indicators and collecting data on environmental innovation.

Innovations in the Environmental Policy System: Voluntary, Collaborative and Information-Based Policies in the United States and the Netherlands

Theo de Bruijn and Vicki Norberg-Bohm

1. Introduction

Fundamental changes in production and consumption systems are required in order to meet the needs and aspirations of a growing world population while using environmental resources in a sustainable manner (IHDP Industrial Transformation Science Plan 1999). The necessary industrial transformation goes beyond the notion of eco-efficiency. It is about system innovation, both technological and institutional (see for instance Clark and George 1995; Huber 1995; Schot et al. 1997). The real challenge for the coming years is to redesign industrial systems (McDonough and Braungart 1998). This is a particularly complex task, given the numbers of actors involved, the number of linked, multiple technological and social options, the innovations and new practices that need to be undertaken and the uncertainties that have to be considered (Roome 2001). Over the past 15 years, a number of innovations in environmental policy have been undertaken in efforts to meet this challenge.

When environmental degradation emerged as a priority for government action in the early 1970s, most industrialized countries enacted media-specific legislation based on direct regulation resulting in a set of ambient, emission, and technology standards that were enforced through permitting systems. Although direct regulation has been a powerful tool for adjusting industrial behavior, it has been criticized for being incapable of addressing the challenges of sustainability. Its main shortcomings, from the perspective of stimulating industrial transformation, are (i) shifting pollution from one media to another rather than eliminating pollution, (ii) constraining innovation, and (iii) reinforcing the adversarial relationships between the public, private and non-profit sectors. This last point is relevant because, given that much of the capacity to undertake a transformation lies within industry, the private sector will need to be engaged proactively in environmental protection in order to move down the path to sustainability.

The United States and many European countries have developed new approaches to overcome these shortcomings, including both market-based approaches (economic incentives) and the voluntary, collaborative and information-based approaches examined in this chapter. This latter group of programs represents an attempt to engage industry in significant environmental improvements through dialogue, consensus-building and voluntary action rather than the imperatives of direct regulation or the incentives of market-based approaches. The rationale for these approaches lies not only in the shortcomings of direct regulation, but perhaps more importantly in the complexity and severity of environmental problems, which necessitate a redefinition of the scope and methods of environmental policy.

This chapter draws on a larger project, which included twelve innovative environmental policies[1]. Overall, we see rather successful programs as well as relative failures. Our standard for evaluation is the extent to which the programs could stimulate substantially improved environmental performance, with a particular emphasis on stimulating the private sector to invest in technological innovation to achieve this goal. In this regard, we are particularly interested in the ability of these programs to create niche markets for radical innovation. The notion of radical innovation is, however, often ex-post (Geels 2002). A radical innovation may start small and gradually, when the actors involved recognize the potential, develop into something large. What makes an innovation into a radical one in the context of industrial transformation is its ability to change the technological trajectory in a direction that results in significant environmental improvement. Niche markets play an important role in this respect, providing steppingstones for the maturing and diffusion of radical new technology (Dosi 1982; Kemp et al. 1998; Nelson and Winter 1982). In niche markets, alternative technologies can be developed and tested that may later change the direction of technological development (Kemp et al. 1998, 2001). Niche markets develop as a result of actors willing to look for novel solutions. An important goal of environmental policies in the pursuit of sustainable development therefore is to motivate "first movers", i.e. firms willing to take the risk of investments in entrepreneurial and technological innovations that substantially reduced environmental impacts. Once a new product or process is developed and demonstrated, it can be passed from the leading to the lagging firms. The motivation of first movers, therefore, is of special interest in the context of environmental policy to promote industrial transformation.

In this chapter we compare three U.S. programs to three Dutch programs in order to understand why seemingly comparable approaches are successful in one context yet fail in another. Drawing on new institutional theory and the literature on technology policy and management, our analysis shows the need for a careful examination of the ways in which any policy innovation can either work within or change the existing regulatory structure. Two factors stand out as being particularly influential - providing incentives for change and "fit" with the national policy style.

The chapter is organized as follows. The following section describes the characteristics of voluntary, collaborative and information-based strategies and discusses the role that they can play in technological innovation. The third section presents three U.S. and three Dutch programs, discussing their basic structure and achievements. The fourth section compares and contrasts these programs, examining the factors influencing success. The paper ends with conclusions on the role of voluntary, collaborative and information-based strategies for promoting techno-

---

1 The project resulted in a workshop report (De Bruijn and Norberg-Bohm 2001) An expanded and revised version of the arguments put forward in this chapter will be appear in an edited volume (De Bruijn and Norberg-Bohm, forthcoming). We are grateful to our authors for their contributions. While we build on their analyses we take responsibility for the views expressed in this chapter.

logical innovation, focusing on their role within the broader environmental system.

2. Fostering Innovation through Voluntary, Collaborative and Information-based Strategies

At first brush, the group of programs in our analysis is defined best by what they are not: neither direct regulation nor economic instruments. But this distinction is more than simply a negation, as these programs were all conceived as an alternative or significant addition to the existing command-and-control system (which has been judged inadequate) and to environmental taxes (which often had limited support from industry or the public). While the policy innovations discussed and compared in this paper had a varying mix and approach to the key characteristics that identify this set of innovations - voluntary action, collaboration, and information generation and disclosure - all of them relied on a combination of these elements for effectiveness. Furthermore, they were created with high expectations of their ability to use these new approaches to stimulate significant improvements in the effectiveness and efficiency of environmental protection.

In terms of voluntary action, some programs were strictly voluntary, allowing firms to choose whether to participate without sanctions for non-participation. Other programs encouraged voluntary action, but provided regulatory back-ups if firms did not step forward voluntarily. A third way in which voluntary action was important is for the information disclosure programs, where disclosure was mandatory, but environmental improvement depended on voluntary action.

In terms of collaboration, many of the programs evaluated in this paper were based on a collaborative model, bringing together a range of stakeholders for goal setting and/or implementation. In contrast, information disclosure falls more within the traditional relationships between regulators and firms, but in the United States resulted in increased collaboration as many firms found it in their own best interest to develop community advisory boards in the wake of the disclosure of new information.

Turning last to information-based approaches, even the programs that did not mandate disclosure relied on increased information flows for transparency in negotiation and implementation, as well as for capacity building.

Looking across this set of programs, it was not only the ways in which they combined the elements described above, but rather the broader context in which these programs operated that determined success or failure. Firms do not respond to a specific program in isolation; rather their response depends largely on the environmental policy system as a whole. Thus, the context of a program is as relevant to its successful implementation as the design of the program itself. Programs that are not designed to fit with and complement the other elements of a nation's environmental policy system are likely to be less successful (Bressers and Klok 1988; Jänicke and Weidner 1996; De Bruijn, forthcoming 2003). In relation to context, these cases suggest two important elements for successful policy change:

the need for adequate incentives for change and "fit" with the national policy style.

Although some leading firms are now considering environmental sustainability in product and process design, historically speaking, and currently for most businesses, there has been a preference for incremental innovation and end-of-pipe solutions to environmental problems (Gottlieb 1995; Hoffman 1997). Firms always face large uncertainties and thus risks for investments in radical innovation. These risks can be even greater for environmentally sustainable innovations because of the role that government policy plays in market creation and a lack of core capabilities in this area. The innovative approaches in this chapter can only be effective to the extent that they decrease these uncertainties and risks, creating opportunities for firms to profit from their investments in superior environmental performance and technological innovation. To accomplish this, these programs must either change the competitive environment of firms, or change firms' perception of their competitive environment.

These innovative approaches can succeed in doing this in two ways. First, they may enhance the ability for first movers to profit from investments in superior environmental performance. Researchers in environmental management and policy have identified a handful of pathways through which this can occur (DeSimone and Popoff 2000; GEMI 1999; Hoffman 2000; Reinhardt 2000). The types of innovative programs examined in this paper have the potential to enhance the ability of firms to pursue three of these pathways: product differentiation, cost savings through increases in resource efficiency (e.g. pursue "win-win" approaches), and improved management of environmental risk.

Second, they can provide the characteristics through which regulation is known to elicit an innovative response, including strictness, reduction in uncertainty, flexibility, and information generation[2]. Unfortunately, experience suggests that strictly voluntary programs (a characteristic shared by many of the cases discussed below) are best suited to providing firms with flexibility in implementation (i.e. finding the best solution for reaching goals) and information generation. Strong incentives are found only external to many of these innovative programs. These policy innovations are thus likely to be most effective when used synergistically with or as a complement to other policy approaches that provide incentives (or disincentives) to action.

Turning next to the question of fit, to succeed is not simply a matter of changing legislation, but rather requires that these new approaches set in motion a process that changes organizational structures, expertise, and working routines in government and firms. Existing institutions, however, limit the range of options (March and Olsen 1989). Thus, while change is possible, it takes a good deal of pressure to produce that change and the range of possibilities for change is constrained by the institutional context (Peters 1999). As a consequence, while the programs in this chapter might represent promising alternatives to current national

---

[2] This summary of the characteristics of regulation that stimulate innovation come from a number of studies, including Ashford 1993, Ashford and Heaton 1983, Porter and Van der Linde 1995.

systems of environmental policy, it is also the existing systems that constrain their design and implementation, and ultimately their effectiveness.

Policy style, defined as a "system of decision-making, different procedures for making social decisions" [3], creates the parameters into which new policies must fit (Van Waarden 1995). Misfits occur when a new program assumes a certain style that is contrary to the existing policy style. Consultation and collaboration, central characteristics of many of the programs, are likely to flourish better in a more corporatist context characterized by pragmatic bargaining and consensus building between administrative and societal actors than in a more adversarial system. Conversely, information disclosure may be a stronger impetus for change in societies with well developed interest groups that have access to the media, courts and other points of pressure, i.e. in more adversarial cultures. Trying to implement a program that does not fit the current national policy strategy requires more than a slight adjustment of current practices, rather it can require changing the very nature of institutions, which in many cases requires changes in legislation. It is near impossible for single, small programs that do not fit the national style in environmental policies to catalyze such fundamental changes.

In sum, there are numerous ways in which these innovative policy mechanisms, often in conjunction with other programs and policies, can stimulate firms to become first movers, and in this way initiate a process of industrial transformation. By incorporating a winning combination of the characteristics of stringency, uncertainty reduction, flexibility and information generation, these mechanisms can create an external environment in which firms choose to solve environmental challenges through radical innovation. By incorporating elements that support corporate efforts in product differentiation, eco-efficiency and environmental risk reduction, these programs can help firms create an internal environment in which technological innovation is a profitable business strategy. However, programs will be less likely to succeed when they do not fit with and complement the other elements of a nation's environmental policy system. Furthermore, efforts to make fundamental changes in the existing environmental policy system face great resistance from existing institutions and interests. Because the programs in this chapter are only one part of the environmental policy system, their effectiveness ultimately depends on the policy context.

---

3 This definition is derived from Richardson (1982) who further speaks of 'standard operating procedures' for the government's approach to problem solving and the relationship between the government and other actors in the policy process (idem: 13). A policy style shows itself foremost in the pattern of interaction between administrative and societal actors, which can either be formal and closed (interventionist style) or can be characterized by pragmatic bargaining, consensus and transparency (mediating style) (Knill and Lenschow 1998).

3. Cases of Policy Innovations in the USA and the Netherlands

In this section we describe several relatively recent policy innovations from the United States and the Netherlands based upon voluntary, collaborative and information-based strategies. These innovations include industry sector collaborations, programs that sponsor and promote the implementation of environmental management systems and environmental disclosure programs.

3.1 Industry Sector Collaboration

The first category concerns programs that engaged industry sectors rather than individual firms. The basic rationale of these programs is that a collaborative, consensus-based approach is expected to result in more effective and efficient solutions for setting and achieving challenging long-term environmental goals.

The Common Sense Initiative (CSI) is the prominent example in the United States of a sector-based, collaborative approach to environmental policy (Coglianese and Allen 2001). It brought together representatives from six industrial sectors and sought to forge a consensus within each sector over innovations in environmental management and policy. CSI operated from July 1994 through December 1998, seeking to take advantage of the in-depth knowledge by firms and organizations within specific industrial sectors to develop "cleaner, better, cheaper" approaches to environmental control. Its goals were to overcome the problems of the media-specific, adversarial command-and-control system by fine-tuning environmental regulation to the specific circumstances of different industrial sectors. By bringing together industry, environmental groups, and other interested parties within a sector, the agency sought agreement on new and better ways of achieving environmental performance goals. Under CSI six subcommittees (one for each sector) in which representatives from industry, NGOs, labor unions, and governments were represented, had to make recommendations to an overarching council. The council renewed these, and sent the recommendation it endorsed to EPA for implementation. The tangible results of this innovative initiative have been at best quite modest. CSI clearly had an ambitious vision (Fiorino 1999) and some within the agency believed it held the potential for much flexibility. But by the end of the 4-year initiative, only about 4 of the approximately 30 subcommittee recommendations that were endorsed by the CSI Council and submitted to EPA for policy change have led to actual revisions in EPA regulations. Moreover, relatively few of the project accomplishments, according to the agency's own reports, have achieved technological innovations. The contribution of CSI to industrial transformation, therefore, is extremely limited. This voluntary, consensus-based program was not able to motivate first movers.

The Dutch Target Group Policy, introduced in 1989 as part of the first National Environmental Policy Plan, is the central element in the current Dutch system of industrial environmental regulation (Hofman and Schrama 2001). Instead of setting technology-forcing standards unilaterally the approach builds on close collaboration with industry. Through negotiations between sectors of industry, the Ministry of the Environment, and regional and local governments, agreements are sought concerning the contribution of specific industrial sectors to the goals laid out in the National Environmental Policy Plan. These goals aim for 50-90 percent emission reductions for specified pollutants. Since 1989 many agreements have been reached, including 11 broad-based agreements with sectors of industry. While some of the agreements are quite demanding, sectors may opt-out if technology does not develop at a pace that enables them to reach the agreed upon goals. In the program, links have been built with government-sponsored technology development programs in order to stimulate the R&D needed for the technology development that will be necessary to reach the Target Group's stringent long-term goals. Furthermore, the Target Group implementation is linked to the permit system to assure that new technologies get adopted. If companies do not comply voluntarily they will eventually be forced to do so by local regulators. The diffusion of the state-of-the-art technology through these negotiated agreements seems rather successful up till now. It is, however, highly dependent on direct regulation in forcing laggards to adopt new technologies. One advantage of the target-group policy is that industry can time the development and implementation of product and process changes to coincide with investment cycles. Another positive attribute is that in the process of eliminating bottlenecks to significant improvements and of searching for new technological options, actors are more likely to engage ideas from other actors and discover the potential for collaboration. Overall, the Target Group policy has shown some positive developments. Whether it will lead to radical (technological) breakthroughs is still questionable; it has created positive conditions for first movers, which is an important initial step.

3.2 Environmental Management Systems

The second category of programs focuses on the use of environmental management systems (EMS) as part of programs that create capacity and incentives for improved environmental performance. As we will see, the Dutch and the U.S. approach differ considerably.

Tiered systems of environmental regulation are a new approach that agencies have developed to encourage companies to strive toward higher levels of environmental protection. Under a tiered approach, regulators invite facilities to institute programs that go beyond regulatory requirements in return for a range of benefits including increased flexibility in meeting environmental standards. Tiered systems have been adopted in 12 states and by the U.S. Environmental Protection Agency (EPA). StarTrack is an example of using the adoption of EMS as part of a tiered system of environmental regulation (Nash 2001). StarTrack was run by the EPA's Region I (New England) office during the late 1990s. For firms to enter the StarTrack program, they had to have a history of compliance and pollution prevention, an EMS or a commitment to adopt an EMS, and a commitment to continued improvement in environmental performance. As part of StarTrack agreements, facility managers promised to undertake audits of internal management and compliance performance, have these audits certified by an independent third party, and publish performance reports. In exchange, EPA managers promised to forego inspections, offer penalty amnesty, provide faster permitting, and publicly recognize StarTrack facilities as environmental leaders. In the end, StarTrack did little to motivate environmental protection in participating firms. In the view of the private sector managers who took part, the program was mainly a paperwork exercise they undertook to garner EPA recognition of established environmental management practices. StarTrack facilities' environmental performance did improve during their participation in the program, but not as a result of their participation. Facilities that met program entry criteria were managed by people who had already invested in environmental performance improvement, and were committed to continue to do so. The benefits EPA provided program participants were meager, and less than the agency had promised. As a result, StarTrack attracted only a handful of participants and was not a critical factor in motivating improved environmental management capabilities or environmental performance.

In the Netherlands EMS are not formally part of a tiered system of environmental regulation. A Memorandum, published in 1989 by the Netherlands Ministry of the Environment, aimed at having all firms implement an EMS by 1995 (De Bruijn and Lulofs 2001). It was, however, a voluntary program. No sanctions were set in the short run for companies who wouldn't implement management systems, other than stating that they might be subject to more and severe enforcement. No specific benefits were provided either, but later on some flexibility during permit procedures was promised. Although the ultimate goal of the program was to contribute to improving the environmental performance of companies, its main programmatic objectives were to generate mutual trust for government-industry collaboration, to enhance capacity building within industry, and to involve third parties in promoting environmental protection. Instead of dealing with SMEs directly, the Dutch government facilitated the formation of networks in which intermediary organizations, especially trade associations, acted as agents for change. The program design proved to be quite effective; networks were established that engaged in collaboration with individual facilities. The majority of firms carried out some of the activities asked for and several hundred implemented a full scale EMS. The program, therefore, was quite successful, although some firms would probably have implemented an EMS anyway. In terms of improvements in the environmental performance of firms, those that took place can probably be attributed to a large extent to other programs, such as the Target Group Policy. The EMS program can be credited with improving the capability of firms to comply with these other programs. The direct environmental benefits of the program, therefore, seem fairly limited. Its main contribution is increasing the capacity for change in a supporting role for other programs such as the Target Group Policy.

3.3 Environmental Disclosure

Our third category of programs concerns mandatory environmental information disclosure. The central reasons why these programs may contribute to improved environmental performance are twofold. First, they may create external pressures on firms through negative publicity about emissions and the ability of interests groups to use information on emissions to pressure for change. Second, they may stimulate firms to develop new information on their emissions, thus identifying previously unknown win-win opportunities.

The U.S. Toxics Release Inventory (TRI) is the prominent example of an environmental disclosure program (Graham and Miller 2001). The TRI was created in 1986 as part of the Emergency Planning and Community Right-To-Know Act. After several amendments during the 1990s, the TRI now requires most medium and large-scale manufacturing firms to provide facility level data on releases of 602 chemicals to all media (air, water and land), as well as on-site and off-site storage, treatment, disposal, recycling and energy recovery. It also requires firms to report qualitatively on source reduction activities and to provide a production index, so that changes in releases and transfers can be related to changes in production. The TRI is heralded as a major success, and an important contributor to a 45 percent reduction in releases of listed chemicals by 1998. The TRI, however, cannot be given credit for this entire decline. A variety of regulations enacted since 1986, as well as other factors, have influenced firms' decisions to reduce toxic emissions. Furthermore, the environmental significance of this decline requires interpretation. Relatively few facilities cut releases by reducing waste at the source; rather recycling increased substantially, although of course positive examples of preventive action do exist. Also, releases declined at a much more rapid rate in early years, raising questions about the long-term impact of this policy approach. Despite these caveats, the TRI was clearly path-breaking legislation that has contributed to toxic emission reductions and provided lessons for the information disclosure policies that followed. Overall, TRI shows positive environmental results but its effects on creating first movers or radical innovation is less obvious. Nonetheless, creating information both for external stakeholders and for internal decision-making is a necessary pre-condition for industrial transformation.

The Netherlands introduced legislation on environmental disclosure in 1997. Although the motivation for the legislation specifically referred to the success of the TRI, its design is quite different. At present, 250 firms are required to publish two yearly environmental reports, one for the government and one for the general public. To a large extent, the firms that fall under this obligation belong to the same industrial sectors that are subject to the Target Group Policy. The report for the government is regulated to some extent. Compulsory models, which contain overviews of the substances and processes for which firms need to provide quantitative data, have been developed for eleven sectors of industry. There is no third party verification but the legislation requires the data to be based upon reliable internal systems, such as in ISO 14001. Governments also have the authority to audit the reporting systems and the resulting data. Nonetheless, official evaluation studies state that the quality of the data is inadequate. As a result, the reports are hard to compare. For the public report, the law does not provide specific information requirements, nor does it require third party auditing, although 10 percent of the firms have done so voluntarily. In short, firms have considerable discretion in the public report. The vast majority of the 250 firms publish a public report but they are rarely read. The effects of the program on the firms environmental performance have not been evaluated fully yet, but given the low profile of reporting in the Netherlands, the expectations should not be too high. The contribution of the Dutch program to industrial transformation is therefore limited.

4. Understanding the Successes and Failures

As is predictable from the discussion in section 3, the Netherlands and the United States differed substantially in the success with which they used the policies described above. Industry sector collaboration, adoption of environmental management systems, and information disclosure, while similar categories of programs, were actually designed quite differently in the two countries. The ability to design innovative programs in these categories that fit with the policy system and that contained and/or were linked to incentives for action also varied considerably in these two countries.

The two industry sector collaborative programs share some important characteristics but also show key differences that resulted in quite opposing outcomes. In the case of the Dutch Target Group program, government established long-term targets that were translated into relatively clear goals. This facilitated implementation, with industry and governments jointly developing plans for meeting these targets. Thus, in this case, strong incentives were internal to the program. Furthermore, the program had a good-fit with the existing national policy style that can be characterized by a long tradition of collaboration, dialogue and consensus building (Liefferink 1997; Bressers and Plettenburg 1997). Notwithstanding the accomplishments of the program, there is some concern that the long-term goals will not be met, despite the linkage with R&D programs.

The U.S. Common Sense Initiative (CSI) lacked the features that created success in the Dutch case. This voluntary, consensus-based approach was largely a failure, providing little incentive for first movers or for industry to reveal valuable information that would improve the government's steering capacity. Several aspects of the program and the larger policy environment contributed to this failure. First and foremost, CSI did not fit well with the adversarial and inflexible U.S. policy regime. The adversarial system creates high risks for firms to voluntarily reveal information about their environmental behavior, and the limited flexibility for implementation meant that the program did not have legal authority to implement innovative solutions proposed during the collaborative process. Furthermore, the program's lofty goals were not translated into clear targets. There were no government-imposed targets (or the threat to impose targets) and the voluntary, consensus-based approach was not able to achieve this.[4].

Turning next to EMS, the manner in which the Netherlands and the United States promoted the use of EMS differed substantially. The Dutch program on EMS is relatively successful for two reasons. First, it is part of a long-term strategy through which firms are encouraged, in a stepwise fashion, to develop and improve their capabilities. Specifically, the strong long-term goals in the Target Group policy provided a signal to firms that they would need to develop the ability to reach these goals. Second, the implementation of this program, through close collaboration between trade associations and the government, fits the corporatist structure of the Dutch society. Nonetheless, caution must be taken in attributing improvements in environmental performance to this program; although the program introduced EMS quite successfully, its contribution to actual improvements of the companies' environmental performance remains questionable. Since the program itself holds only process and no performance requirements the drivers for change had to be found externally.

In contrast, in the United States, EMS has found its way into government policy as part of voluntary programs to create "beyond compliance" and "superior" environmental performance. In this role, it has been used as one indicator of a firm's commitment to being an environmental leader. For example, the StarTrack program challenged firms to voluntarily improve environmental performance in exchange for flexibility in meeting existing environmental requirements, fast-track permitting, reduced monitoring, and recognition. On the whole, the program did not provide adequate benefits to stimulate proactive responses by firms. Recognition as a "StarTrack" company was not particularly valuable, and the EPA was not able to provide the other promised benefits, in large part due to a poor fit with the U.S. regulatory system. Furthermore, there were not clear targets, as "beyond compliance" and "superior performance" were not defined at program inception, and remained sources of controversy amongst stakeholders[5].

In contrast to the two types of programs discussed above, information disclosure programs have had a bigger impact in the United States than in the Netherlands. The TRI was effectively implemented for a number of reasons: information disclosure was required by law, the requirements for disclosure were clearly specified, and the database gained a reputation for accuracy and legitimacy. The TRI also fit well within the U.S. policy system, which has a tradition of information

---

4 The United States had better success with several other more narrowly targeted collaborations, including the Energy Star program (Paton 2001) and R&D collaborations (Norberg-Bohm and Margolis 2001). In these cases, while collaboration remained voluntary, government had a more active role in goal-setting, the projects did not challenge the underlying standards-based regulatory regime, and the programs were able to provide and/or link to a set of incentives for private-sector participation.

5 Efforts are underway to overcome these shortcomings in a national program called Performance Track. If it is successful, then this initial program could be viewed as a step along the way. Performance Track will also have difficulty providing regulatory flexibility due to the U.S. environmental regulatory system. This same problem appeared in Project XL (Marcus et al. 2001).

disclosure, and of protecting competitively sensitive information while facilitating public access to information. Furthermore, the United States has a well-developed set of stakeholders, at national and local levels, which use information as part of their political strategy - creating economic and political pressure for change.

This context is very different in the Netherlands. The Dutch program has less specific requirements for disclosure, especially for the public report. Most problematical though is that the reports receive scant use outside of a limited set of professional groups such as academics. The active monitoring and pursuit for change of a firm's behavior by citizen's groups is not part of the Dutch culture. This reduces the chances for effective information disclosure programs.

5. Conclusions

Taken as a whole, the six programs examined in this chapter as well as the other programs we evaluated in the overarching project demonstrate more success than failure. But even successes are limited if measured against the radical innovation needed for industrial transformation. To the extent that programs have stimulated innovation, it was more often incremental than radical. While there is evidence of private sector leadership, there is concern that it may be one-off rather than on going, and focused on near-term opportunities rather than longer-term and more difficult targets.

The previous section explored the differences and similarities in the design and implementation of these new approaches. It showed the importance of the integration of the new approaches with the rest of the environmental policy system. Programs that do not fit with and complement other elements of the environmental policy system will likely be less successful. While all the programs represent efforts to overcome the limitations of the "command-and-control" system, many with ambitions of creating new regulatory regimes, changing the existing approach to environmental regulation is a tall order. Yet, change is possible. In the Netherlands, over the course of a decade the basic policy approach has been changed quite fundamentally. Through the late 1980s, the Dutch government relied almost exclusively on direct regulation supplemented by some taxes for water pollution. Today the target group policy, with its emphasis on collaboration and negotiation, stands central. In the Netherlands, the government was able to draw on the strong neo-corporatist traits of the Dutch society when changing the core features of its policy system. Both representatives of industry and governmental actors were willing to look for a way out of the traditional regulatory system, although for different reasons. Without a broader context of collaboration and both parties willing to work towards an alternative system, this change would have been very difficult.

The United States had less success with its industry sector program, as well as many other voluntary programs that bumped-up against the rigidity of the current laws and enforcement culture of the EPA (Marcus et al. 2001; National Academy of Public Administration 2001). In contrast, information disclosure in the United States has created more environmental progress, as it fit well with a culture that values access to information and has well-developed interest groups that can use this information to press for change within the adversarial environmental policy system.

The key lesson is not that all countries should follow the Dutch example for industry sector collaboration and EMS, or the U.S. example for information disclosure. While these programs have contributed significantly to environmental improvement, they have all been implemented in ways that suggest both strengths and weaknesses. Rather, the lesson from this exercise is that there is a need for a careful examination of the ways in which any policy innovation can either work within or change the existing regulatory structure. Notwithstanding the potential advantages of voluntary, collaborative and information-based approaches, these cannot be effective unless designed to work synergistically with the larger policy system. In some cases this will require legislative changes; in others a carefully design package of programs and policies that can build capability and provide incentives for action. We conclude, therefore, that there is not one way for environmental policies to stimulate the fundamental innovations necessary for industrial transformation. It is the environmental policy system as a whole that must respond to this enormous challenge.

Voluntary, collaborative and information programs can play a useful role in such a comprehensive strategy but only if they are carefully designed to fit with and complement the other elements of a nation's environmental policy system. In the end, the real question therefore is not whether the new approaches should be used, but rather *how* they should be used. Regardless of goals, there will remain a role for direct regulations and market-based approaches as part of an overall strategy - they will be needed to create sufficient pressures to push industry along the path toward sustainability.

The IPPC Directive and Factors Influencing the Economic and Environmental Performance of Firms and Plants in the Cement, Non-Ferrous Metals and Pulp and Paper Sectors in the EU[1]

David Hitchens, Frank Farrell, Josefina Lindblom and Ursula Triebswetter

1. Introduction

The IPPC Directive is concerned more with the diffusion of presently available environmental techniques than with the innovation process, which is a main consideration of this book. The aim of the IPPC Directive is to achieve a high level of protection of the environment as a whole by preventing or reducing the pollution emanating from industrial installations directly at source, through the use of best available techniques as described in BAT reference documents (BREFs). Member States had to bring their national legislation into line with the IPPC Directive by the end of October 1999 and the Directive must be applied to existing plants by October 2007.

2. BAT Reference Documents

The Directive does not require BREFs, the BREFs facilitate the requirements of the Directive e.g. by allowing competent authorities to be informed of developments in BAT. The BREFs consist of "vertical" sector specific and "horizontal" subject specific BREFs, the first is concerned with identifying BATs appropriate to a particular industry, the second is concerned with cross cutting issues such as economic and cross media impacts, which are relevant to all sectors.

---

1 The study was carried out under the auspices of DG Enterprise by the Institute for Prospective Technological Studies (IPTS), Seville, Spain. The full report is based on a synthesis report and a number of industry studies and appendices. These are referred to in this article. The European IPPC Bureau (EIPPCB) is also located at IPTS and this has faciliated meetings with the individual BREF authors, access to background material underlying the BREF documents, attending Technical Working Group (TWG) meetings and so on. The investigation has been helped and encouraged by IPTS and in particular by Luis Delgado and Per Sørup. Inputs and help with the research have been received from Michalis Vasilopoulos. The team has also liased with European industry associations and their environmental committees and various national associations. Expert advice has also been received from research associations and company personnel.

The directive forbids authorities to prescribe the use of any specific BAT in permits issued to operators. Instead, permits must contain conditions, such as emission limit values, which are sufficient to ensure that BAT requirements are met taking account of the particular characteristics and circumstances of the installation. This flexible approach recognizes the fact that different techniques can be combined to achieve equivalent environmental performance.

In this way IPPC also promotes innovation. "The starting point of the IPPC approach is that continuous process innovation, in combination with resource management and enforcement of environmental quality standards, will lead to both sustainable development and economic growth." (Gislev 2000)

While BREFs are interested in identifying BAT, and to do so they consider many possible techniques, both process integrated and "end-of-pipe" techniques, operating practices, inspection routines, maintenance systems, process control methods etc. from which to draw BAT. They also identify novel pollution prevention and control techniques that are reported to be under development and may provide future cost or environmental benefits, thus attempting to show where the frontier lies for clean technology and other environmental developments in any particular industry.

3. Competitiveness Effects of BAT

Identifying BAT, from the pool of techniques requires, *inter alia*, that they are **'available`**, this means that they are developed on a scale which allows:

"implementation in the relevant industrial sector, under economically and technically **viable** conditions, taking into consideration the costs and advantages, whether or not the techniques are used or produced inside the Member State in question, as long as they are reasonably accessible to the operator"

Viability has for the purposes of this research been defined as the ability of the firms or plants to maintain *competitiveness* in the long run after the adoption of BAT, while sectoral competitiveness is defined as the industry maintaining the level of sectoral output as hitherto. In this study a range of indicators are used to measure competitiveness. These include output measures of performance – profitability, productivity and growth – and input measures of performance, namely: physical and human capital, R&D spending etc. (Measurement of competitiveness is debated Krugman (1996), alternative measures are discussed in Hitchens (2001)). What are the expected influences on firm competitiveness of environmental regulation (OECD 1993)?

3.1 Negative Impacts

The impact on competitiveness, or firm output and employment, will be greater where demand is sensitive to price increases (is price elastic); where firms face strong competition (from countries where regulation is less stringent); where the environmental compliance costs rise and the differential cost penalty relative to domestic and external competitors is greater; where margins and profits are tight; and where environmental costs rank high among the threats facing the firm. The reverse circumstances will lessen the impact of regulation.

3.2 Positive Impacts

However, environmental regulation may also have positive competitiveness implications for some sectors and may encourage firms to develop more resource-efficient methods and to reduce costs. Environmental regulation can yield competitiveness benefits through (1) stimulating innovation (2) improving efficiency (3) creating comparative advantages (4) spinning off new production activities and advantages.

4. Previous Work in the Area

What do previous studies in the area tell us about the impact of regulation on competitiveness and about factors that facilitate or inhibit the adoption of environmental initiatives? At current levels of regulation there does not seem to be any serious trade-off between competitiveness and environmental protection:

Jaffe, Peterson, Portney and Stavins (1995), in a survey article, concluded "... there is relatively little evidence to support the hypothesis that environmental regulations have had a large adverse effect on competition". Or, as Porter and van der Linde (1995) put it, "... it is striking that so many studies find that even the poorly designed environmental laws presently in effect have little adverse effect on competitiveness".

Why do environmental policies have negligible effects on competitiveness? Probably the most important reason is that the cost of complying with regulation is a small fraction of total costs, sufficiently small to be overridden by differences in labour costs, exchange rate variations and so forth (OECD 1993). Second, although stringency varies between countries, the differential in compliance costs between major trading partners is unlikely to be large. (In the present case IPPC is also concerned with an incremental change.)

Technological improvement helps to compensate for increases in the severity of regulations; firms may be starting from a position with some super-normal profits (this means they have some capacity to absorb increased costs); there may be partial substitution away from the more expensive factors of production and, lastly, environmental compliance costs are typically less than one per cent of total costs.

At the same time, it would be unwise to generalise from the evidence available that regulations in general boost the international competitiveness of a region/nation.

It is very important to note that at least up until now, the power of environmental regulation to do a great deal of harm or good to company competitiveness within the EU has been limited. This is not to imply that further upward pressure on standards of regulation is unlikely to have much by way of trade-off with competitiveness. To the extent that there is a trade-off between environmental outcomes and company competitiveness, we would ideally wish to know how strong this is. Additionally, if the political judgement is that environmental outcomes should be attained even at the cost of diminished competitiveness, then the aim would be to design policies where this cost is minimized.

4.1 Hypotheses Tested

The research focused on a set of hypotheses (Hitchens 2001; Vassilopoulos 2001a, 2001b). The basic hypothesis is that the implementation of BAT could place firms at a competitive disadvantage and lead to the loss of markets, particularly to countries with less stringent regulation. The regulated firm needs to redirect resources from other profitable opportunities, costs and prices rise, and markets and customers may be lost.

On the other hand the implementation of BAT, although it may represent a short-term cost and burden to the firm, could push firms on to a higher growth path by forcing them to make product and process changes that yield higher competitiveness (Porter 1990; Porter and van der Linde 1995).

In fact the relationship between BAT and competitiveness is likely to be two-way: the fact that the firm is competitive may lead to the early implementation of environmental initiatives while at the same time environmental initiatives are expected to have consequences for the competitiveness of firms. Competitive firms may have strengths in R&D, skills, modern equipment and other factors which reduce the cost of compliance with BAT relative to less competitive firms in the industry.

This study seeks to capture both these negative and positive factors which influence the costs or benefits arising from an adjustment to the implementation of BAT.

4.2 Methodology

The general approach (Hitchens et al 1998, 2000) is based on case studies and examines the actual experience of plants which have adopted BAT compared with 'matched' representative plants in the industry. There are three stages:

1. Tests are made of the competitiveness performance of plants which have adopted BAT.

2. Tests are undertaken to identify special or unique factors associated with BAT plants based on a matched plant analysis.
3. Plants at risk of closure following the implementation of IPPC are identified.

BAT is itself a variable, hence a BAT plant is defined as one which has adopted most of the elements of BAT as stated in the BREF. The representative plant will have some BATs and the assumption is made that they will require the installation of all BATs. This is a strict assumption. The BREF preface states the need to take account of local considerations, while the BREF itself is concerned with generic BAT.

Questionnaires[2] were constructed in order:

1. To ask managers about the economic effects of adopting environmental initiatives. For each BAT, managers are asked a series of questions on the impact of that initiative on plant performance. Answers were backed up where possible by records and annual audited accounts.
2. To measure the impact of compliance costs or the implementation of initiatives on overall firm performance, as judged by a comparison between matched plants.
3. To relate the importance of environmental costs to other factors influencing the firm's competitive performance. Respondents were asked to specify the competitive advantages and disadvantages they faced, including those arising from environmental regulation and costs, again backed up by evidence where possible.
4. To analyse the influence of human and physical capital, R&D and plant size on compliance costs and ease of implementation of BAT.
5. In the case of the pulp and paper industry an additional (macro) approach based on the expected investments required to meet 80% of BAT requirements by all EU plants and the consequences for plant closure and displacement by imports was undertaken[3].

The methodology has been applied to three sectors for which finalized (after submission and acceptance by DG Environment) versions of BREFs were available at the outset of the research. These are pulp and paper manufacture; cement; and non-ferrous metals production processes. Across the three industries there are important variations in the cost and economic impacts of the BAT elements recommended.

The main approach for all three studies is based on individual plant case studies where plants with and without BAT are compared. The application of this methodology for each industry has differed slightly.

---

2 See, for example, Lindblom et al (2001).

3 By subcontract to Jaakko Pöyry Consulting, Helsinki. Major world-wide consultants to the industry.

1. The cement study (Wagner et al. 2001; Wagner 2001) compares the average performance of plants in different countries having different degrees of environmental stringency, and therefore different mixes of BAT and emission standards. The study sought to compare the average environmental and economic performance of plants in country A with counterparts in countries B, C, D etc.. BAT and non-BAT were defined by emission achievements alone.
2. (a) In the cases of non-ferrous metals (Farrell and Hitchens 2001) and pulp and paper (Lindblom et al. 2001), attempts were made to compare the economic and environmental performance of individual plants with and without BAT, irrespective of their European country of origin (though the sampling is restricted to particular Member States of the EU).
   The measurement of BAT, and therefore the identification of BAT plants, has differed across the three industries. In the cement sample a BAT plant is identified as a plant with low emissions. In non-ferrous metals it is measured by the strength of the BAT input, with total BAT input equal to the sum of individual BAT strengths. In pulp and paper, BAT is measured by the number of BAT implemented by mills and the resultant emissions i.e. by both environmental inputs and outputs.
   (b) In the pulp and paper sector the impact of BAT on international competitiveness was measured by comparing the economic performance of representative competitor plants in country A (outside the EU) with BAT plants in the EU.

BAT plants were selected from a number of sources including industry sources (trade associations) and various directories. There is no census showing a list of BAT used by plants from which a random sample could be drawn. Selected BAT plants were matched with representative plants in the industry, identified from the same sources. They were matched by size and product produced.

In total about 16% of European cement plants in target EU countries were included in the overall sample. In non-ferrous metals, for the selected metals, 45% of EU plants are included and 69% in target EU countries. In pulp, 25% of mills in target EU countries are included and in white line chipboard this is 34%.

4.3 Industry Findings

4.3.1 Cement

A sample of 37 plants in 4 EU countries– Germany, Italy, Spain and the UK were drawn. All processes i.e. dry, semi-dry/semi- wet and wet were considered, with dry accounting for 29 of the plants sampled. This article reports on findings for the dry process only.

Emissions to air are the dominant environmental impact and primary and secondary measures to reduce dust, NO_x and SO_2 are important.

The economic experience of undertaking individual BATs.
Primary measures have a positive cost and quality implication, they have positive paybacks and positive implications for competitiveness (as defined by a reduction in costs). Investments in primary BAT are a feature of new, modern and updated processes. Similarly, energy saving and process improvements have the same positive outcomes.

Secondary BATs for NO_x, SO_2, dust and noise have either mixed, neutral or negative effects. It is from this set of BAT that negative competitiveness implications can arise.

Implications of environmental protection and performance on the overall economic performance of cement plants.
The approach to this question has been tackled by comparing plants between each of the four Member States (taken as an external EU competitor). Environmental performance of these plants differ according to differences in regulatory stringency. The analysis that follows is illustrated by the case of the dry process. Table 1 shows that Germany and Italy lead in terms of environmental stringency and numbers of BATs adopted, Spain and the UK lag.

Economic performance of sample plants
Performance measured by productivity was lower for German plants than that for plants in the other countries considered, while Italian sample plants with the second largest number of BATs, have the highest productivity performance of plants in the countries compared. Low physical productivity at German plants is partly explained by higher than average excess capacity and also a wider variety of products produced than in the other countries compared.

Investment, on the other hand, is higher in Germany (per tonne) than at counterpart plants in other EU countries sampled. Similarly, profitability at German plants owned by three major cement companies was reported to be similar to that in plants owned by the company in other EU countries, despite low capacity utilisation and strong environmental regulation in Germany.

The analysis therefore concluded that the strong environmental performance observed at German plants did not negatively affect their economic performance, despite low productivity and low capacity utilisation. Hence, it is argued that while more BAT are associated with the more stringent regulation recorded there, they do not have a negative impact on the overall economic performance of plants sampled.

Table 1. Average Number of Measures Implemented (within the last 10 years) per Plant Classified by Country

Type of BAT	Germany	Italy	Spain	UK
Average no. of general primary measures per plant				
Expert system	0.8	0.6	0.6	0.2
Automatic quality control	1.0	0.7	0.6	0.2
Precalciner	0.5	0.9	0:0	0.8
Modern clinker cooler	0.5	0.3	0.4	0.2
New or modernised mill	0.6	0.3	0.6	0.8
Raw material storage closed	0.8	0.3	0.2	0.2
Clinker closed	0.9	0.7	0.6	0.6
Paving, fugitive dust	1.0	0.7	0.4	0.2
Optimal fuel feeding	0.1	0	0	0.2
Optimal burning process	0.3	0	0	0.2
Continuous measurement	1.0	0.7	0.4	0.8
Average no. of general primary measures per plant	**7.5**	**5.2**	**3.8**	**4.4**
Average no. of NO_x primary measures per plant				
MSC	0.3	0.1	0	0
low-NO_x burner/ flame cooler	0.9	0.6	0.6	0.6
Sum of NO_x primary measures	**1.2**	**0.7**	**0.6**	**0.6**
Sum of total primary measures per plant	**8.7**	**5.9**	**4.4**	**5.0**
Average no. of secondary measures per plant				
SNCR	0.4	0	0	0
Absorbent addition	0.5	0.1	0	0
wet scrubber	0	0	0	0.2
Sum of secondary measures	0.9	0.1	0	0.2
Sum of all measures per plant	9.6	6.0	4.4	5.2
No. of plants	8	7	5	5

Important factors that influence plant competitiveness and relative importance of environmental regulations and costs.

Competitive advantages and disadvantages were mainly connected with product quality and range, raw material quality, plant location relative to the market and transport costs. More stringently regulated German and Italian plants did report environmental standards and costs as a major competitive disadvantage, primarily as a consequence of secondary measures for NO_x and SO_2. Despite these additional costs, the plants were profitable, in part as a consequence of costs counterbalanced by the use of cheaper alternative fuels.

4.3.2 Non Ferrous Metals (NFM)

The non ferrous metals (NFM) considered in this study are international commodities. They compete on world markets and face world prices. The study focused on the metals: aluminum, copper, lead and zinc.

48 plants are included in the study and they represent between 57% and 100% of the relevant population of plants in the target countries. The main countries included are Spain, Germany, UK and Italy (plants were also visited in Sweden and Belgium).

The NFM BAT as indicated in the BREF were simplified to ten BAT factors and it is the impacts of these factors on competitive performance that are considered.

The principal methodology for assessing environmental inputs used is a derivative of the Operator Performance and Risk Assessment methodology employed in the UK. This methodology ranks BAT on a five-point scale on the basis of strength of input. A score of 4 signifies BAT and across 10 factors a score of 40 indicates a BAT plant (provided a minimum of 4 is scored for each factor). The major weakness in the application of this methodology, recognised in the study, is that individual BAT factors are not given an appropriate weight but merely summed arithmetically.

The focus of the environmental impact is on air, particularly fugitive emissions. Ninety per cent of plants visited had adequate air and water abatement plant. Most require low cost improvements e.g. modern filter materials and/or extensive maintenance. The main impact of IPPC is on improving fugitive emissions and there are significant differences in emissions between plants.

Two tests of the economic impact of BAT were undertaken for this sample of plants[4].

1. Based on plant visits, environmental inputs, as described above, are related to physical productivity, energy use and use of raw materials. Analyses show a positive correlation between high BAT scores and physical labour productivity. There are also positive correlations found between BAT and reductions in energy use and yield of metal (material productivity). The close association between BAT and production efficiency has meant that there was no average relationship in this sample between individual Member State stringency of regulation and BAT performance, as in the cement case. It is argued that labour productivity, metal yield and energy reduction are all closely related to improvements in profitability (see figure 1). Capital costs as measured by technical age (also correlated with productivity) are achieved at low expense provided that

[4] In addition, based on published data, a comparison was made of the financial performance and environmental performance of the majority of primary copper plants world wide. This indicated that those plants in areas with high environmental performance i.e. EU and Japan, were more efficient and had lower costs than plants in other world areas or regions. This relationship was also found to be true for profitability, but these data are not considered fully reliable.

improvements are undertaken progressively (over time). In fact IPPC drives to improve efficiency in terms of use of raw materials and energy and is consistent with this.

2. An analysis of perceived competitive advantages and disadvantages following BAT implementation, reported by managers, indicated 6 responses where environmental cost were a disadvantage. All six related to installations which were not using all or most of the BATs and were less efficient producers. No plants that were achieving BAT standards reported any European or worldwide competitive disadvantage arising from environmental protection. Some claimed that there were competitive advantages linked to the use of BAT.

4.3.3 Pulp and Paper

Detailed work undertaken on the pulp and paper industry was based on two products: kraft pulp and white line chipboard. Two broad methodologies were used - the first centres on a set of case studies matching BAT plants with representative plants in the industry. The sample includes 26 mills located in Europe (Sweden, Finland, Austria, Germany, France, Italy, Spain and Portugal) and 10 strong pulp competitor plants in North and South America.

The second method is based on a macro study of the additional investment required by EU mills, given their cost and environmental performance. External threats from mills in North and South America and Asia are considered on the basis of their environmental and economic performance relative to the EU.

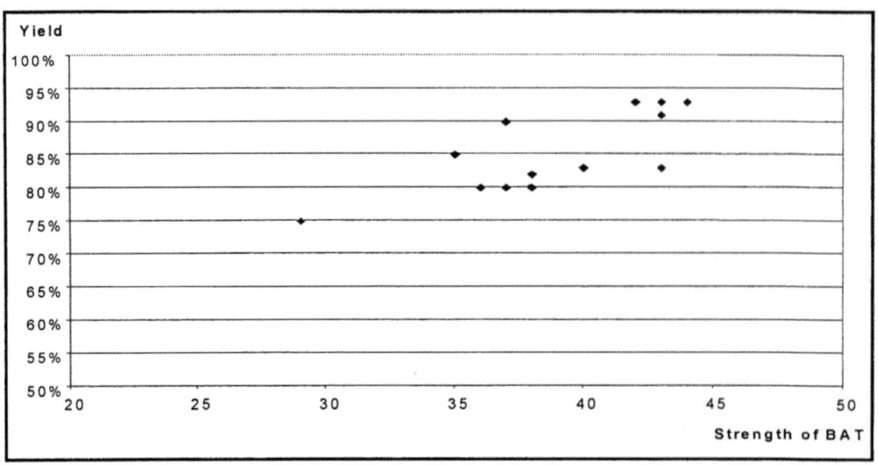

Fig. 1. Secondary Aluminum – Strength of BAT and Yield

5. Findings

5.1 Micro-analysis

Sample mills were divided into three categories, A, B and C, on the basis of their environmental performance. 'A' mills had most BATs and low emissions, those with fewer BATs and lower emissions were classified as B and C mills. Mills were matched on the basis of product and size.

Comparisons of BAT 'A' plants with matched plants
On three tests of performance, findings show that pulp A (strong BAT/emission performers) mills have stronger economic performance. White line chipboard A mills also have strong economic performance but are less distinguished from their B counterparts. Key findings were as follows based on the matched comparisons:

Productivity
Pulp: Physical labour productivity and sales per head are higher for A mills *White line chipboard:* Lower productivity at A mills on average (but there is also a wide variability in productivity performance indicating no significant difference between A and B mills).

Costs
Pulp: Costs per tonne were lower for A mills compared with B and C mills. *White line chipboard:* A mills incurred lower costs per tonne than their counterpart B mills.

Volume growth
Pulp: Volume growth (measured for the last five years) is greater for A mills than B/C mills.

White line chipboard
Show similar growth rates between A and B mills.

5.1.1 International Competition in Pulp

International competitors were sampled in Canada and Brazil in the pulp sector only.

Pulp: In Brazil and Canada competitors sampled were matched with A mills in the EU. Differential environmental costs were not the source of competitive advantage; indeed European markets were claimed to be a driver for improving the environmental performance of those plants in Canada and especially Brazil. The competitive threat from these mills is based on factors other than environmental costs e.g. quality and differential labour and raw material costs.

5.1.2 Competitive Advantages and Disadvantages Arising from Environmental Costs and Performance Relative to Other Factors

- *Pulp*: Environmental strength was recognized as a competitive advantage arising from environmental certification more than performance. No "A" mill reported environmental costs as a competitive disadvantage.
- *White line chipboard:* Of the five "A" mills, 2 claimed a competitive advantage from environmental performance and one a competitive disadvantage.

5.1.3 Impact on Profitability of Individual BAT

Was there difference in the economic experience of implementing BAT between related to the environmental performance of individual plants? It was notable that while, in pulp production, end-of-pipe techniques have a negative impact on profitability. For clean technologies some BATs are reported to have negative impacts in B an C mills while having positive economic impacts at A mills. There was also some evidence to suggest that when investments in combinations of BATs were made, these lowered the cost of any individual BAT.

5.1.4 Suppliers' View on Impact of Bat on Competitiveness

Four interviews with major producers of BAT equipment took place. These companies produce a wide range of pulp and paper technologies. Their view is that there is no effect from BAT on competitiveness except when firms have to make a step change in technology. However, if the mill has been investing continuously there are no difficulties. They emphasized the positive relationship between environmental performance and productivity performance and therefore competitiveness.

5.2 Macro-analysis

As part of the study an analysis was made of the environmental and economic performance performance of all plants in Europe, Asia, Latin America and N America (Jaakko Pöyry Consulting 2001). The approach involved (a) an estimate of the percentage of plants with above and below average environmental performance and costs of production (approximated from a mix of variables and expert opinion) in each region and (b) an estimate of investment requirements to meet 80% of BAT needs, assuming a stringent implementation of BAT, at EU plants.

Results show that in Europe 60% of the output of the kraft pulp sector is produced in mills that already reach above-average environmental and economic performance. These plants are not vulnerable as a consequence of IPPC. In fact those plants with above average performance on one variable, environmental performance or economic performance, were not considered vulnerable. However those plants with below average environmental and economic performance are vulnerable, having high costs and poor environmental standards, this amounted to 10%

of pulp output in the EU (similar analyses for white line chipboard and copy paper yielded figures of 15% and 10%) is vulnerable. Endangered plants also included the better performing plants which were known to have cash flow and other financial problems. Taking these into account and allowing for the improvement in performance and survival of some poor environmental and economic performers, the analysis suggested that 15% of pulp plants and 15% and 20% of white line chipboard and paper plants would close. These sectors are dominated by old and small plants which have undertaken little environmental investment in the past.

This capacity is also shown to be vulnerable to displacement by products produced by strong environmental and economic performers in the Americas and Asia, but also by strong performers in Europe.

Factors associated with environmental performance and the cost of compliance: factors associated with BAT implementation in the three sample industries:
The matched plant analysis indicated a set of factors which were associated with those plants which had adopted a most BATs as compared with average performers in the industry. These are summarised for the three industries in table 2. The table shows:

1. In many cases and for certain industries there are factors that correlate with environmental and BAT performance by individual plants. Such plants are competitive and have high productivity, are modern or technically up to date, are growing, or have high quality human capital inputs (including skills, management and R&D).
2. Past continuous investment in environmental initiatives is important in determining the size of investment required for the implementation of BAT. Past environmental investment can be related to the plant location and the history of regulation in that region.
3. Ownership can be important for reasons of economies in finance, use of human capital and, where necessary, plant rationalisation.

Having environmental management systems is a neutral factor in influencing environmental performance and the take-up of BAT. Why this is the case requires further investigation.

Table 2 lists, under 'favouring' factors, those identified as important attributes of 'BAT'plants. Those listed as 'neutral' are factors expected to influence implementation, although evidence indicates that they are unimportant in this study. Those that have a negative effect are those factors for which the study has accumulated evidence to signify this (i.e. it was not simply the lack of the attribute in the non-BAT plant but it was also shown to be an obstacle). The table summarises factors for the three industries, hence e.g. age of technology may be an important factor in one industry but not in another.

Table 2. Factors Influencing the Implementation of BAT (Summary for three Sectors)

	Favouring	Neutral	Factors that may have negative effects
Regulatory Framework	X	X	
Plant characteristics			
New plant	X		
Plant size	X	X	X
Technology		X	
Current technology	X	X	X
Technical age	X	X	X
Process control	X		
Original plant age		X	X
Plant performance characteristics			
Labour productivity	X	X	X
Current price/cost relationship/profitability		X	
Volume growth	X	X	X
Production costs	X	X	
Energy efficiency	X		X
Existing competitive disadvantages			X
Environmental characteristics			
Prior investment in env. protection and rate of investment	X		X
Current environmental performance			X
Environmental management		X	
Plant inputs			
R&D	X		X
Skills	X	X	
Innovation	X	X	
Price of inputs	X		
Other			
Location		X	
Ownership	X	X	X

5.3 Importance of R&D and Innovation

For two of the industries under consideration, R&D and innovation were important to improved environmental performance.

5.3.1 Non Ferrous Metals

During the interviews and site visits it was evident that a group of "Front End" factors that relate to how a process is managed, developed and controlled is very important in differentiating between good and poor performers at both an economic and environmental level. These factors include the technical age of the process and a number of "skill" factors that include training, innovation, operator competence, management and supervision and the elements of maintenance.

There is a strong influence of technical age on economic and environmental performance, the successful development of furnaces and processes, which reduces technical age is in turn strongly influenced by the level of innovation used within a company. Process development allowed environmental performance by minimising gas volumes and process variations. A large number of process improvements have been made by companies using high levels of innovation and skills to solve problems.

5.3.2 Pulp and Paper Industries

In kraft pulp it was shown how A mills are actively involved in R&D while most of the B mills are doing less but still some. C performers in all size groups have undertaken considerably less R&D efforts. In white Line Chipboard the "A" mills have more research and development on site than the "B" mills and they claim that R&D affects their environmental performance in a positive way. In copy paper "A" mills were shown to explicitly use R&D to solve BAT related issues.

5.4 Implementing BAT

For each industry, analysts emphasised the importance of prioritizing environmental initiatives, a careful timing of those initiatives and time to undertake them. Special consideration is recommended for those initiatives expected to yield a positive economic return. The detail for each industry is given below.

5.4.1 Cement

For dry technology the analysis shows little competitive risk arising from implementation of BAT primary measures. As indicated above, secondary measures are a different matter. The most important is a possible requirement for a reduction of NO_x to 500 mg/m^3. However, additional costs associated with secondary measures were importantly offset (in Germany but also at many Italian plants) by the use of cheaper (alternative) fuels. There are special problems for small, old and independent kilns in adjusting to BAT investment requirements and simultaneously remaining competitive in the short run.

Sufficient time for planning investments is important in the cement industry. Installations have a life cycle of about 20-30 years and require heavy investments. As major changes in the equipment are usually expensive (e.g. the improvement of EPs, implementing MSC or a precalciner into existing equipment) in comparison to a completely new installation, it will be economically beneficial if investments are planned in anticipation of future environmental requirements.

Time is also important to develop, test and evaluate new methods. Depending on technology, type of raw materials, type and quality of fuel etc. environmental measures might lead to different outcomes. Notification of stricter emission levels in advance allows time to find means for improving processes in the most economical way. For reduction of most emissions, experiences are published and easily accessible. In particular, international companies have the advantage of being able to transfer experience among their plants.

The length of time to react to the provisions also depends on the present state of the plant. No plant will be able to move from a low to a high BAT-associated emission level immediately. Plants which are more backward may need more time, but even these plants should be required to state their plans and how and when they will achieve the BAT-associated emission levels.

5.4.2 Non Ferrous Metals

The responses and observations show that in general, existing plants in the sectors studied can incorporate BAT relatively easily, provided that they use innovation and planning to prioritise the work needed. Vulnerable plants are those with few BAT factors which will need to implement more major improvements and without any economic benefit.

Based on the experience of the best performers, the most effective route for most plants in this sector to comply with IPPC will be to improve elements of existing plants by developing the way that the technology is used e.g. by using better methods of controlling and optimising the process. For poor performers, as for good ones, improvement to the "front end" of the process (e.g.process control) is of primary importance, followed by development of the process itself. Skills and the way that they are implemented and directed can be used by medium and poor performers to improve both of these elements. For example, in cases where under-designed gas collection is the main issue, process improvements can reduce gas volumes to a level that is acceptable, although some plants need to up-rate fan sizes and possibly the size of abatement plant.

The improvement of skills is an area where many companies have had success by adapting established systems. Many of these improvements relate to management issues. The study presents a methodology for ranking required BAT and the means of improvement. Using such a methodology it is possible to identify the areas that need to be improved, the techniques that are available to give the improvement and the influences and obstacles involved. The factors can then be used to establish the priorities for a particular site and a timetable for improvement.

5.4.3 Pulp and Paper

A number of factors are listed which are expected to affect the cost of compliance with environmental regulation and the implementation of BAT. These include R&D, skills, innovation, age of technology, the degree of product specialisation and the distance from home markets.

The macro analysis stresses that the competitive impact will be a function of (i) the ease of implementing BAT at a relatively low cost of compliance, (ii) the potential for specialisation to absorb the costs of compliance, (iii) the extent of international exposure for the product and (iv) for poor environmental and economic performers, the need to make large environmental investments.

The results of the study pointed to the necessity of the planning and timing of the BAT investments.

For example, pulp mills on many occasions reported that a reduction in emissions in small steps is cheaper for them. The possible speed of upgrading differs case by case and that there may be the possibility of combining environmental investments with other investments (for capacity increase or quality improvement). These opportunities play a major role in determining whether or not a company should invest in a single jump or choose a stepwise approach.

6. Conclusions

There is no evidence that BAT hindered those companies using BAT and achieving good environmental standards from remaining competitive both nationally and internationally, but it does not follow that early implementation of BAT by other firms or plants in the industries studied would similarly have little or no impact on their competitive performance. There are plants that would have technical difficulties in implementing all BATs. There was some evidence to suggest that more competitive, or best practice firms, with a capacity for innovation including workforce skills, R&D efforts, but also with up to date equipment and methods of production, can more readily absorb the costs of BAT competitively.

In non ferrous metals the study shows that in most cases environmental improvements arise from the "front end" of the process followed by development of the process itself and that skills and innovation are important. In cement ease and cost of the take up of BAT were influenced by the degree of previous regulation, modernity, technology, size, skills and form of ownership. In pulp and paper the economic impact of BAT on individual mills is tightly linked to the mills past competitive performance and technical characteristics, especially, mill size, age, productivity level, growth and R&D capacity.

Hence because IPPC requires permits to contain conditions, such as emission limit values, which are sufficient to ensure that BAT requirements are met taking account of the particular characteristics and circumstances of the installation. The flexible non-prescriptive approach also promotes innovation.

Back-Casting for Environmental Sustainability: From STD and SusHouse towards Implementation

Philip J. Vergragt

1. Introduction

In the recently published Dutch National Environmental Policy Plan 4 (NEPP-4 2001) it is stipulated that the solution of big and persistent environmental problems requires "system innovations". According to this policy document, system innovations sometimes require a social transformation process (or transition) of more than one generation. In order to achieve this a new policy instrument will be created: "transition management". In the NMP-4 transition management is proposed for problems related to energy and transportation (the greenhouse effect), loss of biodiversity and natural resources, and agriculture.

The aim of this paper is to review some of the developments in the last decade, including the development of the 'back-casting' concept, that have lead to the appearance of 'system innovation' on the political agenda. Also, the paper will reflect on lessons learned, on unsolved issues, and it will attempt to formulate recommendations for government policy.

Transitions are loosely defined as gradual continuous processes of societal change in which society changes structurally (Rotmans et al. 2000; Kemp and Rotmans 2001). A transition is the result of connected developments in several societal domains: culture, technology, economics, ecology, institutions, behavior, and worldviews. The distinction between transition and system innovation is not very clear: it appears that transition emphasizes the time dimension of the process, while system innovation emphasizes its systemic character. In this paper we will consider them more or less synonymous.

The idea that environmental (or sustainability) problems are deeply rooted in structural aspects of society is of course not new. In the 70-ies the first environmental movements were anti-capitalistic; in their view the capitalist mode of production cause both environmental and development problems. In order to solve these, nothing less than a marxist revolution was necessary. This view is echoed in the present anti-globalization movement.

The idea that a national government is at the cradle of a system innovation thus raises a lot of fundamental questions. The first is if system innovations can be managed at all, or if they are more or less autonomous processes. Transitions are visible everywhere in society: the ICT revolution, the globalization process, the graying of the Western-European population, the international migration streams, the biotechnology revolution. These transitions are indeed complex and multidimensional; they are the result of many developments in society and in technology, and they can hardly be influenced from a central point, let alone be managed. In relation to sustainability, the question arises if and how these dominant quasi-autonomous transitions can be influenced in the direction of sustainable development.

An interesting question in this context is the role of the national governments in transitions. National governments may be part of the problem rather than of the solution: in many governments there is a heavy entrenchment of policy practices and bureaucratic cultures and structures, which may act as impediments for desired transitions towards sustainability. System innovations will also affect the government system, and thus the question of 'who manages' is an interesting one. Another relevant question is if the international nature of transitions enables a national approach.

There is also the question about the relationship between technological and societal transitions. In technology dynamics a lot of attention is given to the socially contextualised character of technological innovations, and the degree in which they can be influenced. A lot can be learned from technology dynamics literature; however, in the transition discussion it is sometimes unclear if we are talking about a technological transition (a sustainable energy system, a sustainable transportation system) or about a societal transition (towards reduction of energy use by changing behavior, towards a different mobility culture)

The fact that transitions and system innovations are now on the political agenda is partly a spin-off of earlier developments in the nineties. For this reason we first review in this paper an important Dutch innovative program: the Sustainable Technological Development program (STD)[1]. This program was the first to call for deep "leapfrog' technological, cultural and structural changes in society in order to address sustainability issues on a global scale. It introduced the concept of 'back-casting', which is 'looking back from a desirable or unavoidable future'. Next we will review the project "Strategies towards the Sustainable Household" (SusHouse), because in this project the role of the consumer and of the demand side of innovations was stressed. From the STD program and the SusHouse project lessons may be learned for system innovation. We will subsequently address the role of the government and the role of private enterprises in collaboration with other stakeholders.

The general conclusion of this paper will be that in order to achieve system innovation, the role of the government should be to formulate and legitimize the direction to be taken towards sustainable development: the government should set long-term objectives but should abstain from managing too closely specific processes. Social experiments should be undertaken in multi-stakeholder setting and on a small and medium-seize scale in order to foster learning processes among stakeholders, and in order to explore directions to be taken. The government should stimulate these social experiments, and it has a role to play in the upscaling of successful experiments and in providing the relevant incentives and infrastructures. Private enterprises should be part of social experiments, which will enable

1 In Dutch known as DTO program: Duurzame Technologische Ontwikkeling.

them to innovate in sustainable technologies, products and services, in close communication with consumer demands and with requirements of sustainable development.

2. The STD Program

In 1993 five Dutch Ministries launched the Sustainable Technological Development program (Jansen et al. 1992; Vergragt et al. 1993). This program was based on the report of the Brundtland committee (WCED 1987) which introduced the often-cited notion of sustainable development:

"Sustainable development is development that meets the needs of the present generations without compromising the ability of future generations to meet their own needs".

In this definition three basic elements stand out: the fulfillment of (basic) needs; equity between the developed and the now underdeveloped world (between 'North' and 'South'); and solidarity with future generations. Moreover, sustainable development stresses the interwovenness of three aspects: economic development, ecological protection, and social priorities like quality and quantity of labor, health and safety at work, anti-discrimination.

The mission of the STD program (1993-1997) was "to explore and to illustrate how, together with policy makers, technology developers, and opinion leaders, by looking backwards from a sustainable future vision, processes of sustainable technological development can be initiated and kathalysed" (Jansen and Vergragt 1992). In this mission the notion of "back-casting" (Goldemberg et al. 1985) was applied: looking back from a sustainable future vision. Other elements of the program are also visible in the mission: the necessity of stakeholder collaboration, the concept of "illustrative processes", and the focus on collective learning processes.

In the beginning the focus was on the identification of leapfrog technologies that could potentially reduce the environmental impact of activities by a factor 20 in 50 years. The idea of the factor 20 was derived from the Holdren and Ehrlich (1974) IPAT equation[2]. If in the next 50 years the world's populations would increase by a factor 2, and if the welfare of the world's population goes up by a factor of five (a condition for equity), the environmental burden per unit of need fulfillment should go down by a factor of 10-20 in order to reach a sustainable society. A sustainable society is more than just a pollution-free society; it also includes social equity, quality and quantity of labor (Ashford et al. 2001), and sustainable economic development.

The STD took the (basic) needs as a starting point, and concentrated on the following "areas of need": Nutrition, Water, Shelter (housing), Mobility, and needs

[2] The presently popular version of the IPAT equation is I=PxAxT: the environmental impact (I) equals the product of population size (P), the degree of affluence (A) per person, and the environmental impact from technology (T) used to produce one unit of affluence.

for Materials and Chemicals. For each of these areas of needs, future visions have been created together with stakeholders. From these future visions as a starting point, proposals have been developed for "Illustration Processes" in order to start leap-frog innovation processes. (Vergragt and Jansen 1993). All together, 16 Illustration Processes have been carried out, for instance 'Novel Protein Foods', 'Multiple Land Use', 'Sustainable Offices', 'the Municipal Water Chain', and C1 chemistry. Each of these illustration processes brought together stakeholders from the entire 'area of need", thus not only technology developers and business, but also consumers, environmental organizations, and government agencies.

Looking back from now, these 'back-casting' processes were the first intuitive steps to explore the possibilities of evoking 'system innovations'. Many methodologies have been tried out, in order to investigate which were most successful. Leading principles were 'learning by doing' together with stakeholders, thus the initiation and carrying out of collective learning processes together with stakeholders. Another principle was the idea of illustration and communication: not to start activities for transforming an entire area of need, but small scale experiments endorsed by science and technology in order to illustrate and communicate possibilities.

During and at the end of the program a methodology (see table 1) emerged that involves seven steps from problem recognition towards implementation (Weaver et al. 2000).

Table 1. The STD Methodology

Develop long-term vision	Develop short-term vision	Implementation
1. Strategic problem orientation and definition	4. Explore solution options	6. Set up cooperation agreement-define roles
2. Develop future vision	5. Select among options: set up action plan	7. Implement research agenda
3. Back-casting		

STD illustration processes

Nutrition:

- Novel protein foods
- High-tech agroproduction
- Integral crop conversion
- Multiple land use

Transport/Mobility:

- Underground freight transport
- Information technology for transport systems management
- Demand-responsive public transport

Buildings and urban spaces:

- Sustainable public housing
- Sustainable offices
- Urban restructuring

Services provided by water:

- The municipal water chain

Services provided by materials/ chemicals:

- C1 chemistry
- Fine chemistry
- Structural materials from natural fible composites

In the first block, long term orienting activities are carried out. A problem orientation has to be carried out in order to define the system under investigation and its boundaries, and the dimensions of the problem under study. Future visions are important but should be always open for adjustments as results from learning processes. In the second block, the result of the back-casting exercise is to explore, generate, and select options and develop an action plan for implementation. In this phase often a definition study is carried out. In the third block, the actual implementation is carried out. This may take the form of a research project, a policy project, or a social experiment. Not shown in this scheme are the feed-back loops between all stages.

Although the STD program initially focused on technology for addressing the factor 20 challenge, it soon became clear that non-technological factors (called *cultural* and *structural* aspects) were at least as important as barriers and conditions for implementations. Often technologies are more or less available but the barriers are institutional, economical, and especially cultural.

In the STD program it became clear that long-term oriented thinking was one of the most essential conditions for implementation. At the same time it became clear that this was also one of the main bottlenecks. Most industrial companies are more concentrated on short-term profits, and long-term oriented R&D is diminished over the last decade. One of the main challenges for STD is to create synergy between the long and the short term: How to create a vision and a strategy for the long term that is endorsed by stakeholders, and at the same time creating short-term spin-offs that make it attractive for private companies.

Another lesson learned is that during illustrative processes it is extremely difficult to maintain a long term perspective. The dynamics of each project, and especially in multi-stakeholder processes, is such that the short-term objectives become easily dominant. The long-term vision recedes behind the horizon and after a while is not leading any more for the realization of the short-term illustration.

This means that there needs to be a mechanism to attune the long-term vision and the operational goals on a regular basis.

In several cases it proved that the collaboration between existing institutions for carrying out illustrative processes was not strong enough to survive over a prolonged period of time. In several cases towards the end of the STD program new institutions have been created for the continuation of the initiatives undertaken. It can be discussed if this fits into the STD strategy; STD's aim was more to transform existing institutions rather than adding more institutions to the already crowded institutional landscape. But on the other hand the creation of institutions guarantees continuity and creates a channel for funding.

The challenge of sustainable development is a global challenge and can never be solved on the national level alone. The STD Program has been quite active in building international networks for knowledge dissemination and for dialogue, for instance with developing countries[3].

Summarizing, the STD program has set in motion a 'learning by doing' perspective on sustainable development: it has generated a number of illustrative processes and involved a number of stakeholders into long-term thinking; it has developed a methodology and has explored the interactions between technology, culture, and structure. Most importantly, it has operationalised the concept of sustainable development from a fuzzy phrase into tangible activities that could be recognized by "hard" technology developers, and it has developed an operational approach. Still, sustainable development has not yet become a central activity of innovators and policy makers. It takes time and effort to diffuse these new notions of cooperative sustainable innovation and development deeply into society. But after more than 10 years it is now echoed in the NEPP-4 document in the form of transition and system innovation.

3. Strategies towards the Sustainable Household (SusHouse)

In the STD program most of the activities were concentrated on the supply side: on technology developers and policy makers, together with intermediate institutions and knowledge providers such as Universities and technological institutes. In the "Strategies towards the Sustainable Household" (SusHouse) project (Vergragt 2000; Vergragt and Green 2001) the focus was more on the consumers and on the demand side of innovations. Although there have been other projects concentrating on the consumers and the potential of behavioral changes (Schmidt et al. 1999; HOMES 1999), the SusHouse project added to that by applying the STD methodology in order to create future visions of a future sustainable household. The SusHouse project, an international project sponsored by the EU's Environment and

[3] On 26 March 1997 an international workshop was organized, together with the TU Delft and the Institute for Social Studies "The Sustainable Technological Development Approach: Potentials and Pitfalls for Developing Countries.

Climate program (1998-2000)[4], concentrated on three household 'functions": Clothing Care (Vezzoli 2000), Shelter (heating, cooling, and lighting) (Pfeiffer 2000), and Nutrition (shopping, cooking, and eating) (Green and Young 2000)[5]. Together with stakeholders from the entire chain (including consumers and housewives) creativity workshops have been organized in which stakeholders created future visions of the sustainable household. The challenge was to deviate as far as possible from the current entrenchment, using feasible technologies but also extreme behavioral and cultural changes.

A factor 20 efficiency improvement by 2050 requires not only that we considerably change our production processes, but also our consumption patterns taking into account that these are strongly interconnected and interdependent. Other reasons for the focus of the SusHouse project on sustainable households and sustainable consumption include:

- There is a considerable environmental burden and resource usage in consumption and household activities (e.g. Noorman and Schoot Uiterkamp 1998).
- The direct environmental burden of households has been increasing considerably during the last decades. (Noorman and Schoot Uiterkamp 1998).
- Households and their members are important social actors for achieving sustainability. They are responsible for 'demand' and could stimulate the growth of sustainable or 'green' demand.
- Together with sustainable technological innovation, cultural changes will be necessary for sustainable development. From this point of view, households and their members are also important actors.

The methodology developed and evaluated in the *SusHouse* Project (see figure 1) has been derived from the STD methodology, and more specifically from the STD Sustainable Washing Project (Vergragt and Van der Wel 1998). However, the *SusHouse* Project has made substantial changes, namely:

- less emphasis is laid upon technology as the main agent for sustainable development; rather, a combination of *technological, social, and cultural changes* is envisaged.

---

4 The SusHouse project was a collective endeavor of six research groups in five European countries which are: Technology Assessment Group/Design for Sustainability group, Delft University of Technology, (the Netherlands); Szeged College of Food Industry (Hungary); Dept. of Industrial Design, Politecnico di Milano (Italy), Avanzi (Milano, Italy), CROMTEC, Manchester School of Management, UMIST (UK), Lehrstuhl Markt und Konsum (Universitat Hannover, Germany). This research has been supported by the EU DG 12 Environment and Climate RTD Programme, Contract no. ENV4-CT97-446.

5 The full reports on these functions are published on a CD-ROM, together with the final report, the methodology reports, and the country reports; see Vergragt 2000. CD-ROMs are available from the author.

- more emphasis is laid upon the *participation of non-governmental stakeholders* in the process.
- a *design orientation* is chosen, rather than a policy-making orientation. So, the aim of the *SusHouse* Project was *to develop and test a methodology* that would:
- enable companies, governmental policy organizations and NGOs to carry out their own analyses of household functions.
- identify possible product, system and social innovations which offer business opportunities and policy initiatives now.
- develop scenarios for sustainable household functions using industry-consumer-government creativity groups.
- develop methods of assessing the viability of these ways of sustainable household function fulfillment.

The development of the methodology was backed up with case examples of imaginative scenarios developed during the Project (Young and Vergragt 2000). For the fulfillment of functions of the sustainable household; the scenarios were subject to environmental assessment (Bras-Klapwijk 2000), economic analysis (Young and Simms 2000) and consumer acceptance analysis (Bode 2000)[6], and have been 'endorsed' by the social partners in the project.

The project's approach was briefly as follows. (Vergragt and Green 2001). With the help of experts from different stakeholder groups, and with the help of brainstorm techniques, the project research group formulated normative scenarios of sustainable fulfillment of these household functions by the year 2050, including technological, cultural and institutional innovations. The scenarios were evaluated as to how much they decreased the overall environmental burden, whether they were economically credible, and whether they were acceptable to European consumers. After this it was possible in a second set of stakeholder workshops to define the trajectory leading towards this sustainable future.

An essential element of the *SusHouse* methodology is the creation of 'micro' normative scenarios for a sustainable household function. These scenarios are based on the creativity workshops and the ideas generated by the stakeholders. The scenarios are based on the following general notions:

- Technological innovations are necessary but insufficient to bring about factor 20 sustainability improvement.
- A shift from products to services may offer new options for changes towards sustainability.
- Sharing household activities offers a potential for sustainability gains.

The scenarios are intended to generate visions of sustainable household function fulfillment that differ radically from the present. This is why workshop participants were asked to focus on the year 2050 to envision futures that might breach

---

6 These methodology reports are also published on the SusHouse CD-ROM; see previous footnote.

current trends. Such visions may open up new ways of thinking, researching, designing and acting *in the present* (or, at least in the next few years) and thus offer a way out of the present consumption deadlock. The project developed the concept of the *Design-Orienting Scenario* (DOS), as opposed to the more common notion of the Policy-Orienting Scenario (POS) (Manzini and Jegou 2000). A Design-Orienting Scenario is defined at the micro-level of the (future) household, rather than the whole society of economy, and is supposed to create inspiration for 'designers', whether in industry, government, universities or NGOs, to design products, services and social arrangements that might help to realize steps towards these scenarios .

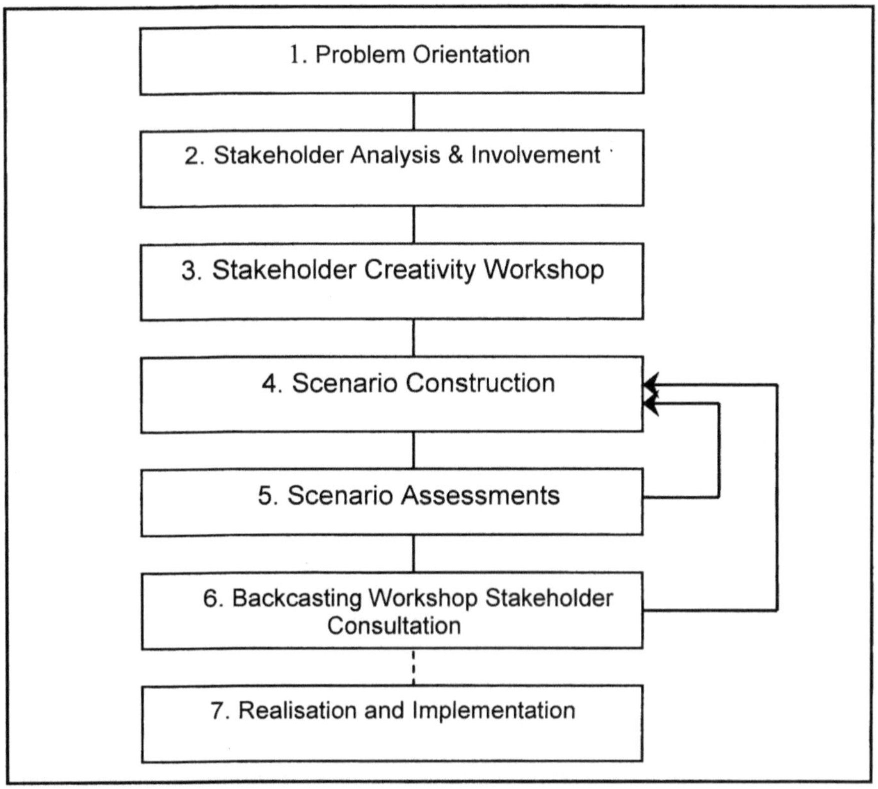

Fig. 1. SusHouse Methodology

A DOS should contain the following elements:

- Various "Proposals" developed as concrete products and/or services
- A global "Vision" picturing the effect of the implementation of the Proposals and their possible impact

- The "Essential Characteristics" explaining the main effects and benefits the DOS is expected to have in terms of sustainability, economics and user acceptance
- A story board, describing "a day in the life...." for the household function in the year 2050

Each of the 'Design orienting Scenarios' devised in the *SusHouse Project* was assessed with respect to three criteria: environmental improvement; economic viability, and consumer acceptance.

The last stage of the process was to reconvene the stakeholders in a backcasting or implementation workshop, in which steps towards implementation have been investigated. In this workshop both new business coalitions have been investigated, research agendas have been constructed, and policy agendas have been drafted (Quist et al. 2000). However, the jump towards actual implementation of these projects appeared to be very large.

As an example we present here one of the scenarios of the Clothing Care function (Knot 2000):

Example of Clothing Care DOS: Eternally Yours

Vision. Clothes are similar to jewels: clothes are precious and durable goods that have high emotional value to the user. Clothes are functional and comfortable, but also have an important function in reflecting a personal style of living and life-stage (personality and identity). In this sense, clothes are in fact part of the person, and regarded as a 'second skin'. Personal style is far more important than fashion. Clothes are closely related to ceremony: the purchase of new clothing is linked to memorable events (changes in life-stage), and is in itself a very special and festive occasion. People own and use a limited amount of clothing, which is used intensively for a long time. Clothes are made to measure: unique pieces, mass-individualized pieces, personally finished semi-products. Cleaning and maintenance is organized as a service, which is paid for at purchase (service contracts). The need for cleaning and maintenance is minimized and optimized because of the use of durable materials and designs, anti-dirt fabrics, dirt-indicators, local stain-removers, care while wearing, dark colours, etc.

Product-service proposals. The scenario involves many different products and services:

- Made to measure high quality clothing, mediated by information technology: unique pieces, mass-individualized pieces, semi-finished pieces;
- Adaptable and repairable fabrics and designs;
- Flexible and multi purpose fabrics and designs;
- Dirt indicators and anti-dirt fabrics;
- Service contracts;
- All-round clothing centers, co-ordinating all business activities concerning clothing and clothing care (manufacturing, maintenance, adaptation, waste

processing), and offering many other services related to body care and recreation;
- Fiber-rights or –quota.

To help gain an understanding of how the DOSs differ on the conceptual level, and how they relate to one another and to the present situation, they were clustered on a matrix presented in figure 2, with the two axes thus:

- *social/collective* (members of the household will tend to collaborate as a social community) versus *individual* (members of the household will behave as separate individuals); and,
- *do-it-yourself* (technical infrastructure enables the members of the household to fulfil the functions on their own) versus *service* (technical infrastructure involved in the functions tends to provide the household with finished, ready-to-use products or services).

The 18 DOSs can be clustered into five groups: ***Care Socializing, Care Outsourcing, High Care, Soft Care, and Easy Care.*** The five clusters can be characterized thus:

- The **Easy-Care** household is characterized by high-tech equipment helping users in their daily life.
- The **Care Outsourcing** household actually involves a certain 'deconstruction' of the household as it is traditionally conceived as a place for the fulfillment of domestic functions.
- The **High-Care** household is based on a lifestyle in line with 'natural' models.
- The **Care Socializing** households are based on a certain level of community life, of collective resources, of sharing of products and services.
- **Soft Care** describes a household characterized both by a high attention/active involvement of the household members in the fulfillment of domestic tasks and a highly sophisticated system assisting them in these tasks.

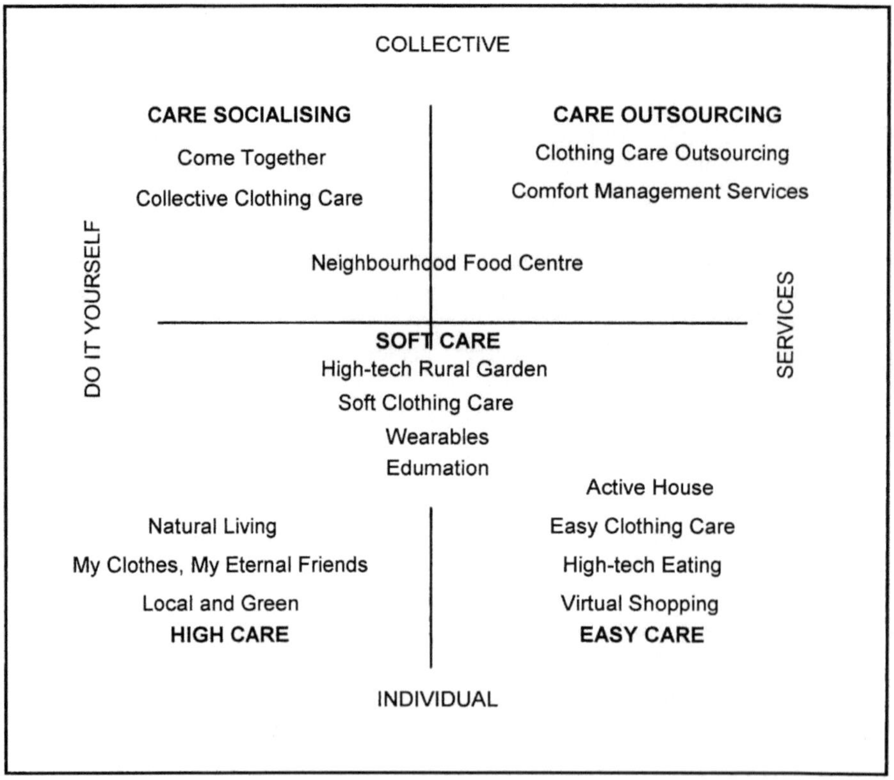

Fig. 2. Integrated Vision of Household DOSs (Manzini and Jegou 2000)

The SusHouse project can be seen as an experiment in methodology development for system innovation with an emphasis on the demand side of the production-consumption chain. It has taken the STD methodology as a starting point, and adapted it to become a working back-casting methodology oriented at system changes in a multidimensional (technological, structural, cultural, behavioral) way. Although it has been developed for functions in the household, and more generally at the consumption side of the production-consumption system, it can be adapted to other societal systems as well. The methodology calls for stakeholder collaboration, vision development, and back-casting towards experiments in the present. The methodology has been tested on an experimental level; the bottleneck lies in the implementation of niche experiments in the short term, and in upscaling from successful experiments toward large-scale innovations (system innovations). The government could facilitate here by fulfilling boundary conditions in order to facilitate working experiments.

4. The Role of the Government: New Governance Models

From the STD program and the SusHouse project a clear message stands out: future visions that are shared among stakeholders are a necessary but not sufficient condition for achieving system innovation towards sustainability. Another lesson learned is that stakeholder management is very important: understanding the culture and the interests of stakeholders, trying to understand their motives for collaboration, and understanding in which phase of the process they can play a role. Further we have learned that even on a small scale level these type of innovations are extremely time-consuming and costly, especially because they deviate from existing development paths and explore new cultural and structural options.

In the beginning of this paper we stipulated that the role of the government should be a restricted one. Here we argue that the role of the government is indispensable because of the long time horizon, the complexity of the processes, and the need for an actor that guard the general direction of sustainable development. However, the government has also to reflect upon its own functioning, and has to develop a concept of governance that is suitable for influencing transitions.

In a paper to be published in Dutch, Van de Graaf et al. (2002) explore the contours of a governance policy concept for steering system innovations or transitions. In this paper they argue that it is time to develop a third generation environmental policy, after the second generation based on stakeholder orientation and social learning. The boundaries for this second-generation environmental policy are reached because of structural impediments in society: the physical infrastructure, social conventions, existing regulations, available knowledge and the knowledge infrastructure. They take as example the car and the mobility system. The car system has become heavily entrenched in society because the developments in car technology have become heavily intertwined with developments in the infrastructure and in societal culture (Sachs 1984; Mom 1997). The challenge is how to formulate a policy that addresses this cultural and structural entrenchment without falling into the trap of planning by blueprints? Van de Graaf et al. (2002) argue that not a great planned attack on the existing system, but system changes may be brought about by "*....concrete contextual practices that, to a certain extend, do not follow the existing rules....*". Various practices may reinforce each other and eventually may lead to a system innovation. For the government there are two roles: *foster innovative practices*, and *foster mutual reinforcement* of these practices.

The recommendation that the government should foster innovative practices was also made in the SusHouse project (Vergragt 2000) and even in the earlier SMEC project (Social Management of Environmental Change (Irwin et al. 1994). Often the development of these practices is done by small innovative firms or by citizens initiatives, and they are the result of 'slumbering reverse salients' (Moors 2000): the growing understanding that certain problems on the system level may prove to be insoluble without system changes. The difficulties of access by the increasing car traffic jams could be an example here: sooner or later the traffic jams and the lack of accessibility may prove not to be soluble within the present car mobility system, and new systems need to be developed.

Mutual reinforcements of innovative practices may be achieved by developing connective infrastructures, regulation on a more general level, technologies that fulfil needs in various contexts, and research programs aimed at investigation of knowledge gaps. It may be added that forms of network management aiming explicitly at connecting innovative practices may be useful here.

Future visions may play an important role here for contextualizing and connecting individual innovative practices, and to provide them with a meaning that goes beyond the innovative practice itself.

In an advice to the Ministry of the Environment about the follow-up of the STD program, Diepenmaat and Te Riele (2001) advice the establishment of two interactive layers: a strategic layer and a practice layer. The strategy layer creates the boundary conditions and sets the stage. They concentrate on long-term signals and develop visions; and they interact with the practice layer. The practice layer consists of changing coalitions between five types of stakeholders: Government, Companies, Knowledge infrastructure and advice, Intermediates, and Citizens/Consumers. They collaborate on specific innovative issues. It is important that each of the five stakeholders is present in each of these projects. This is a change with respect to the present situation in which government and companies dominate together with the knowledge infrastructure. The policy agenda for transitions includes among others network formation, collective vision formation, organization of consistency within the government, growing importance of intermediate organizations, research for endorsing social sustainability experiments, and co-existence of different time scales. They advice a new institution at some distance from the government, somewhat similar to the SER (Social-Economic Council, responsible for the famous Dutch Polder Model).

The recommendations by Diepenmaat et al. point into the same direction as the analysis and recommendations by Van de Graaf et al.; they point at the importance of bottom-up experiments and citizen initiatives (Irwin et al.), and they assign specific roles to the government. It is to be hoped that the government listens very carefully to these recommendations, because they are based on ten years of experiments in the STD program and in projects like SusHouse, and they are endorsed by a deep policy analysis.

5. Towards Implementation in Coalitions with Business

At the end of the day, system innovations need to be implemented by private companies that innovate successfully, which means bring new products and services successfully to the market. In the past ten years we have seen a shift from cleaning up production processes towards the design of environmental friendly products (ecodesign), and then towards the design of sustainable product-services. Presently methods are being developed how to develop sustainable services

(Brezet 2001), and how to innovate together with several companies (the Kathalys method) (Van der Horst et al. 2001).

It has to be stressed that the bulk of the industrial companies is still in earlier stages of these developments. Although the implementation of clean technologies in business is quite far in the Netherlands, the implementation of ecodesign, and especially the development of eco-efficient services stays behind and needs more attention. Beyond these, innovative practices towards system innovation are quite new and hard to organize with companies. What is necessary here is (social) business experiments in which short-term business success and consumer acceptance is achieved in the context of an explicitly formulated long-term vision.

As an example we take the Wash-in project. This project is both a spin-off from the Sushouse project and an innovative design by an industrial design student.

"The Wash In (Van den Bremen 1999) is a new concept for integrated clothing maintenance. It offers not only laundry services, but also various upgrading services for clothes and textiles, contributing to their life extension: taking-in and selling used clothes, reparation and adaptation, re-coloring. The collective laundry processes can be environmentally more efficient than washing at home due for example to the larger scale processes, the professional equipment, faster technology replacements, decreased use of equipment and professional handling. The Wash In uses green electricity from own generation (wind, solar) or from energy providers. The water for the laundry process is filtered rainwater, that is collected on the premises; also less softener is needed because of that. Experimental full-enzymatic detergents are used, making bleaching agents unnecessary, and allowing lower washing temperatures. The use of hotfill washing machines makes it possible to heat the water more efficiently, outside the machine.

The upgrading services for life-extension can however be at least as important for environmental savings. For example, more than half of the total energy and almost all of the water that is used for "being clothed" is not due to the washing, but due to the production and distribution of the clothes (Knot 2000). The laundry services that are offered by the Wash In are washing, drying, ironing, aqua-clean, dry-clean, 'hand-wash'. There is choice between self-service, full service and fast-service. It is not necessary to wait until the laundry is ready. The laundry can be dropped of and poicked up at any time, at the wash In or at service points. The transport of the laundry to and from the Wash In or service points is as much as possible integrated in existing activities and facilities (train stations, petrol stations, childcare, supermarkets).

The Wash In is meant as trendy, modern and fresh. It offers its large choice of services on a 24-hour basis, allowing clients to fit the service in their different daily lives in a flexible way. In this way, the Wash In is designed to attract new target groups for collective laundry processes (compared to the current launderettes): single person and two-income households who are able and willing to spend money but have little time. And although the Wash In is presented as comfort service rather than as a 'green alternative', it can play a role in environmental education concerning washing and clothing related themes inconspicuously. Furthermore, in cases the Wash In may fulfil social functions, like local and green employment for deprived, a neighborhood meeting place (coffee-corner), and a neighborhood information point.

Like the product-service systems in the SusHouse scenarios, the Wash In is meant as a new, attractive arrangement with environmental advantages. It proposes a whole clothing maintenance concept rather than a mere 'green' laundry concept, combining elements that have not been combined before and that can strengthen each other. An example is the combination of laundry and second hand services. About one third of the clothes that people possess sits unused in their wardrobes. The handing in of clothes for re-use becomes easier in the Wash In concept. No extra efforts have to be made (like finding a second hand boutique, washing and ironing the clothes, bringing the clothes to the boutique), and people may get something back for it (clothes exchange points, service points). Also the selling of used clothes becomes different. The users-group of used clothes may enhance through this concept. The Wash In shows a sphere of hygiene and freshness and it is evident that the clothes are thoroughly clean. Furthermore the 'second chance' collection may be an extra 'fun' element for the clients.

The Wash In concept is furthermore related to the SusHouse Clothing Care scenarios in the sense that similar strategies and principles can be recognized:

- The service-strategy, which is central in Wash In, is present in almost all SusHouse scenarios, the most clearly in the Outsourcing scenario.
- The principles of caring for your things and life-extension which are the strongest in the SusHouse Eternally Yours scenario, is also recognizable in the Wash In concept: your clothes are professionally taken care of and get exactly the treatments they need. They are worth it to have them repaired or re-colored.
- The principle of enhancing the use-intensity by successive use, which is central in the SusHouse Chains of Users scenario, is present in the secondhand service of the Wash In. This 'second chance collection' of the Wash In is however also connected to the SusHouse Outsourcing concepts: it may grow into a professionally managed collection in which the clothes change from user to user.

By 'selling' these principles through being on the market, the Wash In may facilitate the upcoming of Wash In concepts or other services, extended or adapted with other, longer term (SusHouse-like) ideas." (cited from Knot and Vergragt 2002)

The problems of implementation of the Wash-in concept as a business concept illustrate the problems of starting up innovative practices or social experiments. The economic success of the business is unknown in advance, and thus private parties are reluctant to invest in this stage. The project in its present form does not fit in existing financing schemes. First a business plan needs to be written but, within a consortium of parties, no party is the prime mover or the most interested stakeholder. Without a product champion parties wait for each other to act. A complicating factor is that extra money is necessary for necessary monitoring activities, like consumer behavior and environmental gain. The project needs to be set up as an experiment, in a flexible way, in order to enable learning processes underway; this does not fit in existing business practices.

These points illustrate on a small scale the problems encountered in small scale social experiments that may the stepping stones of eventual system innovations. The government's approach, advocated in the previous section, should be such that social experiments like these get an easy chance for implementation without extensive paper work and waiting times, and enable research in action about the success and failure factors, the consumer acceptance, and the environmental gain. Another problem to be tackled is how to organize temporarily coalitions and how to protect knowledge that each of the partners brings into this temporary coalition.

Much experience needs to be obtained in the setting-up, managing, monitoring, and evaluation of these small-scale innovative initiatives, before one can even start to think about transition management on a large scale.

6. Conclusions and Recommendations

In this paper I have essentially looked back upon ten years of experience with back-casting in the Netherlands, and looked forward towards implementation of social experiments in the direction of system innovation for sustainable development. Ten years ago the early adopters of the concept of back-casting in the Netherlands (Jansen 1991; Vergragt 1992) stipulated that the setting of long-terms goals was essential, in order to mobilize creativity and commitments in society. The concept of back-casting, originally seen as looking back from a desired or unavoidable future (a sustainable global society in which 12 billion people will live) has eventually developed into a methodology of incorporating future visions in short-term oriented social experiments and innovative projects. Conceptually, back-casting has also indicated that sub-optimal short-term oriented choices should be avoided if they hamper desirable long-term developments. It calls for keeping many options open as long as possible, and choosing robust options that fit into many future visions (Knot et al. 2001). Back-casting has become as accepted as forecasting, and in the mean time has been transformed into concepts as transitions and system innovations, which always presuppose that they go into a certain direction (towards a sustainable society).

One of the problems is of course that it is not known what a sustainable society will look like. There are many notions about sustainability, but it is fairly sure that in a sustainable society basic needs will be fulfilled, that a certain social equity will be reached (including enough cultural diversity), and that the ecosystem will be safeguarded. In this sense the 'factor 20" should not be taken too literally, but should be seen as a symbol for a sustainable society. Because we do not know exactly what a sustainable society will look like, transitions need to set up *iteratively* and *interactively* in order to allow *flexibility* in the process and to enable adjustments and collective learning processes during the process.

In the last ten years we have learned that technology is not the bottleneck, although it is always advantageous to try to influence technological developments into directions relevant for a sustainable society. In this respect it is important to reinforce the knowledge infrastructure, and to remove the barriers for implementation the result from the separation between the knowledge infrastructure and innovative companies. The present Dutch ICES-KIS-3 investment impulse contains both Sustainability and System Innovation as themes, and this is a very promising sign.

The bottlenecks for system innovation towards a sustainable society are more in the separation of networks, in short term thinking, in the dominance of short-term thinking in financial circles, and in risk-averseness of social actors. Also the consumer behavior of citizens is an impediment; the same citizens that as voters behave environmentally conscious, behave in the opposite way as consumers. Further the present regulatory, bureaucratic, and policy system does not foster societal transitions towards a sustainable society.

What we have learned is that it is important to develop future visions that are endorsed by stakeholders. It is very well possible to bring together actors not only from the production chain but also from other stakeholder circles and develop entirely new ideas about future function fulfillment (see for instance Partidario et al. 2000 about the paint chain). Future visions may be based on a combination of expert knowledge and free brainstorming. It is essential that they are elaborated to such an extent that they are concrete enough to mobilize, and open enough to accommodate a variety of experiments.

Stakeholder management is also identified as an essential aspect. Stakeholders are primarily concerned about their own interests and they have their own view on future developments. By creating interactive networks of stakeholders around issues that are recognized as relevant, stakeholders broaden their view and learn to accommodate other perspectives in their own thinking.

Transitions cannot be managed from a central point and certainly not by the government alone. The government will be subject to the transition process too, in the sense that new forms of policy making need to be developed and new bureaucratic rules and structures need to be developed in the process. There will be a need for a center of facilitation, legitimized by, but on a certain distance from the government, which will operate on two levels: a strategic and a practical level. On the strategic level the overall goals will be guarded, and consensus will have to be created by most important societal stakeholders about the general direction of the process. On the practical levels stakeholders including intermediates and consumers/users around certain specific issues will organize social experiments. This may lead to innovations (new products, services, subsystems, infrastructures, and behavior) that may eventually become part of an overall system innovation.

The role of the government will be to legitimize the transition process: by seeking political support by the political parties; to guarantee its support by the creation of a high-level institution that fulfils the strategic function and facilitates the social experiments; to fund the bottom-up experiments, and to create the conditions for upscaling of successful experiments, to attune the infrastructure developments in line with the sustainability strategy, and, more generally, to reform itself and to take away existing institutional and economic barriers to sustainable development.

Towards Environmental Innovation – A Policy Synthesis

Ken Green

The papers in this book seek to bring together two research communities; namely, those who write from the "innovation systems" perspective, looking at the patterns and dynamics of technological innovation within different national and international contexts and those who are engaged in the specific study of "environmental innovation", looking at the processes whereby new products and processes can be developed which take account of ecological impacts and resource usage. The aim of the book is to see whether new insights could be gained regarding appropriate policies for better environmental innovation and what the possibilities and limitations for policy are when the two perspectives collide and combine.

Studies of environmental innovation over the last ten years have concentrated on a few key issues. Firstly, they have been interested in investigating what stimulates companies to come up with 'green(er)' innovations in either the products they sell or the processes they use to make or distribute them. The main focus has been on whether government-imposed regulations have been and should continue to be a main stimulus for such innovations. The answer has been clear – regulation is the most important stimulus to innovation, though other factors are also influential in some circumstances, such as changes in market demand, effects of supply chain pressures (which is a form of inter-organisational demand change) and culture change within the innovating organisation or the campaigning of NGOs. There has also been much work on the institutional contexts of pollution abatement, the adoption of clean(er) technologies and the redesign of products to incorporate materials with less ecological impact. More recently, focus has begun to move towards thinking about innovation for *sustainability,* rather than just for the more limited aims of reduction of ecological impact of existing products and processes. If this new direction is to come up with some new policy insights, it should be able to benefit from recent work on innovation systems, which sees innovation as part of broader socio-economic processes.

Understandings of the mechanisms of innovation by those involved in 'innovation studies' have changed since the 1960s, when innovation was usually seen as the private actions of individual firms carrying out 'R&D' by exploiting scientific 'discoveries' that emerged from public investment. Nowadays, though individual products and services may appear to emerge from individual firms, the process of innovation is seen as involving many social and economic actors. Indeed, some innovations are strongly affected by the action of actors outside firms, in a process of 'social shaping'. This happens as a result of interactive processes that involve the exchange of information and knowledge. It can be taken as obvious that these interactions are:

- between firms in the same or technologically related industries, in partnerships and alliances;
- between firms within a particular supply chain, who act as suppliers and customers to each other;
- between firms and other organisations that regulate them and lobby them

Together, this collection of relationships means that the dynamics of innovation – where and how it takes place, how it links to sources of scientific and market knowledge – is now understood as having a systemic dimension. Crucial to an understanding of these systems is another element which has, until recently, tended to be ignored by innovation systems researchers: namely, patterns of consumption behaviour. Understandings of consumption of and demand for new products and ways of providing and using them, as opposed to just design, production and supply of those products, is so important that, to the three interactions listed above, we can add a fourth:

- between firms and their customers (which may be other firms or final consumers)

Studies of the innovation process over the last 20 years have not only shown that innovation is *multi-actor* (another way of saying that innovation is "distributed" across an number of firms and other agents), but that it is also *multi-level,* that is the actors involved in/affecting innovation may be supranational ones (such as the European Commission or the World Trade Organisation), national (central government agencies) or regional (development assistance agencies). However, it is important to remember that, whatever the influences on a particular innovation and the conditions for its successful diffusion, innovation is inevitably 'local', in that the entrepreneurial actions necessary to realize it have to take place in some particular geographical location. Such locations however shift historically so that radical new developments emerge from novel combinations of knowledge and firms in unpredictable ways and places. The importance of the generation of novelty requires that we encourage *variety.* Forever changing social, economic, political and, we now realize, resource and ecological, uncertainty requires an equally strong matching generation of new ideas, products, processes and systems of provision. In the 'restless' system that is capitalism, responding to the challenges of ecological problems and resource supply changes gives rise to the need for equivalent restlessness in the provision of innovative solutions.

Innovation System studies has therefore being developing a number of theoretical frameworks and outlooks that, whilst focusing on the importance of the capitalist firm as the major location for innovation in our economies, emphasise that those firms have to be seem as 'embedded' in larger socio-economic structures and subject to a range of influences, from governments and markets.

It would seem that these insights might be of some use in trying to understand how innovation can be steered into more sustainable directions so that we might suggest that studies of innovation systems and studies of the particularities of environmental innovation need to come together to provide a set of ideas for the study of sustainable innovations: in other words

Innovation Systems Studies + Environmental Innovation Studies =
Innovation for Sustainability Studies (for environment and quality of life).

The overarching issue of how to steer innovation in a more sustainable direction can be further sub-divided into four main questions:

1. What are the next big challenges to achieve the shift from 'environment' towards 'sustainability' innovation policy?
2. What kind of innovation policy is needed to foster the sustainability?
3. What kind of research is needed to provide guidance to policy targeted at sustainable objectives?
4. What is the appropriate division of labour between national and supranational institutions?

I will briefly indicate what seem to me to be the possible answers to these questions:

Question One: What are the next big challenges to achieve the shift from 'environment' towards 'sustainability' innovation policy?
It is difficult to deny that the challenge of responding to the prospects of climate change, due to the accumulation of greenhouse gases in the atmosphere over the last 100 years of global economic development, is one of the biggest challenge, if not the biggest. Shifting to more sustainable methods of using energy requires substantial investments in new energy technologies, to lower the carbon-dependency of industrial production; in addition, sustainability in energy use demands all kinds of industrial and product innovations that focus on energy efficiency. But improvements in energy efficiency in the most carbon-using areas of industrial production, however massive those improvements might be, make no sense without reductions in the overall consumption of energy and other resources. This puts the focus increasingly on the scale and patterns of *consumption* of advanced industrialised economies and the growing consumption in industrialising countries of South and East Asia. Innovations in product design can of course go a long way to reduce the environmental impacts of how things are used in consumption in households. But major reductions require the re-thinking of *systems* of provision, with redesign of infrastructures and patterns of social living. Such system changes, are of course, difficult to achieve and depend more on changes in social attitudes and new innovation in new organisational forms for the delivery of household tasks. Nevertheless they also involve product and process innovation, especially in methods of distribution (which are currently heavily dependent on transport systems based on the burning of carbon-based fuels) and of the organisation of household tasks and urban structures. So: the big challenges are those of climate change and systems of household and collective provision (the latter often reduced to 'consumption'.)

Question Two: What kind of innovation policy is needed to foster sustainability?
Given what I have said about the importance of variety generation in responding innovatively to rapidly changing uncertainties, then we need innovation policies that will both encourage this generation and support those initiatives that might seem to have strong chances of success. Encouraging variety needs to be taken seriously by seeking to level the playing field so as not to favour existing 'unsustainable' regimes. This includes reduction of subsidies and tax policies which encourage the continued 'lock-in' to technological regimes that depend on burning of fossil fuels or providing substantial support to technological directions that move such regimes in more sustainable directions. Connected to this is the notion of 'niche management experimentation' in which new ideas are protected from the full rigour of the existing markets (which are built on long-term efficiencies in existing technological regimes). More experimentation will of course lead to more failures as well as more successes – we should expect and permit such failures and learn from them. Needless to say, regulations need to be constructed in such a way that this experimentation is stimulated, with the recognition that successful ideas are as likely (if not more likely) to emerge from multi-actor experimentation out of the mainstream than from conventional sites of Research and Development.

The greater focus on the generation of new ideas should not blind us however to the need for policies that seek to *choose* the most appropriate innovations in pursuit of sustainability goals. It is this notion that is behind current debates about 'transitions' to sustainability, in which the focus is on the management of a number of connected innovations that bring about some 'step change' from one (unsustainable) system of provision to another, more sustainable, system. The issue here is one of getting the *timing* of the innovations correct.

Question Three: What kind of research is needed to provide guidance to policy targeted at sustainable objectives?
Following the answers given to questions one and two, it should be obvious that the kind of research that is needed is that which focuses on innovation for the mitigation of climate change and on new patterns of provision in households and communities. It is in the nature of these topics that they cannot be studied at a national level. The 21st century economic system is increasingly internationalised and environmental impacts – where they originate or where they have their effects - can no longer be seen as confined to particular regions. However, despite the claims of increasing 'globalisation', since innovation relies on novel combinations of knowledge and competences in firms and other actors it is still 'localised'. Innovative solutions to problems of sustainability will therefore arise in unpredictable places and ways; generalising these solutions is therefore especially important. This points to a need for many more comparative studies of variety generation and of the introduction and diffusion of innovations. Coupled with this is the need for more research into these 'unsustainabilities' themselves, including their social and technological causes and explorations of the possible ways in which they might be overcome. Such studies will inevitably be of systems rather that of narrowly prescribed industrial production methods or household practices – in short, we need more work on systems as sets of connected innovations.

Question Four: What is the appropriate division of labour between national and supranational institutions?

Whilst studies of innovation for sustainability – whether successful or not – need an international and comparative focus, the balance between national and supranational policies to support them is still not clear. Indeed it has been made more complicated by the existence of treaties and agencies that work at the global level rather than at the level of, say, Europe or some similar large global region. This stems partly from the 'local' nature of the innovation process: 'variety generation' happens on the local scale but 'selection' has to be on a larger scale. In addition, if sustainability requires systems change – transitions to new combinations of social and organisational arrangements built around new technological applications – then the national is not the appropriate level for action. For many societal functions (transport is a good example), it is hardly possible to imagine transitions to sustainability taking place in one country! This emphasises the point made before about the need for broader studies of innovation – its influences and reasons and for success or failure – studies that now need to include the relationship between the national, the supranational and the global.

References

Abernathy W, Utterback J (1975) A dynamic model of product and process innovation. Omega 3: 3-22

Abernathy W, Utterback J (1978) Patterns of industrial innovation. Technology Review 80(7)

Advisory Committee on Science and Technology (ACOST) (1992) Cleaner Technology. London

Aggeri F (1999) Environmental policies and innovation - A knowledge-based perspective on cooperative approaches. Research Policy 28: 699-717

Aldrich H (1999) Organizations evolving. London

Altvater E, Mahnkopf B (1996) Grenzen der Globalisierung. Ökonomie, Ökologie und Politik in der Weltgesellschaft, Münster

Altvater E, Mahnkopf B (1999) Grenzen der Globalisierung. Ökonomie, Ökologie und Politik in der Weltgesellschaft. Münster

Anastas P, Williamson T (1998a) Frontiers in Green Chemistry. In: Anastas P, Williamson T (eds) Green Chemistry: Frontiers in Benign Chemical Syntheses and Processes. Oxford, pp 1-26

Anastas P, Warner J (1998b) Green Chemistry: Theory and Practice. Oxford

Andersen M, Massa I (2000) Ecological Modernisation: Origins, Dilemmas, and Future Directions. Journal of Environmental Policy and Planning 2(4): 337-345

Andersen M, Liefferink D (eds) (1997) European Environmental Policy, The Pioneers. Manchester

Anderson M, Massa I (2000) Ecological modernisation - Origins, dilemmas and future directions. Journal of Environmental Policy Planning 2(4): 337-345

Andersson M (1998) Nine Lessons from Poland. International Environmental Affairs 10(1): 3-7

Ansoff I (1965) The firm of the future. Harvard Business Review Sept-Oct: 165-178

Arrow K (1962) The Economic Implications of Learning by Doing. Review of Economic Studies 29(80): 155-173

Arthur W (1988) Competing technologies: an overview. In: Dosi G et al. (eds) Technical Change and Economic Theory. London, pp 590-607

Arthur W (1989) Competing technologies, increasing returns, and lock-in by historical events. The Economic Journal 99: 116-131

Ashford N (1993) Understanding Technological Responses of Industrial Firms to Environmental Problems: Implications for Government Policy. In: Fischer K, Schot J (eds) Environmental Strategies for Industry: International Perspectives on Research Needs and Policy Implications. Washington, pp 277-307

Ashford N (1994) An Innovation-Based Strategy for the Environment. In: Finkel A, Golding D (eds) Worst Things First? The Debate Over Risk-based National Environmental Priorities. Washington, pp 275-314

Ashford N (2000) An Innovation-Based Strategy for a Sustainable Environment. In: Hemmelskamp J, Rennings K, Leone F (eds) Innovation-Oriented Environmental Regulation: Theoretical Approach and Empirical Analysis. Heidelberg, pp 67-107

Ashford N et al. (1979) Environment, Health, and Safety Regulation, and Technological Innovation. In: Hill C, Utterback J (eds) Technological Innovation for a Dynamic Economy. Cambridge, pp 161-221

Ashford N, Heaton G (1983) Regulation and Technological Innovation in the Chemical Industry. Law and Contemporary Problems 46(3): 109-157

Ashford N, Miller C (1998) Low-Level Exposures to Chemicals Challenge Both Science and Regulatory Policy. Environmental Science & Technology 32(21): 508A-509A

Ashford N, Ayers C, Stone R (1985) Using Regulation to Change the Market for Innovation. Harvard Environmental Law Review 9(2): 419-466

Ashford N, Hafkamp W, Prakke F, Vergragt P (2001) Pathways to Sustainable Industrial Transformation: Cooptimising Competitiveness, Employment and Environment. Cambridge

Ayres I, Braithwaite J (1992) Responsive Regulation: Transcending the Deregulation Debate. New York

Ayres R, Ayres L (1996) Industrial Ecology. Cheltenham

Balzejczak J, Edler D, Hemmelskamp J, Jänicke M (1999) Environmental Policy and Innovation - an International Comparison of Policy Frameworks and Innovation Effects. In: Klemmer P (ed) Innovation and the Environment. Berlin

Bartholomew S (1997) National Systems of Biotechnology Innovation: Complex Interdependence in the Global System. Journal of International Business Studies 2

Bassala G (1988) The Evolution of Technology. Cambridge

Bätcher K, Böhm E, Tötsch W (1992) Untersuchung über die Auswirkungen geplanter gesetzlicher Beschränkungen auf die Verwendung, Verbreitung und Substitution von Cadmium in Produkten. Karlsruhe

Baumol W (1982) Contestable Markets: An Uprising in the Theory of Industrial Structure. American Economic Review 72: 1-15

Becker F (2002) Water Benign Process Innovations and Environmental Regulations. A Case Study of German Chemical Firms during the 1990s. Stuttgart

Becker M, Ashford N (1995) Exploiting Opportunities for Pollution Prevention in EPA Enforcement Agreements. Environmental Science & Technology 29(5): 220A-226A

Beise M (1999) Lead Markets and the International Allocation of R & D. (Paper prepared for the 5th ASEAT Conference "Demand, Markets, Users and Innovation: Sociological and Economic Approaches", Sept. 14-16, Manchester)

Beise M (2001) Lead Markets. Country Specific Success Factors of the Global diffusion of Innovations. New York

Berger P, Luckmann Th (1966) The Social Construction of Reality: a Treatise in the Sociology of Knowledge. New York

Bergmann J, Brandt G, Körber K, Mohl E, Offe C (1969) Herrschaft, Klassenverhältnisse und Schichtung. In: Adorno T (ed) Spätkapitalismus oder Industriegesellschaft, Verhandlungen des 16. Deutschen Soziologentages. Stuttgart, pp 67-87

Berkhout F (2002) Technological regimes, path dependency and the environment. Global Environmental Change 12: 1-4

Berkhout F, Almemark M, Lindfors L, Holm-Mueller K, Hanhoff-Stemping I, Smith A, Stripple H (2000) Sustainability, Competitiveness and Technical Change (SCOTCH). Final Report to the European Commission. Brighton

Bernauer T (2000) Staaten im Weltmarkt: Zur Handlungsfähigkeit von Staaten trotz wirtschaftlicher Globalisierung. Opladen

Binder M, Jänicke M, Petschow U (eds) (2001) Green Industrial Restructuring, Intenational Case Studies and Theoretical Interpretations. New York

Binswanger R (1978) Induced Innovation: Technology, Institutions and Development. Baltimore

Blazejczak J, Edler D (2000) Elements of innovation-friendly policy regimes - an international comparative study for the paper industry. In: Hemmelskamp J et al. (eds) Innovation-Oriented Environmental Regulation. Heidelberg

Bode M (2000) Consumers' Acceptance Analysis of Scenarios. Final Report SusHouse Project. Delft

Boden M, Miles I (2001) London and New York, Conclusions: Beyond the Service Economy. In: Boden M, Miles L (eds) Services and the Knowledge-based Economy. London and New York, pp 247-264

Botcheva L (2001) Expertise and International Governance: The Role of Economic Assessments in the Approximation of EU Environmental Legislation in Eastern Europe. Global Governance 7(3)

Botcheva L (2001a) International Markets, Industries, and the Environment: Cross Country Evidence from Eastern Europe. (Proceedings of the Ninth International Conference of the Greening of Industry Network. January 21-24, Bangkok, Thailand)

Braczyk HJ, Cooke P, Heidenreich M (eds) (1998) Regional Innovation System: The Role of Governances in a Globalized World. London

Bras-Klapwijk R (2000) Environmental Assessment of Scenarios. Final Report SusHouse Project. Delft

Braun E (1994) Promote and Regulate: The Dilemma of Innovation Policy. In: Aichholzer G, Schienstock G (eds) Technology Policy: Towards an Integration of Social and Ecological Concerns. Berlin, New York, pp 95-124

Breschi S, Malerba F (1997) Sectoral innovation systems: technological regimes, Schumpeterian dynamics and spatial boundaries. In: Edquist C (ed) Systems of Innovation. London, pp 130-153

Bressers H, Plettenburg L (1997) The Netherlands. In: Janicke M, Weidner H (eds) National Environmental Policies: A Comparative Study of Capacity-Building. Berlin, pp 109-132

Bressers J, Klok P (1988) Fundamentals of a theory of policy instruments. International Journal of Social Economics 15(3-4): 22-41

Brezet J, Bijma A, Ehrenfeld J, Silvester S (2001) Design of Eco-efficient Services. (DES-) report of the DfS program. Delft

Brickwedde F (ed) (1997) Umwelt und Arbeit - Innovationen als Motor des Strukturwandels. Bramsche

Brown H, Angel D (2000) Environmental Reforms in Poland: Lessons for Industrializing Economies. Environmental Science and Technology 34(18): 3849-3857

Brown H, Angel D, Derr P (2000) Effective Environmental Regulation: Learning from Poland's Experience. Westport

Butter M (2002) A three-layered policy approach for system innovations. (Paper for the first workshop of the BLUEPRINT network, 23-24 January 2002. Brussels)

Caldart C, Ashford N (1999) Negotiation as a Means of Developing and Implementing Environmental and Occupational Health and Safety Policy. Harvard Environmental Law Review 23(1): 141-202

Cameron D (1978) The Expansion of the Public Economy: A Comparative Analysis. American Political Science Review Vol. 72(4)

Campbell J, Hollingsworth J, Lindberg L (1991) Governance of the American economy. Cambridge

Carlsson B, Jacobsson S (1997) Diversity creation and technological systems; a technology policy perspective. In: Edquist C (ed) Systems of Innovation. London, pp 266-294

Carlsson B, Stankiewicz R (1991) On the nature, function and composition of technological systems. Journal of Evolutionary Economics Vol 1: 193-118

Central Bureau of Statistics GUS (1996) Environmental Protection 1996. Warsaw

Central Bureau of Statistics GUS (1998) Annual Statistical Report 1998. Warsaw

Central Bureau of Statistics GUS (1998a) Environmental Protection 1998. Warsaw

Christensen A, Tangen C (1999) The Finnish Combined Heat and Power Sector. In: Koefoed A, Eikeland P, Midttun A (ed) Green Energy-Industrial Innovation: A comparative study of green energy transformations in Northern Europe. Research Report no 7. Sandvika

Christensen C (2000) The Innovator's Dilemma. When New Technologies Cause Great Firms to Fail. Harvard

Christoff P (1996) Ecological Modernization, Ecological Modernities. Environmental Politics 5(3): 476-500

Clark S, Georg S (1995) From Greening to Sustaining: Transformational Challenges for the Firm. (Summary Report of the third international conference of the Greening of Industry Network, Copenhagen)

Cleff T, Rennings K (1999) Determinants of Environmental Product and Process Innovation. European Environment 9: 191-201

Coglianese C, Allen L (2001) Building Sector-Based Consensus: A Review of the U.S. EPA's Common Sense Initiative. (Paper presented at the workshop "Voluntary, Collaborative and Information-Based Policies: Lessons and Next Steps for Environmental and Energy Policy in the United States and Europe", May 10-12 2001, Harvard University)

Coglianese G (1997) Assembling Consensus: The Promise and Performance of Negotiated Rulemaking. Duke Law Journal 46: 1255

Cohen M, Levinthal D (1990) Absorptive capacity: a new perspective on learning and innovation. Administrative Science Quarterly Vol 35: 128-152

Cohen M, Levinthal D (1989) Innovation and learning: the two faces of R&D. Economic Journal Vol 99: 569-96

Colborn T, Dumanowski D, Myers J (1996) Our Stolen Future. New York

Cole D (1997) Instituting Environmental Protection: From Red to Green in Poland. New York

Conrad J (1998) Environmental Management in European Companies. Amsterdam

Constant I (1980) The Origins of the Turbojet Revolution. Baltimore

Cooke P, Schienstock G (2000) Structural Competitiveness and Learning Regions. Enterprise and Innovation Management Studies 1(3): 265-280

Coriat B (1995) Organizational Innovations: The Missing Link in European Competitiveness. In: Andreasen L, Coriat B, De Hertog F, Kaplinsky R (eds) Europe's Next Step: Organizational Innovation, Competition and Employment. London, pp 3-32

Cowan R, Gunby P (1996) Sprayed to death: path dependence, lock-in and pest control strategies. Economic Journal 106: 521-542

COWI Consulting Engineers And Planners (1997) Study on Voluntary Agreements Concluded Between Industry and Public Authorities in the Area of the Environment. Final Report to the European Commission DG-XIII. Soeborg

David P (1985) Clio and the economics of QWERTY. American Economic Review 75: 332-337

David P (1985) Clio and the Economics of QWERTY. Economic History 75: 227-332

David P (1987) Some new standards for the economics of standardization in the information age. In: Dasgupta P, Stoneman P (eds) Economic policy and technological performance. Cambridge, pp 206-239

David P (2000) Understanding Digital Technology's Evolution and the Path of Measured Productivity Growth: Present and Future in the Mirror of the Past. In: Brynolfsson E, Kahin B (eds) Understanding the Digital Economy: Data, Tools, and Research. Cambridge

Davies C, Mazurek J (1998) Pollution Control in the United States. Evaluating the System. Washington D.C.

De Bruijn T. (forthcoming) Multi-level Governance between the European Union and its Member States: The Importance of Policy Style. Achieving Sustainable Development: the Challenge of Goverance Across Social Scales

De Bruijn T, Norberg-Bohm V (2001) Voluntary, Collaborative, and Information-Based Policies: Lessons and Next Steps for Environmental and Energy Policy in the United States and Europe. Cambridge

De Bruijn T, Lulofs K (2001) The Dutch Policy Program on Environmental Management: Policy Implementation in Networks. (Paper presented at the workshop "Voluntary, Collaborative and Information-Based Policies: Lessons and Next Steps for Environmental and Energy Policy in the United States and Europe", May 10-12 2001, Harvard University)

De Bruijn T, Norberg-Bohm V (eds) (forthcoming) Industrial Transformation: Environmental Policy Innovation in the United States and Europe, Cambridge

Dearing A (1999) Technology Co-operation for Sustainable Development. In: OECD (ed) Special Issue on Sustainable Development, STI Review 25. Paris, pp 161-176

Dearing A (2000) Sustainable Innovation: Drivers and Barriers. In: OECD (ed) Innovation and the Environment. Paris

DeSimone L, Popoff F (2000) Eco-Efficiency: The Business Link to Sustainable Development. Cambridge

Deuten J, Rip A, Jelsma J (1997) Societal Embedment and Product Creation Management. Technology Analysis & Strategic Management 9(2): 219-236

Diekmann A, Preisendörfer P (2001) Umweltsoziologie. Eine Einführung. Hamburg

Diepenmaat H, Te Riele H (2001) Boven het Klaver bloeien de Margrieten; een maatschappelijk netwerk voor innovaties richting duurzaamheid. (For the Minsitry of the Environment, mimeo)

Dorfman M, Muir W, Miller C (1992) Environmental Dividends: Cutting More Chemical Waste. New York

Dosi G, Freeman C, Nelson R, Silverberg G, Soete L (eds) (1988) Technical Change and Economic Theory. London

Dosi G (1982) Technological Paradigms and Technological Trajectories: A Suggested Interpretation of the Determinants and Directions of Technical Change. Research Policy 11: 147-162

Dosi G (1988) The nature of the innovative process. In Dosi G et al (eds) Technical Change and Economic Theory. London, pp 221-38

Downing P, Lawrence J (1986) Innovation in Pollution Control. Journal of Environmental Economics and Management 13: 18-29

Dresel T, Blättel-Mink B (1997) Ökologie im Unternehmen. In: Renn O und Blättel-Mink B (ed) Zwischen Akteur und System. Die Organisation von Innovation. Opladen, pp 235-255

Drezner D (2001) Globalization and Policy Convergence. The International Studies Review 3(1): 53-78

Dryzek J (1997) The Politics of the Earth: Environmental Discourses. Oxford

ECN (2000) Energietechnologie in het spanningsveld tussen klimaatbeleid en liberalisering, Energie Centrum Nederland, ECN-C-00-020. Petten

Edquist C (ed) (1997) Systems of Innovation Systems of Innovation: Technologies, Institutions and Organisations. London

Edquist C (1997) Systems of Innovation Approaches - Their Emergence and Characteristics. In: Edquist C (ed) Systems of Innovation: Technologies, Institutions and Organisations. London

Edquist C, McKelvely M (eds) (2000) Systems of innovation: Growth Competitiveness and Employment. London

EEA (European Environmental Agency) (1997) Environmental Agreements: Environmental Effectiveness, Environmental Issues. EEA series 3(1)

Eikeland P(1999) The case of bioenergy industry development in Sweden. In Koefoed A, Eikeland P, Midttun A (ed) Green Energy-Industrial Innovation: A comparative study of green energy transformations in Northern Europe. Research Report no 7 1999. Sandvika

Eikeland P(1999) The case of bioenergy industry development in Finland. In Koefoed A, Eikeland P, Midttun A (ed) Green Energy-Industrial Innovation: A comparative study of green energy transformations in Northern Europe. Research Report no 7 1999. Sandvika

Elzen B (1999) Inventory of Market Acceptance Factors, Enschede. (Deliverable 5 of EU-Project UTOPIA. Contract No. UR-97-SC-2076)

Environmental Law Institute (1997) Research and Development Practices in the Environmental Technology Industry. Washington

Environmental Law Institute (1999) Innovation, Cost and Environmental Regulation: perspectives on business, policy and legal factors affecting the cost of compliance. Washington

EPA (US Environmental Protection Agency) (1991) Pollution Prevention: Progress in Reducing Industrial Pollutants (EPA 21P-3003). Washington

EPA (US Environmental Protection Agency) (1994) Technology Innovation Strategy (EPA 543-K-93 002). Washington

Erdmann G (1993) Elemente einer evolutorischen Innovationstheorie. Tübingen

Esty D, Porter M (2000) Measuring National Environmental Performance and Its Determinants. In: Harvard University/World Economic Forum (eds) The Global Competitiveness Report 2000. New York/Oxford, pp 60-75

European Commission (2000) Green Paper Towards a European strategy for the security of energy supply. Brussels

European Commission (2000) European Commission Regular Report on Progress Towards Accession. Brussels

European Commission (2001) Environment 2010: Our future, Our choice, The Sixth Environment Action Programme. Proposal for a Community Environment Action Programme 2001-2010. Brussels

European Communities (2002), Decision No 1513/2002/EC of the European Parliament and of the Council of June 27, 2002 concerning the sixth framework programme of the European Community for research, technological development and demonstration activities contributing to the creation of the European Research Area and to innovation (2002 to 2006). Official Journal, L 232/5, 29 August, 2002.

Farrell F, Hitchens D (2001) The Impact of Best Available Techniques (BAT) on the Competitiveness of European Non-Ferrous Metals Industry. Seville

Finkel A, Golding D (eds) (1994) Worst Things First? The Debate Over Risk-based National Environmental Priorities. Washington

Fiorino D (1999) Rethinking environmental regulation: perspectives on law and governance. The Harvard Environmental Law Review 23(2): 441-469

Fischer K, Schot J (eds) (1993) Environmental Strategies for Industry - International Perspectives on Research Needs and Policy Implications. Washington

Fransman M (1990) The Market and Beyond. Cooperation and Competition in Information Technology in the Japanese System. Cambridge

Freeman C (1982) The Economics of Industrial Innovation. London

Freeman C (1987) Technology Policy and Economic Performance. London

Freeman C (1992) The Economics of Hope. London

Freeman C (1997) Innovation systems: City-state, national, continental and sub-national. (Mimeo, Paper presented at the Montevideo Conference, University of Sussex, SPRU)

Freeman C, Perez C (1988) Structural crises of adjustment: business cycles and investment behaviour. In Dosi G, Freeman C, Nelson R, Silverberg G, Soete L (eds) Technical Change and Economic Theory. London

Freeman C, Soete L (1997) The Economics of Industrial Innovation. London

Freeman C, Clark J, Soete L (1982) Unemployment and Technical Innovation. London

Freeman R (1984) Strategic Management: A Stakeholder Approach. Boston

Friedrich Ebert Stiftung (1991): Wirtschaftspolitische Diskurse Nr. 19: Chemiestandort Ostdeutschland, Bonn.

Frosch R, Gallapoulos N (1989) Strategies for manufacturing. Scientific American 261(3): 144-52

Fukasaku Y (1998) Public/private partnerships for developing environmental R&D. In: OECD (ed) STI Review 23. Paris

Fukasaku Y (1999) Stimulating Environmental Innovation. In: OECD (ed) STI Review 25, Special Issue on Sustainable Development. Paris, pp 47-64

Fukasaku Y (1999) Technology foresight and sustainable development. Futures Research Quarterly 15(3)

Garud R, Karnoe P (2000) Path Creation as a Process of Mindful Deviation. In: Koski J, Marttila S (eds) Proceedings of the Conference on Knowledge and Innovation. Helsinki

Geels F (2002) Understanding the Dynamics of Technological Transitions; A Co-evolutionary and Socio-technical Analysis. Enschede

Geels F, Kemp R (2000) Transities vanuit socio-technisch perspectief. (Background document for chapter 1 of the report "Transities en transitiemanagement" Rotmans et al. (2000). Eindhoven)

Geels F (2002) 'Technological transitions as evolutionary reconfiguration processes: A multi-level perspective and a case-study. Research Policy 31: 1257-1274

GEMI (Global Environmental Management Initiative) (1999) Environment: Value to Business. Washington

Georgescu-Roegen N (1971) The Entropy Law and the Economic Process. Cambridge
Giddens A (1990) The Consequences of Modernity. Cambridge
Gislev M (2000) European innovation and exchange of information about BAT' in The Sevilla process: A driver for environmental performance in industry, Proceedings Federal Environmental Agency. Berlin
Glinski P (1998) Polish Greens and Politics: A Social Movement in a Time of Transformation. In: Clark J, Cole D (eds) Environmental Protection in Transition: Economic, Legal and Socioeconomic Perspectives on Poland. Brookfield
Goldemberg J, Johansson T, Reddy A, Williams R (1985) An end-use oriented global energy strategy. Annual Review of Energy 10: 613-88
Gottlieb R (ed) (1995) Reducing Toxins: A New Approach to Policy and Industrial Decisionmaking. Washington
Gouldson A, Murphy J (1998) Regulatory Realities: the Implementation and Impact of Industrial Environmental Regulation. London
Grabher G (1993) The Weakness of Strong Ties: The Lock-in of Regional Development in the Ruhr Area. In: Grabher G (ed) The Embedded Firm. On the Socio-economics of Industrial Networks. London, pp 255-277
Graedel T, Allenby B (1995) Industrial Ecology. New York
Graham M, Miller C (2001) Disclosure of Toxic Releases in the United States. (Paper presented at the workshop "Voluntary, Collaborative and Information-Based Policies: Lessons and Next Steps for Environmental and Energy Policy in the United States and Europe", May 10 -12 2001, Harvard University)
Green K, Young W (2000) The Shopping, Cooking and Eating Function. Final Report SusHouse Project. Delft
Green K, McMeekin A, Irwin A (1994) Technological trajectories and R&D for environmental innovation in UK firms. Futures 26(10) 1047-1059
Greenspan Bell R (2000) Building Trust: Laying the Foundation for Environmental Regulation in the Former Soviet Bloc. Environment 42(2): 20-32
Greenspan B, Bromm S (1997) Lessons Learned in the Transfer of U.S.-Generated Environmental Compliance Tools: Compliance Schedules for Poland. Environmental Law Reporter; News and Analysis June 97: 10296-10303
Grin J, Van de Graaf H, Vergragt P (2003) Een derde generatie milieubeleid: een sociologisch perspectief en een beleidswetenschappelijk programma. Beleidswetenschap
Groenewegen P, Vergragt P (1991) Environmental Issue as Threats and Opportunities for Technological Innovation. Technological Analysis and Strategic Management 3(1): 43-55
Grubler A, Nakicenovic N, Victor D (1999) Dynamics of Energy Technologies and Global Change. Energy Policy 27:247-280
Grupp H (1998) Foundations of the Economics of Innovation. Cheltenham
Gujarati D (1995) Basic Econometrics. New York
Gunningham N (1999) Integrative Regulation: A Principle-Based Approach to Environmental Policy. Law and Social Inquiry 24(4): 853-896
Habermeier K (1990) Product use and product improvement. Research Policy 19: 271-283
Häfele W (ed) (1990) Energiesysteme im Übergang unter den Bedingungen der Zukunft. Landsberg/Lech
Hajer M (1995) The Politics of Environmental Discourse - Ecological Modernization and the Policy Process. Oxford

Harrison K (1999) Talking with the Donkey: Cooperative Approaches to Environmental Protection. Journal of Industrial Ecology 2(3): 51-72

Hart S, Milstein M (1999) Global Sustainability and the Creative Destruction of Industries. Sloan Management Review 41(1): 23-33

Hemmelskamp J (1996) Umweltpolitik und Innovation. ZEW-Discussion Paper no. 96-23

Hemmelskamp J (1997) Environmental Policy Instruments and their Effects on Innovation. European Planning Studies 5(2): 177-193

Hemmelskamp J (1998) Environmental Taxes and Standards: An Empirical Analysis of the Impact on Innovation. In: Hemmelskamp J, Rennings K, Leone F (eds) (2000) Innovation-oriented Environmental Regulation. Heidelberg, pp 303-436

Hemmelskamp J, Rennings K, Leone F (eds) (2000) Innovation-Oriented Environmental Regulation. Heidelberg

Héritier A, Mingers S, Knill C, Becka M (1994) Die Veränderung von Staatlichkeit in Europa - Ein regulativer Wettbewerb: Deutschland, Großbritannien, Frankreich. Opladen

Hettige M, Huq M, Pergall F, Wheler D (1996) Determinants of Pollution Abatement in Developing Countries. Evidence from South and South East Asia. World Development 24(12): 1891-1904

Hicks J (1932) The Theory of Wages. London

Hicks B (1996) Environmental Politics in Poland: A Social Movement Between Regime and Opposition. New York

Hirschhorn J (1995) Pollution Prevention Comes of Age. Georgia Law Review 29: p 325

Hirst P, Thompson G (1992) The Problem of Globalization: International Relations, National Economic Management and the Formation of Trade Blocs. Economy and Society 12(4) :357-396

Hitchens D (2001) The Implication for Competitiveness of Environmental Regulations in EU. Seville

Hitchens D, Farrell F, Lindblom J, Triebswetter U (2001) The Impact of Best Available Techniques (BAT) on the Competitiveness of European Industry. Seville

Hitchens D, Birnie J, McGowan A, Triebswetter U, Cottica A (1998) The Firm, Competitiveness and Environmental Regulation. Cheltenham

Hitchens D, Triebswetter U, Birnie J, Thompson W, Bertossi P, Messori L (2000) Environmental Regulation and Competitive Advantage, A study of packaging waste in the European Supply Chain. Cheltenham

Hodgson G (1988) Economics and Institutions. Cambridge

Hoechst (1996) Change - Standorte: Chemie zwischen Nachbarn. Frankfurt

Hoffman A (1997) From Heresy to Dogma: An Instiutional History of Corporate Environmentalism. San Francisco

Hoffman A (2000) Competitive Environmental Strategy. Washington

Hofman P, Schrama G (2001) The Dutch Target Group Approach. (Paper presented at the workshop "Voluntary, Collaborative and Information-Based Policies: Lessons and Next Steps for Environmental and Energy Policy in the United States and Europe", May 10-12 2001, Harvard University)

Holdren J, Ehrlich P (1974) Human population and the global environment. American Scientist 62: 282-92

HOMES (1999) (Proceedings of the 2nd Int. Symposium on Sustainable Household Consumption, Groningen-Paterswolde, June 3-4)

Honkasalo A (2000) Finnish Environmental Cluster Research Program. In: OECD (ed) Innovation and the Environment: Workshop Proceedings. Paris

Hoogma R, Kemp R, Schot J, Truffer B (2002) Experimenting for Sustainable Transport. The approach of Strategic Niche Management. London
Hoogma R, Kemp R, Schot J, Truffer B (2001) Experimenting for Sustainable Transport Futures. The Approach of Strategic Niche Management. London
Horst Humphrey A (1997) Socialism, Capitalism and Transition: With Special Reference to Poland. Tilburg
Howard J, Nash J, Ehrenfeld J (2000) Standard or Smokescreen? Implementation of a Voluntary Code. California Management Review 42(2): 63-82
Howes R, Skea J, Whelan B (1997) Clean and Competitive? Motivating Environmental Performance in Industry. London
Huber J (1995) Nachhaltige Entwicklung. Strategien für eine ökologische und Sociale Erdpolitik. Berlin
Hübner K, Nill J (2001) Nachhaltigkeit als Innovationsmotor. Herausforderungen für das deutsche Innovationssystem. Berlin
Hughes T (1983) Networks of Power. Electrification in Western Society 1880-1930. Baltimore
Hunt C, Auster E (1990) Proactive Environmental Management: Avoiding the Toxic Trap. Sloan Management Review 31(2): 7-18
IEA (International Energy Agency) (2000) Energy Policies of IEA Countries: 2000 Review. Paris
IHDP (International Human Dimensions Programme) (1999) Industrial Transformation Science Plan. Bonn
IIASA-WEC (1998) Energy in the 21st Century. Laxenburg
Irwin A, Vergragt P (1989) Re-thinking the relationship between environmental regulation and industrial innovation. Technology Analysis and Strategic Management 1(1): 57-70
Irwin A, Georg S, Vergragt P (1994) The Social Management of Environmental Change. Futures 26(3): 323-334
Islas J (1997) Getting round the lock-in in electricity generating systems: the example of the gas turbine. Research Policy 26: 49-66
Jaakko Pöyry Consulting (2001) The Impact of Best Available Techniques (BAT) on the Competitiveness of the European Pulp and Paper Industry. Helsinki
Jackson T (ed) (1993) Clean Production Strategies. Boca Raton
Jacob K (1999) Innovationsorientierte Chemikalienpolitik. Politische, soziale und ökonomische Faktoren des verminderten Gebrauchs gefährlicher Stoffe. München
Jacob K (2001) Chlorine Production in Germany. In: Binder M, Jänicke M, Petschow U (ed) Green Industrial Restructuring. International Case Studies and Theoretical Interpretations. Heidelberg, pp 187-216
Jacob K, Jänicke M (1998) Ökologische Innovationen in der chemischen Industrie: Umweltentlastung ohne Staat? ZfU 21(4): 519-547
Jaffe A, Newell R, Stavins R (2001) Technological change and the environment. In: Mäler K, Vincent G (eds) The Handbook of Environmental Economics. Amsterdam
Jaffe A, Palmer K (1996) Environmental Regulation and Innovation: A Panel Data Study. NBER Working Paper 5545 April
Jaffe A, Newell R , Stavins R (2000) Technological Change and the Environment. NBER Working Paper 7970 October
Jaffe B, Peterson S, Portney P, Stavins R (1995) Environmental Regulation and Competitiveness of U.S. Manufacturing: What does Evidence Tell Us? Journal of Economic Literature 33(1): 136-63

Jänicke M (1979) Wie das Industriesystem von seinen Mißständen profitiert. Opladen
Jänicke M (1985) Preventive Environmental Policy As ecological Modernisation And Structural Policy. Wissenschaftszentrum Berlin IIUG dp 85-2. Berlin
Jänicke M (1998) Umweltpolitik: Global am Ende oder am Ende global? In: Beck U (ed) Perspektiven der Weltgesellschaft. Frankfurt, pp 332-344
Jänicke M, Weidner H (1996) Summary: Global Environmental Policy Learning. National Environmental Policies: A Comparative Study of Capacity-Building. Heidelberg
Jänicke M, Weidner H (eds) (1997) National Environmental Policies: A Comparative Study of Capacity-Building. Berlin
Jänicke M, Klaus J (2002) Ecological Modernisation and the Creation of Lead Markets, this collection.
Jänicke M, Binder M, Mönch H (1997) Dirty Industries: Patterns of Change in Industrial Countries. Environmental and Resource Economics 9(4): 467-491
Jänicke M, Blazejczak J, Edler D, Hemmelskamp J (2000) Environmental Policy and Innovation: an International Comparison of Policy Frameworks and Innovation Effects. In: Hemmelskamp J, Rennings K, Leone F (eds) Innovation-oriented Environmental Regulation. Theoretical Approaches and Empirical Analysis. Heidelberg, pp 125-152
Jansen J (1991) Duurzaam Denken, Duurzaam Doen. Intreerede. Delft
Jansen J, Vergragt P (1992) Sustainable Development: A Challenge to Technology! (Proposal for the Interdepartmental Research Program 'Sustainable Technological Development', 10th june 1992, Leidschendam)
Jendroska J (1996) Environmental Regulatory Framework in Poland: History and Recent Development. (Proceedings of a September 1995 Workshop on Environmental Health and Safety in Private Enterprises in Poland)
Johnson B (1992) Institutional learning. In: Lundvall B (ed) National Systems of Innovation: Towards a Theory of Innovation and Interactive Learning. London
Johnson B (1996) Systems of Innovation: Overview and Basic Concepts: Introduction. In Edquist C (ed) Systems of Innovation. Technologies, Institutions and Organisations. London, pp 36-40
Jörgens H (1996) Die Institutionalisierung von Umweltpolitik im internationalen Vergleich. In: Jänicke M (ed) Umweltpolitik der Industrieländer. Entwicklung - Bilanz - Erfolgsbedingungen. Berlin, pp 59-111
Judge W, Douglas T (1998) Performance Implications of Incorporating Natural Environmental Issues into the Strategic Planning Process: An Empirical Assessment. The Journal of Management Studies 35(2): 241-260
Kaminski M, Bertelli D, Moye M, Yudken J (1996) Making Change Happen: Six Cases of Companies and Unions Transforming their Workplaces. Washington
Karaczun Z (1997) Policy of Air Protection in Poland. Warsaw
Keijzers G (2000) The Evolution of Dutch Environmental Policy: the Changing Ecological Arena from 1970-2000 and Beyond. Journal of Cleaner Production 8(3):179-200
Kemp R (1994) Technology and Environmental Sustainability: The Problem of Technological Regime Shifts. Futures 26(10): 1023-1046
Kemp R (1997) Environmental Policy and Technical Change: A Comparison of the Technological Impact of Policy Instruments. Cheltenham
Kemp R (1998) Environmental Regulation and Innovation Key Issues and Questions for Research. In: Institute for Prospective Technological Studies (IPTS) (ed) The Impact of EU-Regulation on Innovation of European Industry. Seveilla, pp 12-39

Kemp R (2000) Technology Effects of Environmental Policy - An overview of the effects of past policies and suggestions for improvement. In OECD (ed) Innovation and the Environment: Workshop Proceedings. Paris

Kemp R (2000) Technology and Environmental Policy: Innovation effects of past policies and suggestions for improvement. In: OECD (ed) Innovation and the Environment. Paris

Kemp R (2001) Opportunities from a Green Industrial Policy from an Evolutionary Technology Perspective. In: Binder M, Jämicke M, Petschow U (eds) Green Industrial Restructuring. Berlin, pp 151-169

Kemp R, Soete L (1992) The Greening of Technological Progress: An Evolutionary Perspective. Futures 24(5): 437-457

Kemp R et al. (1994) Technology and the transition to environmental stability: continuity and change in complex technological systems. Final report for SEER research programme of the Commission of the European Communities. Maastricht

Kemp R, Arundel A (1998) Survey Indicators for Environmental Innovation. IDEA paper series No. 8, STEP (Studies in technology, innovation and economic policy)

Kemp R, Loorbach D (2003) Governance for Sustainability Through Transition Management. Paper presented at the Berlin Conference on the Human Dimensions of Global Environmental Change: Governance for Industrial Transformation. Berlin

Kemp R, Rotmans J (2001) The Management of the Co-evolution of Technical, Environmental and Social Systems. (Paper for International Conference:"Towards Environmental Innovation Systems", 27-29 September 2001, Garmisch-Partenkirchen)

Kemp R, Moors E (2002) Modulating Dynamics in Transport for Climate Protection. In: Faure M, Gupta J, Nentjes A (eds) Institutions and Instruments to Control Climate Change: Kyoto and After. Cheltenham

Kemp R, Schot J, Hoogma R (1998) Regime Shifts to Sustainability through Processes of Niche Formation. The Approach of Strategic Niche Management. Technology Analysis and Strategic Management 10(2): 175-195

Kemp R, Truffer B, Harms S (2000) Strategic Niche Management for Sustainable Mobility. In: Rennings K, Hohmeier O, Ottinger R (eds) (2000) Social Costs and Sustainable Mobility - Strategies and Experiences in Europe and the United States. Heidelberg, pp 167-187

Kemp R, Smith K, Becher G (2000) How should we study the relationship between environmental regulation and innovation. In Hemmelskamp J et al. (eds) Innovation-Oriented Environmental Regulation. Heidelberg

Kemp R, Rip A, Schot J (2001) Constructing Transition Paths through the Management of Niches. In: Garud R, Karnøe P (eds) Path Dependence and Creation. London, pp269-299

Kempf A (2000) Selbst die US-Konkurrenz spricht von "Verbund". Stuttgarter Zeitung 250(13)

Kern K (2000) Die Diffusion von Umweltinnovationen. Umweltpolitische Innovationen im Mehrebenensystem der USA. Opladen

Kern K, Jörgens H, Jänicke M (1999) Die Diffusion umweltpolitischer Innovationen. Ein Beitrag zur Globalisierung von Umweltpolitik FFU-Report 99-11. Berlin

King A, Lenox M (2000) Prospects for Industry Self-Regulation Without Sanctions: A Study of Responsible Care in the Chemical Industry. The Academy of Management Journal 43(4): 698-716

Kingdon J (1995) Agendas, Alternatives, and Public Policies. New York

Klemmer P (ed) (1999) Innovation and the Environment. Berlin
Klemmer P, Lehr U, Löbbe K (1999) Umweltinnovationen - Anreize und Hemmnisse. Berlin
Klepper S, Simons K (1997) Technological extinctions of industrial firms: an enquiry into their nature and causes. Industrial and Corporate Change 6: 379-460
Kline S, Rosenberg N (1986) An overview of innovation. In: Landau R, Rosenberg N (eds) The positive sum strategy. Harnessing technology for economic growth. Washington
Kline S, Rosenberg N (1986) An Overview of Innovation. In Landau R, Rosenberg N (eds) The Positive Sum Strategy. Washington, pp 275-305
Knill C, Lenschow A (1998) The Impact of British and German Administrations on the Implementation of EU Environmental Policy. Journal of European Public Policy 5(4): 595-614
Knot M (2000) Eternally yours and Neighbourhood clothing pool. In: Young W, Vergragt P Draft Design-Orienting Scenarios, (DOS), SusHouse Project, Manchester CROMTEC. Manchester
Knot J, Marjolijn C, Van den Ende J, Vergragt P (2001) Flexibility strategies for sustainable technology development. Technovation 21: 335-343
Koefoed A (1999) The case of bioenergy use and industry development in Denmark: In Koefoed A, Eikeland P, Midttun A (ed) Green Energy-Industrial Innovation: A comparative study of green energy transformations in Northern Europe. Research Report no 7. Sandvika
Koefoed A (1999) The case of Danish wind power integration and wind industry development. In Koefoed A, Eikeland P, Midttun A (ed) Green Energy-Industrial Innovation: A comparative study of green energy transformations in Northern Europe. Research Report no 7. Sandvika
Krugman P (1996) Making Sense of the Competitiveness Debate. Oxford Review of Economic Policy 12(3): 17-25
Lanjouw J, Mody A (1996) Innovation and the International Diffusion of Environmentally Responsive Technology. Research Policy 25: 549-571
Lazonick W (1990) Competitive Advantage on the Shop Floor. Cambridge
Lehr U, Löbbe K (2000) The joint project 'Innovation impacts of environmental policy'. In: Hemmelskamp J et al. (eds) Innovation-Oriented Environmental Regulation. Heidelberg
Leibowitz S, Margolis S (1999) Winners, Losers and Microsoft: Competition and Antitrust in High Technology. Oakland
Leonard D (1995) Wellsprings of Knowledge. Building and Sustaining the Sources of Innovation. Boston
Lévêque F (1993) How can environmental policy makers tackle industrial diversity? In: OECD (ed) Environmental Policy and Industrial Competitiveness. Paris, pp 78-85
Lewis M (1963) Technics and Civilisation. San Diego
Lewis S, Henkels D (2000) Good Neighbor Agreements: A Tool for Environmental and Social Justice. Social Justice 23(4)
Lewis S (1993) The Good Neighbor Handbook, A Community-based Strategy for Sustainable Industry. New York
Liefferink D (1997) The Netherlands: A Net Exporter of Environmental Policy Concepts. The Innovation of EU Environmental Policy. In: Liefferink D, Anderson A (eds) European environmental policy. The pioneers. Manchester

Lindblom J, Triebswetter U, Hitchens D (2001) The Impact of Best Available Techniques (BAT) on the Competitiveness of the European Pulp and Paper Industry. Seville
Littmann K (1975) Die Chancen staatlicher Innovationslenkung. Göttingen
Loasby B (1999) Knowledge, Institutions and Evolutionary Economics. London
Luhman N (1982) The Differentiation of Society. New York
Lundvall B (1988) Innovation as an Interactive Process: From User-Producer Interaction to the National System of Innovation. In: Dosi G, Freeman C, Nelson C et al. (eds) Technical Change and Economic Theory. London, pp 349-369
Lundvall B (ed) (1992) National Systems of Innovation: Towards a Theory of Interactive Learning. London
Lundvall B (1998) Why study Naional Systems and National Styles of Innovation? Technology Analysis & Strategic Management 10(4)
Lundvall B (1999) Nation-states, social capital and economic development - a systems approach to knowledge creation and learning. (Mimeo. Paper presented at 'The International Seminar on Innovation, Competitiveness and Environment in Central America: A Systems of Innovation Approach', February 22-23 1999, San José, Costa Rica)
Lundvall B, Borrás S (1997) The Globalizing Learning Economy: Implications for Innovation Policy. (Report based on the preliminary conclusions from several projects under the TSER Programme, DG XII; Commission of the European Union)
Lundvall B, Archibugi D (2001) Introduction: Europe and the Learning Economy. In: Archibugi D, Lundvall B (eds) The Globalizing Learning Economy. Oxford, pp 1-17
Lynn G, Morone J, Paulson A (1997) Marketing and Discontinuous Innovation: The Probe and Learn Process. In: Tushman P, Anderson P (eds) Managing Strategic Innovation and Change. A Collection of Readings. Oxford, pp 353-375
Malaman R (1996) Technological Innovation for Sustainable Development: Generation and Diffusion of Cleaner Technologies in Italian Firms. (Paper presented at the EARE conference, Lisbon)
Malerba F (2002) Sectoral Systems of Innovation and Production. Research Policy 31(2): pp 247-264
Manzini E, Jégou F (2000) The Construction of Design Orienting Scenarios. Final Report SusHouse project. Delft
March J (1991) Exploration and Exploitation in Organizational Learning. Organizational Science 2(1): 71-78
March J, Olsen J (1989) Rediscovering Institutions: The Organizational Basis of Politics. New York
Marchetti C, Nakicenovic N (1979) The Dynamics of Energy Systems and the Logistic Substitution Model. Luxenburg
Marcus A et al. (2001) Cooperative Regulation: Setbacks and Accomplishments of Project XL. (Paper presented at the workshop "Voluntary, Collaborative and Information-Based Policies: Lessons and Next Steps for Environmental and Energy Policy in the United States and Europe", May 10-12 2001, Harvard University)
McDonough W, Braungart M (1998) The NEXT Industrial Revolution. The Atlantic Monthly 282(4): 82-92
McKelvey M (1997) Co-evolution in commercial genetic engineering, Industrial and Corporate Change 6: 503-532
Meeus M, Oerlemans L (1999) National Systems of Innovation. In Steven C, van Waarden F (eds) Innovation and Institutions: A Programmatic Study. Wassenmaar, pp 120-163

Meyer-Krahmer F (1997) Innovation und Nachhaltigkeit im Zeichen der Globalisierung. Ökologisch Wirtschaften 1: 20-22
Meyer-Kramer F (2001) Industrial Innovation and Sustainability - Conflicts and Coherence. In Archibugi D, Lundvall B (eds) The Globalizing Learning Economy. Oxford, pp 177-196
Mintzberg H (1987) Crafting strategy. Harvard Business Review July-August: 66-75
Mol A (1995) The Refinement of Production: Ecological Modernization Theory and the Chemical Industry. Utrecht
Mol A (2001) Globalization and Environmental Reform. The Ecological Modernization of the Global Economy. Cambridge
Mol A, Spaargaren G (2000) Ecological Modernisation Theory in debate: a Review. Environmental Politics 9(1): 17-49
Mol A, Sonnenfeld D (2000a) Ecological Modernisation Around the World: An Introduction. Environmental Politics 9(1): 3-14
Mol A, Sonnenfeld D (eds) (2000b) Ecological Modernisation Around the World: Perspectives and Critical Debates. London
Mol A, Lauber V, Liefferink J (eds) (2000) The Voluntary Approach to Environmental Policy. Joint Environmental Policy-making in Europe. Oxford
Mom G (1997) De geschiedenis van de auto van morgen. Deventer
Moors E (2000) Metal Making in Motion. Technology choices for Sustainable Metals Production. Delft
Müller-Fürstenberger G (1995) Kuppelproduktion. Heidelberg
Mundl A, Schutz H, Stodulski W, Sleszynski J, Welfens M (1999) Sustainable Development by Dematerialization in Production and Consumption. Warsaw
Murphy J (2000) Ecological modernisation. Geoforum 31(1): 1-8
NACEPT (National Advisory Council for Environmental Policy and Technology) (1991) Permitting and Compliance Policy: Barriers to U.S. Environmental Technology Innovation. Washington
NACEPT (National Advisory Council for Environmental Policy and Technology) (1992) Improving Technology Diffusion for Environmental Protection. Washington
NACEPT (National Advisory Council for Environmental Policy and Technology) (1993) Transforming Environmental Permitting and Compliance Policies to Promote Pollution Prevention. Washington
NAPA (National Academy of Public Administration) (2001) Leading Change: Advancing Effective Governance in the 21st Century. Washington
Nash J, Ehrenfeld J (1996) Code Green: Business Adopts Voluntary Environmental Standards. Environment 38(1): 16-20, 36-45
Nash J (2001) Tiered Environmental Regulation: Lessons from the StarTrack Program. (Paper presented at the workshop "Voluntary, Collaborative and Information-Based Policies: Lessons and Next Steps for Environmental and Energy Policy in the United States and Europe", May 10-12 2001, Harvard University)
Nelson R (ed) (1993) National Innovation Systems. A Comparative Analysis. Oxford
Nelson R (1994) Economic growth via the co-evolution of technologies and institutions. In: Leydesdorff L, von Besselaar P (eds) Evolutionary Economics and Chaos Theory. London, pp 21-32
Nelson R, Winter S (1982) An Evolutionary Theory of Economic Change. Cambridge

Nelson R, Rosenberg N (1993) Technical innovation and national systems: Introductory chapter. In Nelson R (ed) National System of Innovation: A Comparative Study. Oxford

Nelson R, Rosenberg N (1993) National Innovation Systems: A Comparative Analysis. New York

NEPDG (National Energy Policy Development Group) (2001) National Energy Policy. Washington

Noorman K, Schoot Uiterkamp T (eds) (1998) Green Households? Domestic Consumers, Environment and Sustainability. London

Norberg-Bohm V (2000) Beyond the Double Dividend: Public and Private Roles in the Transformation to a Sustainable Industrial Society. In: OECD (ed) Innovation and the Environment: Workshop Proceedings. Paris

Norberg-Bohm V, Margolis R (2001) Reaching Environmental Goals through R&D Collaboration: Lessons from the U.S. Department of Energy Programs for Gas Turbines and Solar Photovoltaics. (Paper presented at the workshop "Voluntary, Collaborative and Information-Based Policies: Lessons and Next Steps for Environmental and Energy Policy in the United States and Europe", May 10-12 2001, Harvard University)

North D (1990) Institutions, Institutional Change and Economic Performance. Cambridge

Notestein F (1945) Population, the long view. In: Schultz T (ed) Food for the world. Chicago, pp 36-57

O'Brien M (2000) Making Better Environmental Decisions. Cambridge

OECD (1987) The Promotion and Diffusion of Clean Technologies. Paris

OECD (1992) Technology and the Economy. The Key Relationships. Paris

OECD (1992) Science Responds to Environmental Threats (synthesis report and country studies). Paris

OECD (1993) Environmental Policies and Industrial Competitiveness. Paris

OECD (1993) Pollution Abatement Expenditures in OECD, Environmental Monographs No. 75. Paris

OECD (1994) Capacity Development in Environment. Paris

OECD (1997) Proposed Guidelines for Collecting and Interpreting Technological Innovation Data - Oslo Manual. Paris

OECD (1999) Technology and Environment: Towards policy integration. Paris

OECD (1999) Technology and Sustainable Development, STI Review, No. 25, Paris.

OECD, Eurostat (1999) The Environmental Goods and Services Industry - Manual for Data Collection and Analysis. Paris

OECD (1999) Managing National Innovation Systems. Paris

OECD (1999) The OECD Three-Year Project on Sustainable Development: A Progress Report. Paris

OECD (1999) Voluntary Approaches for Environmental Policy - An Assessment. Paris

OECD (2000) Innovation and the Environment. Paris

OECD (2001) OECD Science, Technology and Industry Scorecard. Paris

OECD (2001) Sustainable Development - Critical Issues. Paris

OECD (2001) Policies to Enhance Sustainable Development. Paris

Ohmae K (1993) The Rise of the Region-State. Foreign Affairs 72: 78-87

Ohta M, Griliches Z (1976) Automobile prices revisited: extensions of the hedonic hypothesis. In Terleckyj N (ed) Household Production and Consumption. New York

Ohta M, Griliches Z, (1986) Automobile prices and quality: did the gasoline price increases change consumer tastes in the USA? Journal of Business & Economic Statistics 4: 187-198
Orru M (1994) The Faces of Capitalism. (Paper presented at the Sixth Annual International Conference on Socio-Economics, HEC. Paris)
Orum P (2002) Working Group on Community Right-to-Know. (Comment for the Toxic Release Ineventory (TRI) Program)
Palmer K, Oates W, Portney P (1995) Tightening Environmental Standards: the Benefit-Cost or the No-Cost Paradigm? Journal of Economic Perspectives 9(4): 119-132
Partidário P, Vergragt, P (2001) Towards leap-frog innovations in the coatings chain; A back-casting study in Portugal and the Netherlands". (Paper accepted for Greening of Industry Conference, 21-25 jan 2001. Bangkok)
Partidário P, Vergragt P (2002) Planning of Strategic Innovation aimed at environmental sustainability; actor-networks, scenario-acceptance, and back-casting analysis with a polymeric coating chain. Futures 34 (9-10): 841-861
Paton B (2001) Converging and Separating Mechanisms in Voluntary Product Labeling Programs. (Paper presented at the workshop "Voluntary, Collaborative and Information-Based Policies: Lessons and Next Steps for Environmental and Energy Policy in the United States and Europe", May 10-12 2001, Harvard University)
Perez C (1983) Structural change and the assimilation of new technologies in the economic system. Futures 15: 357-375
Perez C (1985) Microelectronics, long waves and world structural change: new perspectives in developing countries. World Development 13: 441-463
Peters B (1999) Institutional Theory in Political Science: The 'New Institutionalism'. New York
Petschow U, Hübner K, Dröge S, Meyerhoff J (1998) Nachhaltigkeit und Globalisierung. Berlin
Pfeiffer C (2000) The Shelter Function. Final Report SusHouse Project. Delft
Popper K (1972) Objective Knowledge, An Evolutionary Approach. Oxford
Porter M (1980) Competitive Strategy. London
Porter M (1990) The Competitive Advantage of Nations. London
Porter M (1998) Clusters and the new economics of competition. Harvard Business Review 76(6)
Porter M, Van der Linde C (1995) Towards a new conception of the Environment-Competitiveness Relationship. Journal of economic Perspectives 9(4): 97-118
Porter M, Van der Linde C (1995) Green and competitive - Ending the stalemate. Harvard Business Review 73: 120-137
Quist J, Pacchi C, Van der Wel M (2000) Workshop Organization and Stakeholder Management. Final Report SusHouse Project. Delft
Raymond V (1981) Economic environment of international business. N.J.
Raymond V (1985) Exploring the Global Economy: Emerging Issues in Trade and Investment. Cambridge
Reichel M (1998) Markteinführung von erneuerbaren Energien. Wiesbaden
Reijnders L (1998) The Factor X Debate: Setting Targets for Eco-Efficiency. Journal of Industrial Ecology 2(1): 13-22
Reinhardt F (1999) Market Failure and the Environmental Policies of Firms. Journal of Industrial Ecology (3)1: 9-21

Reinhardt F (2000) Down to Earth: Applying Business Principles to Environmental Management. Boston

Renn O (1997) Die Rolle von Technikleitbildern für technische Innovationen. In: Renn O, Blättel-Mink B (eds) Zwischen Akteur und System. Die Organisation von Innovation. Opladen, pp 271-284

Rennings K (2000) Redefining Innovation - Eco-Innovation Research and the Contribution from Ecological Economics. Ecological Economics 32(2): 319-332

Richardson J (ed) (1982) Policy styles in western Europe. London

Rip A (1995) Introduction of New Technology: Making Use of Recent Insights from Sociology and Economics of Technology. Technology Analysis and Strategic Management 7(4): 417-431

Rip A, Kemp R (1998) Technological Change. In Rayner S, Malone E (eds) Human Choice and Climate Change. An International Assessment Vol 2. Washington, pp 327-399

Ronning G (1991) Mikroökonometrie. Berlin

Roome N (1992) Developing Environmental Management Strategies. Business Strategy and the Environment 1(1): 11-24

Roome N (2001) Policies and Conditions for Environmental Innovation and Management in Industry. (Paper presented at the International Conference Towards Environmental Innovation Systems, September 27-29. Garmisch-Partenkirchen)

Rose R (1993) Lesson-Drawing in Public Policy. A Guide to Learning across Time and Space. Chatham

Rosenberg N (1976) The Direction of Technological Change: Inducement Mechanisms and Focussing Devices. In: Rosenberg N (ed) Perspectives on Technology. Cambridge, pp 108-125

Rosenberg N (1982) Inside the Black Box: Technology and Economics. New York

Rosenkopf T (1998) The co-evolution of community networks and technology: lessons from the flight simulation industry. Industrial and Corporate Change 7: 311-346

Rostow W (1960) The Process of Economic Growth. Oxford

Rothwell R, Zegveld W (1981) Industrial Innovation and Public Policy : Preparing for the 1980s and the 1990s. London

Rotmans J (1994) 'Transitions on the move', Global Dynamics and Sustainable Development. Bilthoven

Rotmans J, den Elzen M (1993) Halting Global Warming: Should Fossil Fuels by phased out? In: Lal M (ed) Global Warming. Concern for tomorrow. New Delhi

Rotmans J et al. (1995) 'TARGETS in Transition', Global Dynamics and Sustainable Development. Bilthoven

Rotmans J, Kemp R, Van Asselt M (2001) More Evolution than Revolution. Transition Management in Public Policy. Foresight 3(1): 15-31

Rotmans J, Kemp R, van Asselt M, Geels F, Verbong G, Molendijk K (2000) Transitions and Transition Management: the case of a low-emission energy supply, ICIS-Report. Maastricht

Rotmans J, Kemp R, van Asselt M, Geels F, Verbong G, Molendijk K (2000) Transities en transitiemanagement. Bilthoven

Roughgarden J (1996) Theory of Population Genetics and Evolutionary Ecology, an Introduction. New York

Royal Commission on Environmental Pollution (2000) A Changing Climate. London

Ruttan V (2000) Technology, Growth and the Environment: an Induced Innovation Perspective. Minneapolis

Sabatier P (1987) Knowledge, Policy-Oriented Learning and Policy Change. An Advocacy Coalitions Framework. Knowledge 8: 17-50

Sabel, C F (1995) 'Bootstrapping Reform: Rebuilding Firms, the Welfare State and Unions', Politics and Society 23, 1: 5-48.

Sachs W (1984) Die Liebe zum Automobil. Hamburg

Sahal D (1985) Patterns of Technological Innovation. Reading

Sahal D (1985) Technological guideposts and innovation avenues. Research Policy 14: 61-82

Särndal C, Swensson B, Wretman J (1992) Model Assisted Survey Sampling. New York

Saviotti P (1996) Technological Evolution, Variety and the Economy. Aldershot

Saviotti P (2002) Empirical analysis and evolutionary theories, presented at the first European Meeting on Applied Evolutionary Economics, to be published as Introduction. In Saviotti P (ed) Applied Evolutionary Economics: New Empirical Methods and Simulation Techniques. Cheltenham (2002)

Saviotti P (2001) Variety, growth and demand. Journal of Evolutionary Economics 11: 119-142

Saviotti P (2001) On the Co-Evolution of Technologies and Institutions. In this volume

Saviotti P (2003) On the co-evolution of technologies and institutions. In this volume

Scharpf F (1999) Governing in Europe: Effective and Democratic? Oxford

Schienstock G (1994) Technology policy in the process of change: Changing paradigms in research and technology policy. In: Aichholzer G, Schienstock G (eds) Technology policy: Towards an integration of social and ecological concerns. Berlin, pp 1-23

Schienstock G (1996) Transformation regionaler Ökonomien: Das Beispiel Baden-Württemberg. In Flecker J, Hofbauer J (eds) Vernetzung und Vereinnahmung. Arbeit zwischen Internationalisierung und neuen Managementkonzepten, Österreichische Zeitschrift für Soziologie, Sonderband 3. Opladen, pp 163-194

Schienstock G (1997) The Transformation of Regional Governance: Institutional Lock-ins and the Development of Lean Production in Baden-Württemberg. In Whitley R, Kristensen P (eds) Governance at work: the social regulation of economic relations in Europe. Oxford

Schienstock G (1999) Transformation and Learning: A Perspective on National Innovation Systems. In Schienstock G, Kuusi O (eds) Transformation Towards a Learning Economy. The Challenge for the Finnish Innovation System, Sitra 213. Helsinki

Schienstock G, Hämäläinen T (2001) Transformation of the Finnish Innovation System: A Network Approach. Helsinki

Schmidheiny S (1992) Changing Course - A Global Business Perspective on Development and the Environment. Cambridge

Schmidt T, Postma A (1999) Minder energiegebruik door een andere leefstijl? Project Perspectief. Rotterdam

Scholz J (1991) Regulatory Enforcement in a FederalistSystem. American Political Science Review 80: 1249-1270

Schot J (1992) The Policy Relevance of the Quasi-evolutionary Model: The Case of Stimulating Clean Technologies. In: Coombs R et al. (eds) Technological Change and Company Strategies. London, pp 185-200

Schot J (1998a) Innoveren in Nederland. In: Lintsen H (ed) Geschiedenis van de Techniek in Nederland. De wording van een moderne samenleving 1800-1890. Volume VI. Zutphen, pp 217-240

Schot J (1998b) The Usefulness of Evolutionary Models for Explaining Innovation. The Case of The Netherlands in the Nineteenth Century. History And Technology 14: 173-200

Schot J, Rip A (1997) The Past and the Future of Constructive Technology Assessment. Technological Forecasting and Social Change 54(2-3): 251-268

Schot J, Brand E, Fischer K (1997) The Greening of Industry for a Sustainable Future: Building an International Research Agenda. Business Strategy and the Environment 6(3): 153-162

Schot J, Lintsen H, Rip A (eds) (1998) Techniek in Nederland in de Twintigste Eeuw. Eindhoven

Schumpeter J (1939) Business Cycles: A Theoretical, Historical and Statistical Analysis of the Capitalist Process, 2 Vol., New York: McGraw-Hill.

Schumpeter J (1912) Theorie der wirtschaftlichen Entwicklung. Berlin

Schumpeter J (1912) The Theory of Economic Development. Cambridge

Schumpeter J (1939) Business Cycles: A Theoretical, Historical and Statistical Analysis of the Capitalist Process. New York

Scott A, Storper M (1992) Regional Development Reconsidered. In: Ernste H, Meier V (eds) Regional development and contemporary industrial response: Extending flexible specialization. London

Skea J (1995) Environmental technology. In Folmer H et al (eds) Principles of Environmental and Resource Economics. Aldershot

Slaughter S (1993) Innovation and Learning during Implementation: A Comparison of user and Manufacturer Innovations. Research Policy 22: 81-95

Smits R, Hekkert M, van Lente H (2001) Intermediairen en transitiemanagement. Nieuwe rollen voor Novem? Utrecht

Socolow R, Andrews C, Berkhout F, Thomas V (1994) Industrial Ecology and Global Change. Cambridge

Solsberg L (1997) Energy Challenges and Opportunities for Action. In: OECD (ed) Sustainable Development. OECD Policy Approaches for the 21st Century. Paris

Spaargaren G, Mol A (1992) Sociology, Environment and Modernity. Ecological modernization as a theory of social change. Society and Natural Resources 5(5): 323-345

Speir J (2001) EMSs and Tiered Regulation: Getting the Deal Right. In: Coglianese C, Nash J (eds) Regulating from the Inside: Can Environmental Management Systems Achieve Policy Goals? Washington, pp 198-219

Statistisches Bundesamt (1998) Konzentrationsstatistische Daten für das Verarbeitende Gewerbe, den Bergbau und die Gewinnung von Steinen und Erden sowie für das Baugewerbe 1995 und 1996. Fachserie 4, Reihe 4.2.3. Wiesbaden

Statistisches Bundesamt (1999) Beschäftigung, Umsatz und Energieversorgung der Unternehmen und Betriebe im Bergbau und im Verarbeitenden Gewerbe 1998. Fachserie 4, Reihe 4.1.1. Wiesbaden

Stodulski W (1999) 10 Years of Transformation in Poland: Environmental Protection. Warsaw

Strasser K (1997) Cleaner Technology, Pollution Prevention, and Environmental Regulation. Fordham Environmental Law Journal 9(1): 1-106

Sturm A, Wackernagel M, Müller K (2000) The Winners and Losers in Global Competition. Why Eco-Efficiency Reinforces Competitiveness: A Study of 44 Nations. Chur

Summerton J (ed) (1994) Changing Large Technical Systems. Boulder

Susskind L, McMahon G (1985) The Theory and Practice of Negotiated Rulemaking. Yale Journal on Regulation 3: 133-165

Swedish Ministry of Environment (1996) Our Environment. Stockholm

Taistra G (2001) Die Porter Hypothese zur Umweltpolitik. ZfU 2: 241-262

Teisman G (2000) Sturen als ontwikkelingsopdracht (Guidance as a development assignment). In: RMNO (ed) Report of discussions with scientists during the NMP4 process. The Hague, pp 64-67

Ten Heuvelhof E, Van Twist M (2000) Nieuwe markten en de rol van de overheid. (ESB dossier, Liberalization of network sectors)

Teubal M (1987) Innovation Performance, Learning and Government Policy. London

Teubal M (1998) Enterprise Restructuring and Embeddedness - An Innovation Systems and Policy Perspective. CRIC Discussion Paper No. 15. Manchester

Teubner G (1983) Substantive and Reflexive Elements in Modern Law. Law and Society Review 17: 239-285

Tuomi I (2001) Theory of Innovation. Changes and Meaning in the Age of Internet. Helsinki

US Department of Commerce (1999) National Trade Data Bank, Poland: Economic Trends and Outlook. Washington

Utterback J (1987) Innovation and Industrial Evolution in Manufacturing Industries? In Guile B, Brooks H (eds) Technology and Global Industry: Companies & Nations in the World Economy. National Academy of Engineering. Washington

Van Asselt M (2000) Perspectives on Uncertainty and Risk: the PRIMA Approach to Decision Support. Dordrecht

Van de Graaf H, Grin J, Vergragt P (2001) Naar een derde generatie Milieubeleid: een mogelijk sturingsconcept voor transities. (submitted for publication in Beleid en Maatschappij, 2002)

Van den Breemen C (1999) Wash-in. Eindhoven

Van de Poel I (2000) On the Role of Outsiders in Technical Development. Technology Analysis & Strategic Management 12(3): 383-397

Van Lente H (1993) Promising Technology. The Dynamics of Expectations in Technological Developments. Twente

Van Waarden F (1995) Persistence of National Policy Styles: A Study of Their Institutional Foundations. Convergence or Diversity? Intenationalization and Economic Policy Response. Gateshead

Vassilopoulos M (2001a) Industrial Competitiveness and Environmental Regulation. Seville

Vassilopoulos M (2001b) Clean Technology Adoption by Firms, Institute for Prospective Technological Studies. Seville

VCI (1995) Guidelines Responsible Care. Daemstadt

VCI (1998) Responsible Care. Darmstadt

Vellinga P, Berkhout F, Gupta J (eds) (1998) Managing a material world. Dordrecht

Vellinga P, Herb N (1999) Industrial Transformation Science Plan. IHDP Report 12. Bonn

Verbong G (2000) De Nederlandse overheid en energietransities: een historisch perspectief. (Background document for the report "Transitions and transition management" Rotmans et al. (2000). Eindhoven)

Vergragt P (1992) Naar een Ecologische Technologie; Technology Assessment en Duurzaamheid. Intreerede. Delft

Vergragt P (2000) Strategies towards the Sustainable Household. Final Report SusHouse Project Delft
Vergragt P, Van Grootveld G (1994) Sustainable Technology Development in the Netherlands: The first phase of the Dutch STD program. Journal of Cleaner Production 2(3-4): 133-139
Vergragt P, Van der Wel A (1998) Back-casting: An Example of Sustainable Washing. In: Roome N (ed) Sustainability strategies for industry: the future of corporate practice. Washington, pp 171-184
Vergragt P, Knot M (2001) Strategieën voor duurzame huishoudens: het SusHouse project. Milieu 16(3-4): 186-193
Vergragt P, Green K (2001) The SusHouse Methodology. Design Orienting Scenarios for Sustainable Solutions, Journal of Sustainable Design
Vergragt, P J, Leo J, Jansen A 1993, 'Sustainable Technological Development; The making of a Dutch long term oriented technology program', Project Appraisal 8(3) pp 134-140
Vezzoli C (2000) The Clothing Care Function. Final Report SusHouse Project. Delft
VN (1997) Critical Trends: Global Change and Sustainable Development. New York
Vogel D (1995) Trading Up. Consumer and Environmental Regulation in a Global Economy. Cambridge
Vogel D (1997) Trading up and governing across: transnational governance and environmental protection. Journal of European Public Policy: 556-571
Vogel D (2001) Is There a Race to the Bottom? The Impact of Globalization on National Regulatory Policies. La Revue Tocqueville 12(1)
Vogel D (2002) The Hare and The Tortoise Revisited: The New Politics of Consumer and Environmental Regulation in Europe. Manuscript submitted to the British Journal of Political Science
Vogt W (1999) Dictionary of Statistics & Methodology. Thousand Oaks
Von Hippel E (1976) The Dominant Role of Users in the Scientific Instrument Innovation Process. Research Policy 5: 212-239
Von Hippel E (1988) The Sources of Innovation. Oxford
VROM (2001) Nationaal Milieubeleidsplan 4: een wereld en een wil, werken aan duurzaamheid. Den Haag
VROM (2000), 'Transities naar Duurzaamheid: Milieubeleid als Transitiemanagement' (Transitions to sustainability: Environmental policy as transition management), Vision document for Knowledge and Technology workgroup, draft 10 April 2000.
VROM-raad (1998), 'Transitie naar een koolstofarme energiehuishouding' (Transition to a low-carbon energy economy, Advice for the Climate Policy Implementation Memorandum, Advice 010, 23 December 1998.
Wagner K (2001) The European Cement Industry. Seville
Wagner K, Triebswetter U, Hitchens D (2001) The Impact of Best Available Techniques (BAT) on the Competitiveness of the European Cement Industry. Seville
Walker W (2000) Entrapment in large technical systems: institutional commitment and power relations. Research Policy 29: 833-846
Wallace D (1995) Environmental Policy and Industrial Innovation. Strategies in Europe, the USA and Japan. London
Walley N, Whitehead B (1994) It's Not Easy Being Green. Harvard Business Review May-June: 46-52
WCED (World Commission on Environment and Development) (1987) Our Common Future. Oxford

WCSD (World Council on Sustainable Development) (1987) Our Common Future (The Brundtland Report). Oxford

Weaver P et al. (1999) Sustainable Technology Development. Sheffield

Weaver P, Jansen L, van Grootveld G, van Spiegel E, Vergragt P (2000) Sustainable Technology Development. Sheffield

Weber K (1999) Innovation Diffusion and Political Control of Energy Technologies. A Comparison of Combined Heat and Power Generation in the UK and Germany. Heidelberg

Weber K (2000) The role of networks for innovation diffusion and system change. CHP in the UK, Germany and the Netherlands. SEIN Research Report No. 14. Bielefeld

Weber M, Dorda A (1999) Strategic Niche Management as a new way of facilitating the introduction of new transport technologies and concepts. The IPTS Report 4(31): 20-28

Weber M, Hoogma R, Lane B, Schot J (1998) Experimenting with Sustainable Transport Innovations. A workbook for Strategic Niche Management. Seville

Weidner H (2002) Capacity Building for Ecological Modernization. American Behavioral Scientist 45(9): 1340-1368

Weidner H, Jänicke M (eds) (2002) Capacity Building in National Environmental Policy. A Comparative Study of 17 Countries. Berlin

Whitefield J (2001) Vital Signs. Nature 29 June

Willems, Van den Wildenberg (2000) Horizonverkenning Programma E.E.T. Den Haag

Wynne B (1995) Technology Assessment and Reflexive Social Learning: Observations from the Risk Field. In: Rip A, Misa T, Schot J (1995) Managing Technology in Society: The Approach of Constructive Technology Assessment. London, pp 19-36

Yakowitz H (1997) Assessing the cost-effectiveness of cleaner production. In OECD (ed) Cleaner Production and Waste Minimisation in OECD and Dynamic Non-Member Economies. Paris, pp 163-178

Young W, Simms J (2000) Economic Analysis of Scenarios. Final Report SusHouse Project. Delft

Young W, Vergragt P (eds) (2000) Strategies towards the Sustainable Household (SusHouse) Project: Design Orienting Scenarios (DOS). Manchester

Ziman J (1978) Reliable Knowledge. Cambridge

Zürn M (1998) Gesellschaftliche Denationalisierung und Regieren in der OECD Welt. In: Kohler-Koch B (ed) Regieren in Entgrenzten Räumen. PVS Sonderheft 29/1998. Opladen

Zylicz T (1993) Environmental policy reform in Poland. In: Sterner T (ed) Economic Policies for Sustainable Development. Dordrecht

Zylicz T (1998) A Survey of Current Issues. In: Tietenberg T, Folmer H. (eds) The International Yearbook of Environmental and Resource Economics 1998-99. Cheltenham

List of Contributors

Ashford, Nicholas A., Prof. PhD
Massachusetts Institute of Technology
Cambridge, MA 02139
USA

Becker, Frank, Dr.
Institute of Economics and Law
University of Stuttgart
Keplerstraße 17
70174 Stuttgart
Germany

Berkhout, Frans, Dr.
SPRU-Science & Technology Policy
Research, University of Sussex Falmer
Brighton BN1 9RF UK
UK

de Bruijn, Theo, Dr.
Center for Clean Technology
and Environmental Policy
University of Twente
Capitool 15, Enschede
The Netherlands

Elzen, Boelie, Dr.
BBT, University of Twente
Capitool A-203
Enschede
The Netherlands

Erdmann, Georg, Prof. Dr.
Chair for Energy Systems
Technical University Berlin
Einsteinufer 25 / TA8
10587 Berlin
Germany

Farrell, Frank
Environment Agency
Government Buildings Burghill Road,
BS10 6BF Bristol
UK

Fukasaku, Yukiko
Directorate for Science Technology
and Industry, OECD
Paris
France

Green, Ken, Prof. Dr.
Institute of Innovation Research
Manchester School of Management
PO Box 88
Manchester M60 1QD
UK

Hemmelskamp, Jens, Dr.
Director Research and
Projectmanagement
University of Heidelberg
Seminarstraße 2
69118 Heidelberg
Germany

Hoogma, Remco
Novem BV
Catharijnesingel 59
3511 GG Utrecht
The Netherlands

Jacob, Klaus, Dr.
Environmental Policy Research Centre
Freie Universität Berlin
Ihnestrasse 22
14195 Berlin
Germany

Jänicke, Martin, Prof. Dr.
Environmental Policy Research Centre,
Freie Universität Berlin
Ihnestrasse 22
14195 Berlin
Germany

Kemp, René, Dr.
MERIT
P.O. Box 616
6200 MD Maastricht
The Netherlands

Koefoed, Anne Louise
Norwegian School of Management
P.O. Box 580
1302 Sandvika
Norway

Lindblom, Josefina
Institute of Prospective Technological
Studies (IPTS)
European Commission
Isla de la Cartuja
41092 Seville
Spain

Midttun, Atle, Prof. Dr.
Norwegian School of Management
P.O. Box 580
1302 Sandvika
Norway

Norberg-Bohm, Vicki, Prof. Dr.
Belfer Center for Science and
International Affairs
Harvard University
79 John F. Kennedy Street
Cambridge, MA 02138
USA

Roome, Nigel, Prof. Dr.
Erasmus Centre for Sustainable
Development and Management
Burgemeester Oudlaan 50
3062 PA Rotterdam
The Netherlands

Rotmans, Jan, Dr.
International Centre for
Integrative Studies
University of Maastricht
P.O. Box 616
6200 MD Maastricht
The Netherlands

Saviotti, Pier-Paolo, Prof. Dr.
INRA
Université Pierre Mendès-France
38040 Grenoble Cedex 9
France

Schienstock, Gerd, Prof. Dr.
University of Tampere
33014 Tampere
Finland

Szejnwald Brown, Halina, Prof. Dr.
International Development,
Community, and Environment
Clark University
950 Main Street
Worcester, MA 01610
USA

Triebswetter, Ursula, Dr.
IFO Institute
Poschingerstr. 5
81679 Munich
Germany

Vergragt, Philip J., Prof. Dr.
Faculty of Design, Engineering
and Production
Delft University of Technology
Jaffalaan 9
2628 BX Delft
The Netherlands

Weber, Matthias K., Dr.
Head of Department Technology
Policy
ARC Systems Research GmbH
2444 Seibersdorf
Austria